METHODS IN MOLECULAR BIOLOGY

Series Editor
John M. Walker
School of Life Sciences
University of Hertfordshire
Hatfield, Hertfordshire, AL10 9AB, UK

For further volumes:
http://www.springer.com/series/7651

Methods in Membrane Lipids

Second Edition

Edited by

Dylan M. Owen

Department of Physics and Randall Division of Cell and Molecular Biophysics, King's College London, London, UK

 Humana Press

Editor
Dylan M. Owen
Department of Physics and Randall Division of Cell and Molecular Biophysics
King's College London
London, UK

ISSN 1064-3745   ISSN 1940-6029 (electronic)
ISBN 978-1-4939-1751-8   ISBN 978-1-4939-1752-5 (eBook)
DOI 10.1007/978-1-4939-1752-5
Springer New York Heidelberg Dordrecht London

Library of Congress Control Number: 2014951432

Printed on acid-free paper

Humana Press is a brand of Springer
Springer is part of Springer Science+Business Media (www.springer.com)

Preface

Cell membranes are important biological structures. Not only do they serve a structural and containment role for the cell but they also regulate transport, intracellular signaling, adhesion, and enzymatic reactions. To fulfill these roles, the cell membrane is composed of a highly complex composite of lipids and proteins which not only serve a direct function but also modify and regulate membrane biophysical parameters; a complexity which is only now being understood. This edition provides protocols for investigating membranes and their component lipids in both artificial membranes, cells and in silico.

The contribution begins with three introductory chapters. Chapter 1 examines the properties of the component lipids themselves, exploring the specific motifs of some of the most common archetypes. Chapter 2 introduces membranes and their biophysical properties and how these relate to their biological role. Finally, Chapter 3 discusses the latest advances in fluorescent probes for studying membranes: Fluorescence has been one of the major tools for examining lipid bilayers in recent years, especially in live cells due to its specific and noninvasive nature, and is a subject of several of the subsequent chapters.

Chapters 4–8 concern sample preparation. The relative simplicity of artificial membranes compared to cellular membranes has been an impasse in comparing in vitro and in vivo results. Recently, however, more realistic systems have been developed. For example, Chapter 4 describes the use of tethered lipid bilayers, Chapter 5 focuses on the use of detergent-resistant membrane methods to extract membrane subsets, Chapter 6 on the formation of artificial vesicles composed of cell membrane lipids, and Chapter 7 on the creation of realistic, asymmetric planar lipid bilayers. It is also useful to be able to modify membrane composition and properties in order to examine the effect on biological function; the subject of Chapter 8.

Once samples have been obtained, a range of physical techniques are available to study membrane composition, properties, and function. Eleven of these are the subjects of Chapters 9–19; for example, bilayer composition can be studied using mass spectrometry (Chapter 9). The packing density of lipids within a bilayer can be studied using environmentally sensitive fluorescent membrane probes (Chapter 10), and the lateral distributions of lipids and proteins within the bilayer can be characterized using newly developed super-resolution fluorescence imaging microscopy methods based on single-molecule detection (Chapter 11), a technique which is complemented by the gold standard in high-resolution imaging—electron microscopy (Chapter 12). Chapter 13 gives a detailed protocol for the use of atomic force microscopy (AFM) to study bilayers whereas Chapter 14 introduces another fluorescence-based method to examine the molecular orientations of lipids. Chapter 15 shows how fluorescence correlation spectroscopy (FCS) can be used to study the diffusion of lipids and proteins within the plane of the bilayer whereas Chapters 16 and 17 give detailed methods on studying lipid phase behavior in artificial systems using X-ray scattering and nuclear magnetic resonance (NMR) respectively. The section concludes with two further fluorescence techniques; measuring lateral mobility in membranes using photobleaching (Chapter 18) and a new method to analyze signals from environmentally sensitive dyes such as those used in Chapter 10 (Chapter 19).

It is important that physical observations be backed up with a strong theoretical framework and comprehensive computer modeling. The volume therefore ends with two examples of in silico analysis: One to model the behavior of cholesterol within a bilayer (Chapter 20) and one to examine cholesterol-dependent phase separation (Chapter 21).

London, UK   *Dylan M. Owen*

Contents

Contributors

G. Ekin Atilla-Gokcumen • *Department of Chemistry, University at Buffalo, The State University of New York, Buffalo, NY, USA*

Svetlana Baoukina • *Center for Molecular Simulation, University of Calgary, Calgary, AB, Canada*

Alison Beckett • *Division of Cellular and Molecular Physiology, Institute of Translational Medicine, University of Liverpool, Liverpool, UK*

Richard K. P. Benninger • *Department of Bioengineering, University of Colorado, Aurora, CO, USA*

Deborah A. Brown • *Department of Biochemistry and Cell Biology, Stony Brook University, Stony Brook, NY, USA*

Sonia Carne • *Victor Chang Cardiac Research Institute, Darlinghurst, NSW, Australia*

Michael Carnell • *Biomedical Imaging Facility, Lowy Cancer Research Centre, University of New South Wales, Sydney, NSW, Australia*

Marek Cebecauer • *Department of Biophysical Chemistry, J. Heyrovsky Institute of Physical Chemistry, The Academy of Sciences of the Czech Republic, Prague, Czech Republic*

Bruce Cornell • *Victor Chang Cardiac Research Institute, Darlinghurst, NSW, Australia*

Charles Cranfield • *Victor Chang Cardiac Research Institute, Darlinghurst, NSW, Australia*

Alex Demchenko • *University of Strasbourg, Strasbourg, France*

Institute of Physics National Academy of Sciences of Ukraine, Kiev, Ukraine

Giavanni Dietler • *Laboratoire de Physique de la Matière Vivante, IPSB, SB, EPFL, Lausanne, Switzerland*

Guy Duportail • *University of Strasbourg, Strasbourg, France*

Ulrike S. Eggert • *Department of Chemistry, King's College London, London, UK; Randall Division of Cell and Molecular Biophysics, King's College London, London, UK*

Richard Epand • *Department of Biochemistry and Biomedical Sciences, McMaster University, Hamilton, ON, Canada*

Ana Garcia-Saez • *Max Planck Institute for Intelligent Systems and German Cancer Research Center, BioQuant, Heidelberg, Germany*

Katharina Gaus • *Lowy Cancer Research Centre, Centre for Vascular Research and Australian Centre for Nanomedicine, University of New South Wales, Sydney, NSW, Australia*

Ottavia Golfetto • *Laboratory for Fluorescence Dynamics, Biomedical Engineering Department, University of California, Irvine, CA, USA*

Enrico Gratton • *Laboratory for Fluorescence Dynamics, Biomedical Engineering Department, University of California, Irvine, CA, USA*

EDUARD HERMANN • *Max Planck Institute for Intelligent Systems and German Cancer Research Center, BioQuant, Heidelberg, Germany*

ELIZABETH HINDE • *Laboratory for Fluorescence Dynamics, Biomedical Engineering Department, University of California, Irvine, CA, USA*

PEICHI HU • *Mork Family Department of Chemical Engineering and Materials Science, University of Southern California, Los Angeles, CA, USA*

SANDOR KASAS • *Laboratoire de Physique de la Matière Vivante, IPSB, SB, EPFL, Lausanne, Switzerland*

ANDREY S. KLYMCHENKO • *University of Strasbourg, Strasbourg, France*

ROB LAW • *Department of Chemistry, Imperial College London, London, UK*

K.R. LEVENTAL • *Department of Integrative Biology and Pharmacology, University of Texas Health Science Center at Houston Medical School, Houston, TX, USA*

I. LEVENTAL • *Department of Integrative Biology and Pharmacology, University of Texas Health Science Center at Houston Medical School, Houston, TX, USA*

GIOVANNI LONGO • *Laboratoire de Physique de la Matière Vivante, IPSB, SB, EPFL, Lausanne, Switzerland*

ALEX MACMILLAN • *Biomedical Imaging Facility, Lowy Cancer Research Centre, University of New South Wales, Sydney, NSW, Australia*

ASTRID MAGENAU • *Lowy Cancer Research Centre, Centre for Vascular Research and Australian Centre for Nanomedicine, University of New South Wales, Sydney, NSW, Australia*

SALEEMULLA MAHAMMAD • *Department of Cell and Molecular Biology, Northwestern University Feinberg School of Medicine, Chicago, IL, USA*

NOAH MALMSTADT • *Mork Family Department of Chemical Engineering and Materials Science, University of Southern California, Los Angeles, CA, USA*

BORIS MARTINAC • *Victor Chang Cardiac Research Institute, Darlinghurst, NSW, Australia*

YVES MÉLY • *University of Strasbourg, Strasbourg, France*

SULE ONCUL • *Department of Biophysics, School of Medicine, Istanbul Medenlyet University, Istanbul, Turkey*

DYLAN M. OWEN • *Department of Physics and Randall Division of Cell and Molecular Biophysics, King's College London, London, UK*

INGELA PARMRYD • *Department of Medical Cell Biology, Uppsala University, Uppsala, Sweden*

IAN PRIOR • *Division of Cellular and Molecular Physiology, Institute of Translational Medicine, University of Liverpool, Liverpool, UK*

JONAS RIES • *Max Planck Institute for Intelligent Systems and German Cancer Research Center, BioQuant, Heidelberg, Germany*

CHARLES RODUIT • *Laboratoire de Physique de la Matière Vivante, IPSB, SB, EPFL, Lausanne, Switzerland*

JOHN SEDDON • *Department of Chemistry, Imperial College London, London, UK*

D. PETER TIELEMAN • *Department of Biological Sciences, Centre for Molecular Simulation, University of Calgary, Calgary, AB, Canada*

ARWEN TYLER • *Department of Chemistry, Imperial College London, London, UK*

RENEE WHAN • *Biomedical Imaging Facility, Lowy Cancer Research Centre, University of New South Wales, Sydney, NSW, Australia*

G.W. ASHDOWN • *Department of Physics, King's College London Strand Campus, London, UK; Division Randall Division of Cell and Molecular Biophysics, King's College London, London, UK*

SEMEN YESLEVSKYY • *Institute of Physics, National Academy of Sciences of Ukraine, Kiev, Ukraine*

Chapter 1

Introduction to Membrane Lipids

Richard M. Epand

Abstract

Biological membranes are composed largely of lipids and proteins. The most common arrangement of lipids in biological membranes is as a bilayer. This arrangement spontaneously forms a barrier for the passage of polar materials. The bilayer is thin but can have a large area in the dimension perpendicular to its thickness. The physical nature of the bilayer membrane will vary according to the conditions of the environment as well as the chemical structure of the lipid constituents of the bilayer. These physical properties determine the function of the membrane together with specific structural features of the lipids that allow them to have signaling properties.

The lipids of the membrane are not uniformly distributed. There is an intrinsic asymmetry between the two monolayers that constitute the bilayer. In addition, some lipids tend to be enriched in particular regions of the membrane, termed domains. There is evidence that certain domains recruit specific proteins into that domain. This has been suggested to be important for allowing interaction among different proteins involved in certain signal transduction pathways.

Membrane lipids have important roles in determining the physical properties of the membrane, in modulating the activity of membrane-bound proteins and in certain cases being specific secondary messengers that can interact with specific proteins. A large variety of lipids present in biological membranes result in them possessing many functions.

Key words Lipid structures, Membrane composition, Membrane physical properties, Membrane biological properties

1 Structure of Membrane Lipids

Biological membranes are composed of many molecular species of lipids. There is a continuing effort to understand the consequences of this diversity. In recent years there has been a rapid advance in our ability to analyze the lipid species in biological membranes through the use of mass spectroscopy, that has developed into the field of lipidomics [1] (*see* also Chapter 9 in the present work). In mammalian membranes the major lipid components can be classified as glycerophospholipids, sphingolipids, and cholesterol (Fig. 1).

Dylan M. Owen (ed.), *Methods in Membrane Lipids*, Methods in Molecular Biology, vol. 1232, DOI 10.1007/978-1-4939-1752-5_1, © Springer Science+Business Media New York 2015

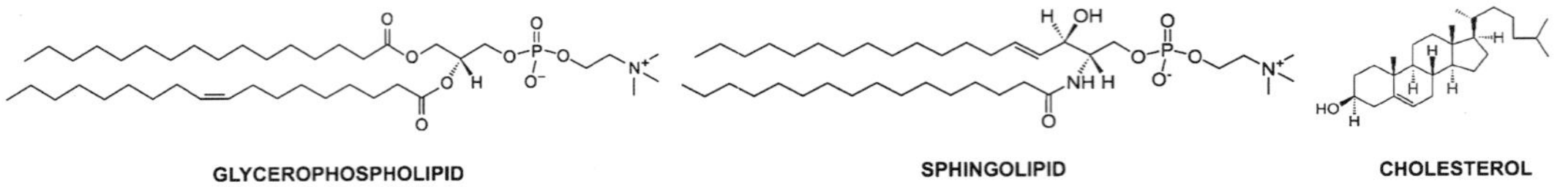

Fig. 1 Chemical structure of principal lipid components of mammalian membranes. The glycerophospholipid shown is 1-palmitoyl-2-oleoyl phosphatidylcholine, the sphingolipid is *N*-palmitoyl sphingomyelin. Structures have been obtained from the Web site of Avanti Polar Lipids (avantilipids.com)

The glycerophospholipids are composed of a glycerol moiety with two fatty acids esterified to the sn-1 and sn-2 positions to make a diacylglycerol precursor of a glycerophospholipid. The most common species have a saturated acyl chain at the sn-1 position and an unsaturated chain at the sn-2 position, although like most things in biology there are exceptions to this generalization. There are also lipid species, the plasminogens, which have ether linkages between the hydrocarbon the glycerol. The sn-3 position of diacylglycerol can be phosphorylated to make phosphatidic acid, structurally the simplest among the glycerophospholipids. The phosphate group of phosphatidic acid can then be coupled with one of several alcohols to produce other glycerophospholipids. The most common among these include the zwitterionic glycerophospholipids phosphatidylcholine (shown in Fig. 1) and phosphatidylethanolamine. The structural difference between these two lipids is that phosphatidylcholine has three methyl groups on the nitrogen, while phosphatidylethanolamine has a free amino group. Despite this similarity in their structure, phosphatidylcholine and phosphatidylethanolamine have very different properties. Biological membranes are arranged predominantly as bilayers, but the lipid composition of each monolayer of the bilayer is different. In biological membranes, phosphatidylethanolamine is located predominantly on the cytoplasmic leaflet of the plasma membrane and has less hydration and more spontaneous negative curvature than phosphatidylcholine. Most model membrane studies have used symmetrical bilayers, but effort is being made to establish systems for producing asymmetric model liposomes (*see* Chapter 7 as an example). The role of bilayer asymmetry in modulating the behavior of cholesterol in model membranes has been approached by molecular dynamic simulations (*see* Chapter 20). Anionic glycerophospholipids are also largely found on the cytoplasmic face of the plasma membrane bilayer. These lipids include phosphatidylserine and phosphatidylinositol as more prominent components. Phosphatidylinositol can also be phosphorylated on the free hydroxyl groups of the inositol to produce several different lipid species that function as secondary messengers. Certain forms of these phosphorylated phosphatidylinositols are found enriched in specific intracellular organelles or in the plasma membrane. Phosphatidylinositols have an acyl chain composition similar to other glycerophospholipids with a saturated acyl chain at the sn-1 position and unsaturated acyl chain at sn-2, but in the case of this lipid class, there is a very high enrichment of 1-stearoyl-2-archidonoyl species (i.e., 18:0/20:4) [2]. Another anionic lipid, cardiolipin, is unique in being found largely in the inner membrane of mitochondria on the leaflet facing the mitochondrial matrix. It is only released from this location during the process of apoptosis or mitophagy [3]. Cardiolipin is abundant in many species of bacteria but is a minor component of mammalian cell membranes. In mammalian cells, the location of cardiolipin in the mitochondria may be a consequence of an early step in evolution. In addition to its unique location in mammalian cells, cardiolipin also has a unique chemical structure, being composed of two phosphatidic acid moieties that are bridged by a glycerol. Cardiolipin is thus an anionic lipid with four acyl chains. In the mammalian heart these acyl chains are remarkably enriched with linoleoyl (18:2) acyl chains, but other organs and other species have more varied acyl chains [4].

Sphingomyelin (shown in Fig. 1) is the most abundant sphingolipid in the mammalian membrane. It is a zwitterionic lipid with a head group similar to that of phosphatidylcholine. However, sphingolipids have no glycerol. They are made by adding groups to the lipid sphingosine. Sphingosine has a free amino group that is protonated at neutral pH, making this lipid one of the few natural cationic lipids, although it is never found in high concentrations in biological membranes. It does however function as a signaling lipid together with its phosphorylated form sphingosine phosphate, which is formed by phosphorylation of the hydroxyl group of sphingosine. The amino group of sphingosine can also be acylated, generally with a saturated acyl chain to produce ceramide. Both ceramide and its phosphorylated form, ceramide phosphate, also function as lipid signaling agents. Attachment of choline to ceramide phosphate produces sphingomyelin, a phospholipid found in abundance in the plasma membrane of eukaryotes. There is particular interest in sphingomyelin because of its apparent role in the formation of membrane domains. There are also glycosylated forms of sphingolipids including cerebrosides and gangliosides. These are present in low concentrations and general have a role in recognition phenomena for other cells or proteins.

The final lipid category is cholesterol. It has a sterol-based structure of four fused rings. In mammalian cells it is present at highest concentration in the plasma membrane. However, it has been suggested that the activity of cholesterol is the same in some intracellular organelles as it is in the plasma membrane [5], but likely because of the presence of sphingomyelin in the plasma membrane that can interact with cholesterol, the concentration of cholesterol in this membrane far exceeds its activity. Other sterols, including ergosterol in yeast and phytosterols in plants, likely play a similar role in modulating the physical properties of the membranes of these organisms [6].

2   Physical Properties of Membrane Lipids

Phospholipid membranes exhibit polymorphic phase behavior. The principal tool for understanding this phase behavior is X-ray scattering (Chapter 16). Model membranes can take up lamellar, hexagonal, or cubic phases. Particularly for the case of cubic phases, of which there are many possible forms, diffraction studies provide one of the few tools to identify the particular form of the cubic phase. Although the predominant phase of lipids in biological membranes is lamellar, conditions or chemical components that favor other phases can provide information about the overall curvature and dynamic properties of a membrane. For example, the Gaussian curvature that can be measured through properties of bicontinuous cubic phases can be related to the mechanism of fusion peptides [7].

An important physical characteristic of a membrane is its "fluidity." However, it should be kept in mind that membrane fluidity is a qualitative concept and there is no single parameter that can describe the motional properties within a membrane. Membrane motions are complex and highly anisotropic. They also vary considerably between the membrane–water interface with its relatively high degree of order, to the center of the bilayer where the disorder is much larger. A more accurate determination of "fluidity" can be achieved using NMR methods (Chapter 17). Fluorescence methods have been useful in determining many aspects of membrane order and polarity (Chapter 3). In particular, use has been made of polarity-sensitive fluorescent dyes (Chapter 10). One such probe is Laurdan, whose use has been developed into a method for probing heterogeneity in the plane of the membrane (Chapter 19). Membrane order can also be probed using fluorescence linear dichroism (Chapter 14). Another aspect of membrane order is given by the rate at which molecules travel along the plane of the membrane. These measurements are also made with fluorescent probes using either fluorescence recovery after photobleaching (Chapter 18) or scanning fluorescence correlation spectroscopy (Chapter 15).

An aspect of membrane structure that has received considerable attention in recent years is the phenomenon of membrane domain formation. It is generally recognized that biological membranes are not uniform, but there are regions in which certain components are sequestered. One aspect of this is a domain enriched in cholesterol and sphingomyelin that has been termed "rafts." Such domains can readily be detected in model membranes where they have micron size dimensions. However, the size and lifetime of such domains and even their lipid composition is more difficult to define in biological membranes [8, 9]. Nevertheless, less direct approaches provide evidence for clustering of molecules in membranes. One criterion that has been a convenient tool for isolating proteins which appear to interact with certain lipids is detergent resistance (Chapter 5). This provides a now well established method for isolating "raft" fractions. The method has limitations [10], but has nevertheless proven to be a valuable empirical tool. Another criterion of "rafts" in biological membranes is that they will disperse upon removal of cholesterol (Chapter 8). This method also has caveats, since it has been shown that cholesterol depletion results in large changes in cell architecture [11]. One of the issues is that cholesterol depletion results in the detachment of the plasma membrane from the cytoskeleton. The loss of the coupling between the plasma membrane and cytoskeleton can by itself lead to loss of domain structure and can therefore not be distinguished from the extraction of cholesterol. The mechanical properties of cytoskeleton-membrane interactions can be studied using atomic force microscopy (Chapter 13). Computer modeling of lipid phase separation is challenging because it occurs on a long time scale for molecular dynamic simulations. Nevertheless, progress is being made in understanding the physical bases of these phenomena (Chapter 21).

The lipids of biological membranes have several important functions. They establish the barrier between one compartment and another. In addition, the nature of the lipids affects the physical properties of the membrane, including intrinsic curvature, "fluidity," and the formation of domains. These physical properties will modulate the behavior of the membrane with regard to stability and leakage as well as the tendency of the membrane to undergo fusion. In addition, the lipid components will also affect the activity of certain proteins. This can occur nonspecifically through modulation of the physical properties of the lipid surrounding the protein or alternatively by binding to a specific site on the protein. Lipids can also drive the formation of membrane domains. Such domains can sequester proteins into a small area to allow for more efficient signal transduction pathways. In addition to these functions, certain membrane lipids are also signaling agents. This is based on their specific molecular structure that allows them to bind to specific protein binding sites.

It is thus important to study the properties of lipid mixtures to understand their behavior and properties in simpler systems as well as their effect in modulating the activity of isolated membrane-bound proteins in vitro. At the same time it is important to extrapolate these findings to biological membranes with their greater complexity of lipid species, their high and diverse content of proteins, relative to most model membrane systems, their asymmetry and the attachment of some membranes to the cytoskeleton. Many biological properties are dependent on membrane phenomenon, so that these studies can lead to significant advances in our understanding of biological systems.

References

1. Myers DS, Ivanova PT, Milne SB, Brown HA (2011) Quantitative analysis of glycerophospholipids by LC-MS: acquisition, data handling, and interpretation. Biochim Biophys Acta 1811: 748–757

2. D'Souza K, Epand RM (2014) Enrichment of phosphatidylinositols with specific acyl chains. Biochim Biophys Acta 1838(6): 1501–1508

3. Patil VA, Greenberg ML (2013) Cardiolipin-mediated cellular signaling. Adv Exp Med Biol 991:195–213

4. Schlame M, Ren M, Xu Y, Greenberg ML, Haller I (2005) Molecular symmetry in mitochondrial cardiolipins. Chem Phys Lipids 138:38–49

5. Steck TL, Lange Y (2010) Cell cholesterol homeostasis: mediation by active cholesterol. Trends Cell Biol 20:680–687

6. Bloom M, Evans E, Mouritsen OG (1991) Physical properties of the fluid lipid-bilayer component of cell membranes: a perspective. Q Rev Biophys 24:293–397

7. Siegel DP (2008) The Gaussian curvature elastic energy of intermediates in membrane fusion. Biophys J 95:5200–5215

8. Frisz JF, Klitzing HA, Lou K, Hutcheon ID, Weber PK, Zimmerberg J, Kraft ML (2013) Sphingolipid domains in the plasma membranes of fibroblasts are not enriched with cholesterol. J Biol Chem 288:16855–16861

9. Lozano MM, Liu Z, Sunnick E, Janshoff A, Kumar K, Boxer SG (2013) Colocalization of the ganglioside G(M1) and cholesterol detected by secondary ion mass spectrometry. J Am Chem Soc 135:5620–5630

10. Lichtenberg D, Goni FM, Heerklotz H (2005) Detergent-resistant membranes should not be identified with membrane rafts. Trends Biochem Sci 30:430–436

11. Kwik J, Boyle S, Fooksman D, Margolis L, Sheetz MP, Edidin M (2003) Membrane cholesterol, lateral mobility, and the phosphatidylinositol 4,5-bisphosphate-dependent organization of cell actin. Proc Natl Acad Sci U S A 100: 13964–13969

Chapter 2

Introduction: Membrane Properties (Good) for Life

Marek Cebecauer

Abstract

Membranes protect cells from the surrounding environment but also provide a means for the optimization of processes such as metabolism, signalling, or mitogenesis. Membrane structure and function is determined by its molecular composition. How lipid species define membrane properties is discussed in this introductory chapter.

Key words Bilayer, Membrane functions, Membrane proteins, Biological roles

In living cells, lipids predominantly form lamellar structures; bilayers. The lipid bilayer represents a two-dimensional (2D) amphipathic structure with hydrophilic surfaces on both sides of a hydrophobic core. The core is largely impermeable for ions and other polar molecules, and to some extent for water molecules. These properties endow lipid bilayers with an ideal structure to function as cellular membranes which protect the cell from the surrounding environment and function in the organization of molecules. Membranes also form the surface of intracellular organelles, such as mitochondria or the nucleus, providing significant compartmentalization of eukaryotic cells. I will focus on membranes of higher eukaryotes such as humans and mice, but the herein described basic principles can be considered (to some level) also for membranes of prokaryotes, fungi, and plants [1–3]. Later chapters of this book describe useful techniques and tools which can be employed in membrane studies. In this introductory chapter I will try to outline a few aspects of membranes which are important for cellular function and are, therefore, in the center of membrane biology. In some cases, future directions of research are suggested.

Lipids define the basic lamellar structure of cellular membranes but proteins also significantly contribute to its form and function. Due to their hydrophilic surface, proteins normally do not integrate into the hydrophobic core of membranes but there are some exceptions. Proteins with hydrophobic helical segments (one or more), called transmembrane domains (TMDs), can integrate into

Dylan M. Owen (ed.), *Methods in Membrane Lipids*, Methods in Molecular Biology, vol. 1232,
DOI 10.1007/978-1-4939-1752-5_2, © Springer Science+Business Media New York 2015

the lipid bilayer. We call these integral membrane proteins. Other examples of membrane proteins are glycosphingolipid-anchored proteins (GPI-AP) of the outer plasma membrane leaflet and intracellular lipid-modified proteins which are anchored to the cytosolic leaflet of cellular membranes. All these proteins are directly influenced by the properties of membrane. Another group is peripheral membrane proteins possessing a specific domain (e.g., pleckstrin-homology (PH) domain) which can associate with the hydrophilic surface of the membrane based on their affinity for the lipid headgroups (e.g., negatively charged phosphatidylinositols). The function of such proteins is also regulated by membrane properties and organization. The list of membrane "associated" proteins can be further extended to include those which interact, directly or indirectly, with membrane components and, in some cases, can assemble into large supramolecular structures important for metabolism, signalling, cell motility, and other cellular processes. The formation and function of these complexes depends on membrane properties. It is therefore important to investigate the relations between individual membrane components and how these cooperate for the optimal local environment.

Only a few lipid species are required to form a lipid bilayer under physiological conditions [4]. Surprisingly, hundreds of different lipid species were found to form cellular membranes [5]. It is now well established that specific membrane lipid compositions can modulate its physical properties and function. In cells, a good example is mitochondria which are surrounded by two membranes, inner and outer. A unique lipid composition of mitochondrial inner membrane (increased presence of negatively charged lipids and large content of cardiolipin (1,3-bis(sn-3′-phosphatidyl)-sn-glycerol)) provides a specific environment for reactions exclusively localized in these metabolic factories of the cell. On the other hand, we do not have detailed information on lipid composition of cellular membranes and their subcompartments. The current development of techniques with dramatically improved sensitivity of such analysis (*see* Chapter 9) can help to better understand why cells use such a variety of lipid species for their function.

In parallel, it is important to investigate how lipid species modulate membrane properties and, as a consequence, how membrane properties regulate cellular processes. Phospholipids of cellular membranes are formed by a polar headgroup and attached acyl chains. Charge, size and the precise conformation of the headgroup as well as its local density can directly influence the association of peripheral molecules with membranes. A nice example of lipid headgroup-determined protein localization was shown for phosphatidylserine (PS) and the C2 domain of lactadherin [6]. PS is the abundant lipid of the cytosolic leaflet of plasma membrane but is present at lower level in other intracellular membranes, endosomes and lysosomes. In untreated cells, fluorescently labelled C2 domain locates specifically to these membranes of murine macrophages. Such localization is exclusively due to its interaction with PS since no changes were observed after depletion of other negatively charged lipids (various phosphatidylinositols). On the other hand, interference with the presence of PS in membranes led to a rapid relocation of C2 domain of lactadherin to the cytosol [7]. Therefore, the PS headgroup with its charge, size and conformation determined the localization of C2 domain of lactadherin. An analogous effect was observed for other cellular proteins including signalling molecules c-Src, Rac1, and K-Ras [6]. In similar manner, the localization of phosphoinositide-binding signalling proteins interacting specifically with headgroups of phosphatidylinositol-4, 5-bisphosphate (phospholipase C-d1; [8]) or phosphatidylinositol-3,4,5-trisphosphate (Grp1; [9]) at membranes are regulated by lipases, kinases, and phosphatases providing mechanism for the control of their activity. An increasing number of available headgroup-specific biosensors [10, 11] and labelled lipids [12, 13] will be important for a better understanding of lipid changes in membranes of living cells [14].

Acyl chains of phospholipids define the rigidity of hydrophobic core of the membrane. Rigidity of membrane can also be described as a lipid packing density, meaning it determines how tightly lipid molecules (their acyl chains) are packed within a defined space of a membrane. Such a property has a dramatic effect on the freedom with which molecules (including proteins) can move and interact and was recognized as important for cells as early as the 1980s [15]. Membrane rigidity can also influence the orientation of membrane molecules, potentially causing change(s) in their conformation. Longer, saturated acyl chains of lipids support higher ordering, whereas increased presence of unsaturated bonds and acyl chain shortening decreases lipid ordering in membranes. This is well studied in model membranes with separated lipid phases. It is important to note here that the implementation of knowledge acquired from model systems to cellular membranes has to be performed with care (*see* Appendix).

In cells, an increased content of phospholipids with long and saturated acyl chains was detected in vesicles of trans-Golgi network (TGN), exo/endosomes, and plasma membrane [16, 17]. Increased rigidity of membranes towards the cell surface was expected (Fig. 1). Indeed, live cell imaging using environmentally sensitive membrane dyes (Laurdan or di-4-ANEPPDHQ; Chapter 10) supports this view with low lipid ordering observed for nuclear membrane and the endoplasmic reticulum (ER) and higher ordering in the plasma membrane and associated membrane structures [18, 19]. This may have consequences: (1) it can provide help for sorting of molecules (e.g., proteins) synthesized at the surface of intracellular membranes (ER) but targeted to the plasma membrane [20], and (2) increased membrane rigidity towards the cell surface can provide

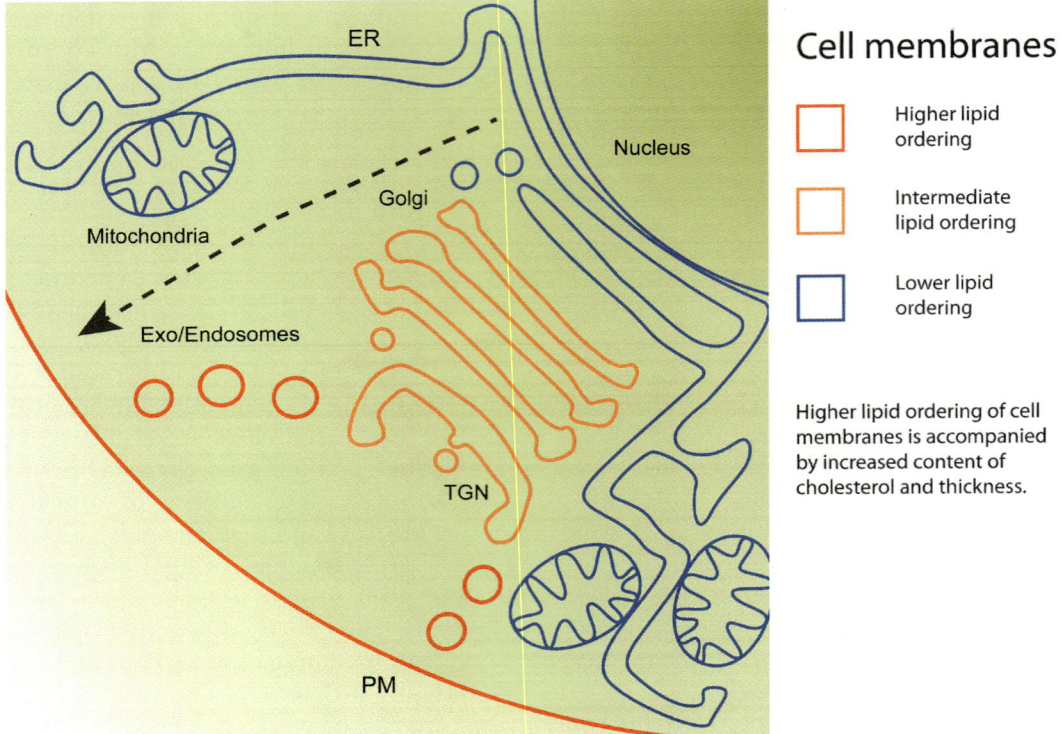

Fig. 1 Membrane order in living cells. Experiments using Laurdan and di-4-ANEPPDHQ fluorescent membrane probes indicate lower lipid ordering of intracellular membranes such as the endoplasmic reticulum (ER) and nucleus but higher ordering at the surface and in endosomal compartments. Somewhat intermediate state is believed for Golgi apparatus and trans-Golgi network (TGN). No clear prediction exists for mitochondria but no similar membrane structure with higher lipid ordering was observed in living cells. *Dashed arrow line* indicates anterograde traffic and sorting of membrane molecules

a specific environment for assembly of larger multimolecular entities [21, 22]. Such assemblies have recently been demonstrated at the surface of immune cells and neurons [23, 24] but also bacteria [25]. Techniques for characterization of multimolecular complexes and protein clusters in cellular membranes are undergoing a renaissance and a rapid development (Chapters 11 and 12). Such studies open new insights into membrane organization of cellular processes such as immune or neurological response, mitogenesis, and cell migration, to name just a few.

Cholesterol is essential component of cellular membranes. Its content is increased in the plasma membrane, endosomes, and membranes of TGN [16, 26]. One can speculate that cholesterol, in addition to its other functions, maintains fluidity of membranes with high content of gel-forming lipids such as phospholipids and sphingolipids with long saturated acyl chains (*see* Appendix). The fluid character of such ordered membranes is essential for the regulation of cellular processes by diffusion. The metabolism and important properties of cholesterol and other sterols in model and cell membranes are described in detail in excellent recent reviews [27–29]. Sphingolipids can be glycosylated to generate glyco-sphingolipids (GSLs) of the outer plasma membrane leaflet. High density of hydroxyl groups in the glyco-structure of GSLs can lead to their aggregation and membrane reorganization (e.g., GM1 and GM3 clustering) adding to the heterogeneity of membranes. For example, GM3 clustering induces caveolin-1 redistribution and inhibition of EGFR activation [30]. Of note, GSLs have high affinity for cholesterol-rich membrane domains and the effect of their clustering can be indirect [31]. Both cholesterol and GSL levels can be modulated in cells by drugs or by the inhibition of metabolic pathways (Chapter 8 and [32]). Such experiments are very useful for our understanding of the importance of these lipids for cell function. But one has to be very careful when interpreting data from such experiments since these treatments are rather systemic [33, 34] and do not reduce cholesterol or GSL levels preferentially from cholesterol- and glycosphingolipid-rich domains, frequently called lipid rafts.

Another property, membrane thickness, also varies due to a lipid packing, acyl chain length, and the presence of cholesterol. This was suggested to have an impact on sorting of integral membrane proteins throughout cellular membranes due to a mismatch between the length of the TMD and membrane thickness. Different sorting of integral membrane proteins due to a mismatch has been observed in cells [35, 36]: proteins with short TMD preferentially reside in the ER having relatively thin membrane and those with longer TMD are sorted to Golgi apparatus and plasma membrane with thicker membrane and higher cholesterol content. On the other hand, membrane protein localization can be also influenced by its affinity for specific lipids [6, 37, 38] and there are proteins with long TMD which were shown to reside in the ER [39]. So it is probably a concert of various protein and lipid features which defines final destination of membrane molecules. It is also important to note here that our knowledge of mechanisms behind sorting and localization of membrane molecules is still very limited and requires further studies in silico, in vitro and in vivo.

Improved imaging techniques and analytical tools (Chapters 14, 15, 18 and 19) recently enabled us to investigate lateral membrane diffusion in living cells. Ensemble (FRAP, FCS, sptPALM) and single-molecule (SPT) techniques provide useful information on mobility of membrane molecules. Even though membrane dynamics studies are still in their early phase, it is already clear that membrane molecules in cells exhibit anomalous diffusion and are frequently constrained in mesoscale (20–200 nm) domains [40–42]. High molecular crowding in cellular membranes is certainly an important factor causing anomalous diffusion but membrane heterogeneities due to the interactions of lipids and proteins are hypothesized to add to this effect. The cytoskeleton further extends the complexity of cell membrane environment by providing

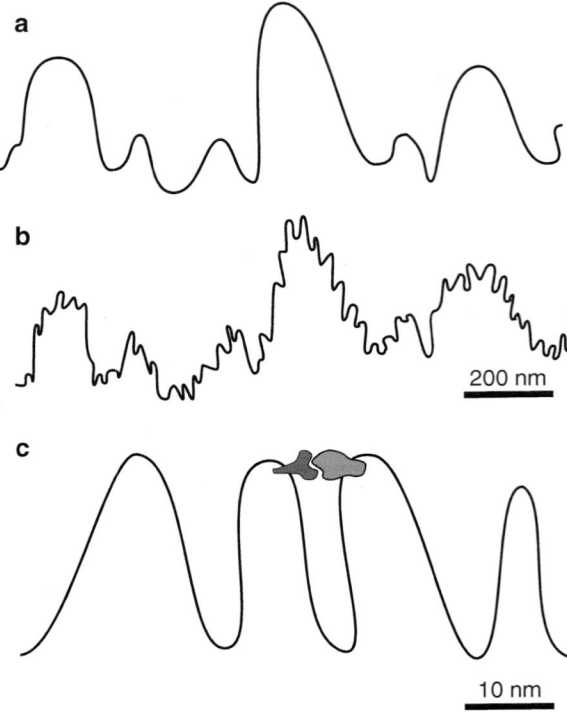

Fig. 2 Plasma membrane ruffling. Surface scanning techniques uncovered large ruffles at the surface of all studied cells. Scheme **A** indicates membrane curvature as observed by AFM or SICM. Fluorescence measured in membrane blebs suggested even more extensive folding of the plasma membrane. Scheme **B** illustrates membrane with low nanoscale ruffles which may support interaction of molecules in *cis* (Scheme **C**). Such intense membrane folding cannot be detected using current surface scanning techniques in living cells

"pickets-and-fences"(as suggested by Aki Kusumi's group; [43]) which were demonstrated to limit the movement of lipids and proteins in cellular membranes [44, 45]. Moreover, the cytoskeleton modulates mechanical forces affecting membranes which can have direct impact on cell function [46, 47]. Mechanical forces of membranes can be studied by atomic force microscopy (AFM; Chapter 13), a method which will certainly provide novel insights on cell membranes. AFM, but also scanning ion conductance microscopy (SICM; [48, 49]) and polarization microscopy [50, 51] can uncover the topology of cell surface which is far from being flat as observed by fluorescence imaging techniques [52]. Such observations have dramatic consequences for the interpretation of single particle tracking and membrane super-resolution image analysis [53]. In addition, if extensive folding and ruffling exists at mesoscale (<100 nm; Fig. 2), this would mean significant curvature of membranes [54], including the plasma membrane. Curvature was previously reported to cause lipid (and protein) sorting and has impact on rigidity of membranes [55, 56], again adding to the heterogeneity of cell membranes. Further studies are required to describe the topology and curvature of plasma and intracellular membranes in higher detail for more general conclusions.

Finally, the exchange of lipid molecules between individual leaflets of a bilayer caused by transbilayer diffusion and transfer was observed in model and cell membranes. This is important due to the fact that a large part of lipid synthesis machinery localizes to the cytosolic leaflet of ER but various lipid species are found in both leaflets throughout the cell membrane compartments [57]. In addition, the distribution of lipids between membrane leaflets varies and in the membranes of the Golgi apparatus and plasma membrane we observe a lipid asymmetry with respect to transbilayer distribution [15]. It is, therefore, important to understand the mechanism(s) responsible for heterogeneous distribution of lipids in cell membranes. In model membranes, which are usually symmetric, transbilayer diffusion was found to be a very slow process in comparison to the pace demonstrated for similar lipids in cell membranes [57]. Asymmetric model membranes are currently available (*see* Chapter 7) but, to my knowledge, transbilayer mobility of lipids has not been investigated yet.

In summary, the organization and local properties of membranes dramatically affects cellular processes and function. The high speed of membrane-associated events frequently makes it difficult to perform all experiments in living cells at the physiological temperature. The current development of advanced imaging techniques and various tools for membrane studies, a good part of which is summarized in this book, will certainly accelerate membrane research. In a few years' time we may know, for example, what is behind the heterogeneities observed in the plasma membrane of all studied cells—is it an artifact, topology, or membrane domains?

Acknowledgements

My work is supported by Czech Science Foundation (P305/11/0459). The author would also like to acknowledge Purkyne Fellowship from The Academy of Sciences of the Czech Republic.

Appendix: Lipid Phases and Acyl Chain Ordering in Model Membranes

Physical properties of membranes are important factors for cell function but due to current technical limitations cannot be studied in living cells with high precision or, in some cases, not at all. Model membranes with well-defined composition (e.g., giant unilamellar vesicles; GUVs) provide an excellent subject for biophysical studies and enable characterization of physical properties with good accuracy and reproducibility. In model membranes, lipids can form at least three phases which differ in freedom with which lipids can move around: the most rigid—solid (So), the most relaxed—liquid disordered (Ld), and somewhat intermediate—liquid ordered (Lo) phase. Phase behavior of lipids depends on the melting temperature (Tm) of lipids in a mixture and the presence of cholesterol. Temperature-gradient experiments with individual lipids and phase diagrams for bi- and ternary lipid mixtures uncovered some interesting complexities in lipid phase behavior and, especially, the importance of cholesterol (and other sterols) for the liquid character of cellular membranes [58, 59]. Cholesterol has rigidifying effect on lipids with low Tm, but it also prevents formation of solid phase by high Tm lipids. Currently, a phase diagram for 4-component lipid mixture was reported [60]. Data generated with even more complex lipid mixtures are difficult to interpret and we may wait another 5–10 years for phase diagram of 5-component mixture. In addition, no classical lipid phases can be expected in non-equilibrium system of living cells. Therefore, a more general property of lipid mixtures, lipid ordering or conformational order of acyl chains, was suggested for studies of complex lipid mixtures and cell membranes in order to better characterize the rigidity of a local membrane environment [61]. Reporters of lipid ordering have been established—fluorescent probes which can sense the presence or absence of water molecules in membranes, solvatochromic membrane dyes. Higher penetration of water molecules to the hydrophobic core indicates reduced rigidity of the membrane. To date, the best membrane order probe is probably Laurdan and its derivative, C-Laurdan (*see* Chapters 3 and 10; [62]). More ordered membranes can be easily distinguished from disordered parts in phase separated monolayers and GUVs [61, 63]. Since solid (or gel)-like phase is not expected to last for long in living cells, there is no problem that these probes cannot distinguish Lo from So phase. In cells, differences in lipid ordering are expected to be less pronounced due to a high complexity of lipid mixture and the presence of proteins. Indeed, data from cell-derived vesicles, GPMVs or cell-derived blebs (Chapter 6), show significant but small difference in values detected by C-Laurdan [64]. Even smaller differences are observed in living cells for intracellular membranes compared to the plasma membrane at physiological temperature [19, 65]. This indicates that organization of membranes in living cells depends more on subtle variation of local membrane composition and properties than on a large segregation of molecules due to their physical properties. It is, therefore, important to focus on a local membrane environments and changes therein. But model membranes represent an excellent tool for studies of molecules in a defined environment and adjusting our methods for future cell experiments. Cell-derived vesicles then function as a bridge between these two worlds, artificial and natural. In addition, in silico simulations provide unprecedented tool to investigate local relations before doing costly experiments (Chapters 20 and 21).

References

1. Goldfine H (1984) Bacterial membranes and lipid packing theory. J Lipid Res 25:1501–1507

2. Cronan JE (2003) Bacterial membrane lipids: where do we stand? Annu Rev Microbiol 57: 203–224

3. Malinsky J, Opekarova M, Grossmann G, Tanner W (2013) Membrane microdomains, rafts, and detergent-resistant membranes in plants and fungi. Annu Rev Plant Biol 64: 501–529

4. Montes LR, Alonso A, Goni FM, Bagatolli LA (2007) Giant unilamellar vesicles electro-formed from native membranes and organic lipid mixtures under physiological conditions. Biophys J 93:3548–3554

5. Ejsing CS, Sampaio JL, Surendranath V, Duchoslav E, Ekroos K, Klemm RW, Simons K, Shevchenko A (2009) Global analysis of the yeast lipidome by quantitative shotgun mass spectrometry. Proc Natl Acad Sci U S A 106: 2136–2141

6. Yeung T, Gilbert GE, Shi J, Silvius J, Kapus A, Grinstein S (2008) Membrane phosphatidyl-serine regulates surface charge and protein localization. Science 319:210–213

7. Yeung T, Terebiznik M, Yu L, Silvius J, Abidi WM, Philips M, Levine T, Kapus A, Grinstein S (2006) Receptor activation alters inner surface potential during phagocytosis. Science 313: 347–351

8. Lomasney JW, Cheng HF, Wang LP, Kuan Y, Liu S, Fesik SW, King K (1996) Phos-phatidylinositol 4,5-bisphosphate binding to the pleckstrin homology domain of phospholi-pase C-delta1 enhances enzyme activity. J Biol Chem 271:25316–25326

9. Klarlund JK, Rameh LE, Cantley LC, Buxton JM, Holik JJ, Sakelis C, Patki V, Corvera S, Czech MP (1998) Regulation of GRP1-catalyzed ADP ribosylation factor guanine nucleotide exchange by phosphatidylinositol 3,4,5-trisphosphate. J Biol Chem 273: 1859–1862

10. Kay JG, Grinstein S (2011) Sensing phosphati-dylserine in cellular membranes. Sensors (Basel) 11:1744–1755

11. Nahorski SR, Young KW, John Challiss RA, Nash MS (2003) Visualizing phosphoinositide signalling in single neurons gets a green light. Trends Neurosci 26:444–452

12. Laketa V, Zarbakhsh S, Morbier E, Subramanian D, Dinkel C, Brumbaugh J, Zimmermann P, Pepperkok R, Schultz C (2009) Membrane-permeant phosphoinositide derivatives as mod-ulators of growth factor signaling and neurite outgrowth. Chem Biol 16:1190–1196

13. Neef AB, Schultz C (2009) Selective fluores-cence labeling of lipids in living cells. Angew Chem Int Ed Engl 48:1498–1500

14. Schultz C (2010) Challenges in studying phos-pholipid signaling. Nat Chem Biol 6:473–475

15. Spector AA, Yorek MA (1985) Membrane lipid composition and cellular function. J Lipid Res 26:1015–1035

16. van Meer G, Voelker DR, Feigenson GW (2008) Membrane lipids: where they are and how they behave. Nat Rev Mol Cell Biol 9: 112–124

17. van Meer G, de Kroon AI (2011) Lipid map of the mammalian cell. J Cell Sci 124:5–8

18. Gaus K, Gratton E, Kable EP, Jones AS, Gelissen I, Kritharides L, Jessup W (2003) Visualizing lipid structure and raft domains in living cells with two-photon microscopy. Proc Natl Acad Sci U S A 100:15554–15559

19. Owen DM, Oddos S, Kumar S, Davis DM, Neil MA, French PM, Dustin ML, Magee AI, Cebecauer M (2010) High plasma membrane lipid order imaged at the immunological syn-apse periphery in live T cells. Mol Membr Biol 27(4–6):178–189

20. Surma MA, Klose C, Simons K (2012) Lipid-dependent protein sorting at the trans-Golgi net-work. Biochim Biophys Acta 1821:1059–1067

21. Gould GW, Lippincott-Schwartz J (2009) New roles for endosomes: from vesicular carriers to multi-purpose platforms. Nat Rev Mol Cell Biol 10:287–292

22. Cebecauer M, Spitaler M, Serge A, Magee AI (2010) Signalling complexes and clusters: functional advantages and methodological hur-dles. J Cell Sci 123:309–320

23. Renner M, Specht CG, Triller A (2008) Molecular dynamics of postsynaptic receptors and scaffold proteins. Curr Opin Neurobiol 18:532–540

24. Varma R, Campi G, Yokosuka T, Saito T, Dustin ML (2006) T cell receptor-proximal signals are sustained in peripheral microclusters and terminated in the central supramolecular activation cluster. Immunity 25:117–127

25. Greenfield D, McEvoy AL, Shroff H, Crooks GE, Wingreen NS, Betzig E, Liphardt J (2009) Self-organization of the escherichia coli chemotaxis network imaged with super-resolution light microscopy. PLoS Biol 7:e1000137

26. Mukherjee S, Zha X, Tabas I, Maxfield FR (1998) Cholesterol distribution in living cells: fluorescence imaging using dehydroergosterol as a fluorescent cholesterol analog. Biophys J 75:1915–1925

27. Nes WD (2011) Biosynthesis of cholesterol and other sterols. Chem Rev 111:6423–6451

28. Olsen BN, Schlesinger PH, Ory DS, Baker NA (2012) Side-chain oxysterols: from cells to membranes to molecules. Biochim Biophys Acta 1818:330–336

29. Maxfield FR, van Meer G (2010) Cholesterol, the central lipid of mammalian cells. Curr Opin Cell Biol 22:422–429

30. Wang XQ, Sun P, Paller AS (2002) Ganglioside induces caveolin-1 redistribution and interaction with the epidermal growth factor receptor. J Biol Chem 277:47028–47034

31. Degroote S, Wolthoorn J, van Meer G (2004) The cell biology of glycosphingolipids. Semin Cell Dev Biol 15:375–387

32. Zha X, Pierini LM, Leopold PL, Skiba PJ, Tabas I, Maxfield FR (1998) Sphingomyelinase treatment induces ATP-independent endocytosis. J Cell Biol 140:39–47

33. Bijl N, van Roomen CP, Triantis V, Sokolovic M, Ottenhoff R, Scheij S, van Eijk M, Boot RG, Aerts JM, Groen AK (2009) Reduction of glycosphingolipid biosynthesis stimulates biliary lipid secretion in mice. Hepatology 49:637–645

34. Mahammad S, Parmryd I (2008) Cholesterol homeostasis in T cells. Methyl-beta-cyclodextrin treatment results in equal loss of cholesterol from Triton X-100 soluble and insoluble fractions. Biochim Biophys Acta 1778:1251–1258

35. Sharpe HJ, Stevens TJ, Munro S (2010) A comprehensive comparison of transmembrane domains reveals organelle-specific properties. Cell 142:158–169

36. Bulbarelli A, Sprocati T, Barberi M, Pedrazzini E, Borgese N (2002) Trafficking of tail-anchored proteins: transport from the endoplasmic reticulum to the plasma membrane and sorting between surface domains in polarised epithelial cells. J Cell Sci 115:1689–1702

37. Haberkant P, Schmitt O, Contreras FX, Thiele C, Hanada K, Sprong H, Reinhard C, Wieland FT, Brugger B (2008) Protein-sphingolipid interactions within cellular membranes. J Lipid Res 49:251–262

38. Contreras FX, Ernst AM, Haberkant P, Bjorkholm P, Lindahl E, Gonen B, Tischer C, Elofsson A, von Heijne G, Thiele C et al (2012) Molecular recognition of a single sphingolipid species by a protein's transmembrane domain. Nature 481:525–529

39. Hennecke S, Cosson P (1993) Role of transmembrane domains in assembly and intracellular transport of the CD8 molecule. J Biol Chem 268:26607–26612

40. Eggeling C, Ringemann C, Medda R, Schwarzmann G, Sandhoff K, Polyakova S, Belov VN, Hein B, von Middendorff C, Schonle A et al (2009) Direct observation of the nanoscale dynamics of membrane lipids in a living cell. Nature 457:1159–1162

41. Ritchie K, Shan XY, Kondo J, Iwasawa K, Fujiwara T, Kusumi A (2005) Detection of non-Brownian diffusion in the cell membrane in single molecule tracking. Biophys J 88:2266–2277

42. Lasserre R, Guo XJ, Conchonaud F, Hamon Y, Hawchar O, Bernard AM, Soudja SM, Lenne PF, Rigneault H, Olive D et al (2008) Raft nanodomains contribute to Akt/PKB plasma membrane recruitment and activation. Nat Chem Biol 4:538–547

43. Ritchie K, Iino R, Fujiwara T, Murase K, Kusumi A (2003) The fence and picket structure of the plasma membrane of live cells as revealed by single molecule techniques. Mol Membr Biol 20:13–18

44. Fujiwara T, Ritchie K, Murakoshi H, Jacobson K, Kusumi A (2002) Phospholipids undergo hop diffusion in compartmentalized cell membrane. J Cell Biol 157:1071–1081

45. Murase K, Fujiwara T, Umemura Y, Suzuki K, Iino R, Yamashita H, Saito M, Murakoshi H, Ritchie K, Kusumi A (2004) Ultrafine membrane compartments for molecular diffusion as revealed by single molecule techniques. Biophys J 86:4075–4093

46. Alsteens D, Garcia MC, Lipke PN, Dufrene YF (2010) Force-induced formation and propagation of adhesion nanodomains in living fungal cells. Proc Natl Acad Sci U S A 107:20744–20749

47. Natkanski E, Lee WY, Mistry B, Casal A, Molloy JE, Tolar P (2013) B cells use

mechanical energy to discriminate antigen affinities. Science 340:1587–1590

48. Korchev YE, Bashford CL, Milovanovic M, Vodyanoy I, Lab MJ (1997) Scanning ion conductance microscopy of living cells. Biophys J 73:653–658

49. Rheinlaender J, Geisse NA, Proksch R, Schaffer TE (2011) Comparison of scanning ion conductance microscopy with atomic force microscopy for cell imaging. Langmuir 27:697–704

50. Benninger RK, Onfelt B, Neil MA, Davis DM, French PM (2005) Fluorescence imaging of two-photon linear dichroism: cholesterol depletion disrupts molecular orientation in cell membranes. Biophys J 88:609–622

51. Lazar J, Bondar A, Timr S, Firestein SJ (2011) Two-photon polarization microscopy reveals protein structure and function. Nat Methods 8:684–690

52. Parmryd I, Onfelt B (2013) Consequences of membrane topography. FEBS J 280:2775–2784

53. Adler J, Shevchuk AI, Novak P, Korchev YE, Parmryd I (2010) Plasma membrane topography and interpretation of single-particle tracks. Nat Methods 7:170–171

54. Drin G, Casella JF, Gautier R, Boehmer T, Schwartz TU, Antonny B (2007) A general amphipathic alpha-helical motif for sensing membrane curvature. Nat Struct Mol Biol 14:138–146

55. Tian A, Baumgart T (2009) Sorting of lipids and proteins in membrane curvature gradients. Biophys J 96:2676–2688

56. Ogunyankin MO, Huber DL, Sasaki DY, Longo ML (2013) Nanoscale patterning of membrane-bound proteins formed through curvature-induced partitioning of phase-specific receptor lipids. Langmuir 29:6109–6115

57. Holthuis JC, Levine TP (2005) Lipid traffic: floppy drives and a superhighway. Nat Rev Mol Cell Biol 6:209–220

58. Veatch SL, Keller SL (2005) Miscibility phase diagrams of giant vesicles containing sphingomyelin. Phys Rev Lett 94:148101

59. Veatch SL, Keller SL (2005) Seeing spots: complex phase behavior in simple membranes. Biochim Biophys Acta 1746:172–185

60. Konyakhina TM, Wu J, Mastroianni JD, Heberle FA, Feigenson GW (2013) Phase diagram of a 4-component lipid mixture: DSPC/DOPC/POPC/chol. Biochim Biophys Acta 1828:2204–2214

61. Parasassi T, Gratton E, Yu WM, Wilson P, Levi M (1997) Two-photon fluorescence microscopy of laurdan generalized polarization domains in model and natural membranes. Biophys J 72:2413–2429

62. Kim HM, Choo HJ, Jung SY, Ko YG, Park WH, Jeon SJ, Kim CH, Joo TH, Cho BR (2007) A two-photon fluorescent probe for lipid raft imaging: C-laurdan. Chembiochem 8:553–559

63. Bagatolli LA, Gratton E (2000) Two photon fluorescence microscopy of coexisting lipid domains in giant unilamellar vesicles of binary phospholipid mixtures. Biophys J 78:290–305

64. Sezgin E, Levental I, Grzybek M, Schwarzmann G, Mueller V, Honigmann A, Belov VN, Eggeling C, Coskun U, Simons K et al (2012) Partitioning, diffusion, and ligand binding of raft lipid analogs in model and cellular plasma membranes. Biochim Biophys Acta 1818:1777–1784

65. Dinic J, Ashrafzadeh P, Parmryd I (2013) Actin filaments attachment at the plasma membrane in live cells cause the formation of ordered lipid domains. Biochim Biophys Acta 1828:1102–1111

Chapter 3

Introduction to Fluorescence Probing of Biological Membranes

Alexander P. Demchenko, Guy Duportail, Sule Oncul, Andrey S. Klymchenko, and Yves Mély

Abstract

Fluorescence is one of the most powerful and commonly used tools in biophysical studies of biomembrane structure and dynamics that can be applied on different levels, from lipid monolayers and bilayers to living cells, tissues, and whole animals. Successful application of this method relies on proper design of fluorescence probes with optimized photophysical properties. These probes are efficient for studying the microscopic analogs of viscosity, polarity, and hydration, as well as the molecular order, environment relaxation, and electrostatic potentials at the sites of their location. Being smaller than the membrane width they can sense the gradients of these parameters across the membrane. We present examples of novel dyes that achieve increased spatial resolution and information content of the probe responses. In this respect, multiparametric environment-sensitive probes feature considerable promise.

Key words Fluorescence, Probes, Dyes, Membrane structure, Membrane dynamics

1 Introduction

In this chapter we review the main trends in the development of fluorescence probes to obtain information about the structure, dynamics, and interactions in biomembranes. The advantages of fluorescence probe methods are manifold, and include extremely high sensitivity extending to the level of single molecules, ultra-high resolution in interaction energies and sub-nanosecond resolution in time. The ability to operate in biological systems of varying complexity can be recognized as a unique feature of fluorescence techniques. The non-destructive nature of these techniques and spatial resolution provided by fluorescence microscopy brought their extended use in the study of live cells.

The versatility of fluorescence techniques is enhanced due to the possibility of recording several parameters that can bring complementary information on the studied systems. Fluorescence intensity can be recorded with resolution in wavelength,

Dylan M. Owen (ed.), *Methods in Membrane Lipids*, Methods in Molecular Biology, vol. 1232,
DOI 10.1007/978-1-4939-1752-5_3, © Springer Science+Business Media New York 2015

polarization, and time. Its recording features a great range of applications for labeling in different assays and formation of contrast in imaging. In sensing techniques, this parameter is used as a quantitative measure of fluorescence quenching, but since its recording cannot be provided on an absolute scale, the introduction of a fluorescent reference may be needed. The resolution in wavelength further extends the possibilities of fluorescence methods. The generation of new bands due to protonation–deprotonation, isomerization, electronic charge transfer, and proton transfer in organic dyes can be used in different fluorescence sensing techniques, and spectral shifts can respond to changes of intermolecular interaction energies. In specially designed dyes such interactions modulate the intramolecular electronic charge and proton transfer of these probes, which enhance the response of these dyes and allow displaying it quantitatively. Fluorescence anisotropy is used to measure the kinetics of fluorophore rotation and thus estimate the viscosity of the microenvironment. Lifetime measurements can provide quantitative information on fluorescence quenching that can be used to study molecular aggregation, lateral diffusion, compartmentalization, or transverse location of labeled species in the membrane.

The combination of these parameters in the fluorescence response and the ability of fluorescent dyes to talk to each other via the electron transfer quenching or via Förster resonance energy transfer (FRET) allows realizing innumerable possibilities in gaining information about the structure and dynamics of biological membranes. These possibilities are further enhanced by the proper design of fluorophores providing the researcher really unlimited choice. The strong standing of organic dyes against new materials such as semiconductor quantum dots is due to these properties but also to their small size and the easy application of well developed methods for their functionalization and introduction into biological systems.

Nevertheless, fluorescence sensing and probing shows some important limitations. In contrast to NMR, Raman and IR spectroscopy, the fluorescence-related information here cannot be obtained and treated on the atomic level. In order to display electronic transitions giving rise to absorption and emission in the visible range, electrons should be delocalized on the molecular scale, and it is not easy (and often impossible) to provide spatial information on the interactions of fluorescent molecules. Therefore, when fluorescent dyes are applied to characterize biological membranes it is often hard to understand at which precise location the probe monitors a given parameter. Smart fluorophore design, location-selective quenching techniques and computer simulations can be used to some extent to provide such information, but high precision still cannot be reached.

Another problem is even trickier to solve. Because of a lack of precise structural information on the atomic level, fluorophore interactions in the studied medium are usually presented in an averaged form, in which the surroundings serve as "dielectric continuum". This allows quantization of the fluorescence response in the tested medium in terms of "micropolarity" or "microviscosity" on the background of polarity or viscosity scales obtained in homogeneous solvent media. One has to be extremely cautious when applying this approach to biological systems, such as biomembranes, featuring high structural anisotropy and heterogeneity on the molecular scale. Being a very difficult task, the switch from a continuous-medium to a molecular-scale interpretation of fluorescence probing data is highly needed. Therefore, present trends involve the design of fluorescence probes of a new generation with smartly located polar and apolar substituents for positioning the fluorophore moiety in desired location and orientation in the membrane. Moreover, rotating, proton-dissociating, and hydrogen bond sensing groups are further inserted in these probes. Though interest in the studies of phospholipid bilayers as biomembrane models tends to decline, the testing of novel smart dyes in these model systems becomes increasingly important. These results can be complemented by molecular dynamics simulations (*see* Chapter 21 in this volume).

During the last few years, as a result of the efforts of many researchers, the information content of the response from fluorophore molecules by the combination of spectral, lifetime, and anisotropy information was found to strongly increase. The response of these parameters can be provided by changes in the ground-state and also in the excited-state and on their combination, by observing the whole-molecular or segmental rotations and the dynamics of fluorescence quenching. The development of instruments capable of recording images in polarized light and fluorescence decays in microscopes, thus combining spatial, anisotropy, and time resolution, have led to a widespread use of FRET imaging in the studies of functional assemblies in cell membranes, so that additional spectroscopic information can be analyzed from both FRET donors and acceptors.

These achievements and tendencies for further development together with selected results of successful applications are the subject of this chapter.

2   The Structures

When fluorescent dyes are used just for labeling, the preferred dyes are those displaying simple decay kinetics, high stability, and invariant fluorescence parameters. The requirements towards reporting (probing and sensing) molecules are quite different.

In this case, the probes should be highly responsive to functionally important changes in the physicochemical parameters of their environment by changes of their fluorescence emission [1].

In the design of fluorescence probes for biomembranes one has to satisfy several requirements. They have to possess high affinity towards the membrane and to its particular sites but not to distort its native structure. The location and orientation should be known and the mechanism of its response understood.

2.1 Dyes Partitioning into Membranes

Biomembrane probes can be classified into several categories. The simplest ones bear no resemblance to lipids, but due to their sufficiently high hydrophobicity can spontaneously partition to the lipid environment. Well-known examples are 1,6-diphenylhexatriene (DPH) or pyrene [2, 3]. These dyes do not exhibit a definite location in the membrane but they are useful since their response (e.g., polarization change in DPH and excimer formation of pyrene) can be correlated with several types of membrane dynamics.

2.2 Dyes as Phospholipid Derivatives

The second category of biomembrane probes is constituted by phospholipid analogs bearing a suitable fluorophore either at their polar head group or their acyl chains. These probes include also sterols with a fluorophore bound to either their side chain or their oxygen atom (e.g., cholesteryl esters) [4, 5]. In order to not perturb significantly the membrane structure, the lipid head group should be labeled by a polar dye, while the acyl chain should be labeled by apolar dyes. However, this limitation was sometimes circumvented. For instance, the 7-nitrobenz-2-oxa-1,3-diazol-4-yl (NBD) group was widely used to label both polar head groups (serine or ethanolamine) and acyl chains [6]. These NBD-labeled lipids were extensively used as fluorescent analogs of native lipids in model and biological membranes to monitor a variety of processes.

The NBD group possesses desirable properties for applications in biomembrane investigation [7]. While its fluorescence is quite weak in water, it becomes bright and exhibits a high degree of environmental sensitivity upon transfer to a hydrophobic medium as the membrane bilayer. Also, its fluorescence lifetime exhibits good sensitivity to environmental polarity [8, 9]. NBD-labeled lipids (Fig. 1) were shown to reasonably mimic endogeneous lipids in a number of studies [10–12] though the polar nature of the NBD moiety can distort the alignment of the acyl chain and alter the solubility of the conjugate, making it prone to spontaneous inter-membrane translocation [13–15]. Moreover, NBD labeling can cause the acyl chain to loop back to the membrane–water interface [16, 17] (Fig. 1).

BODIPY-labeled lipids more closely mirror natural lipids [18–21]. These analogs are conceptually attractive for investigating the hydrophobic region of lipid membranes since the BODIPY

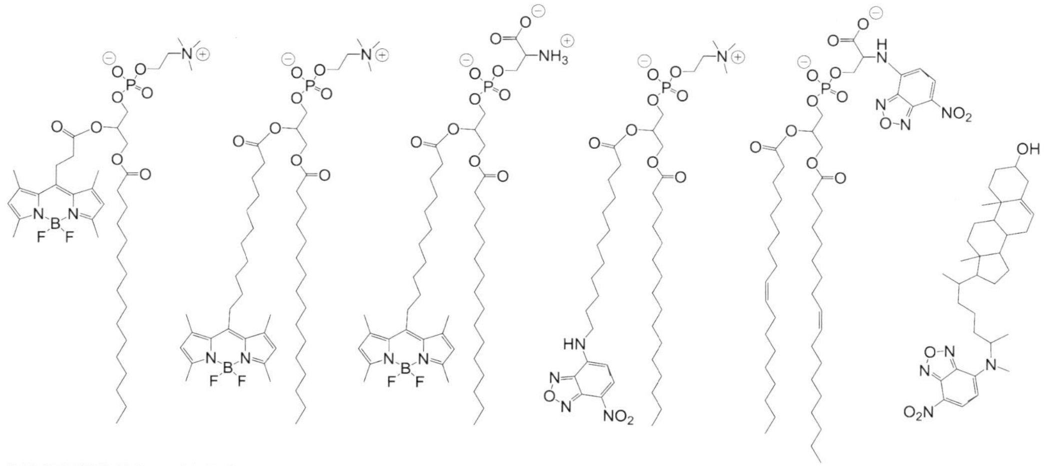

Fig. 1 Examples of BODIPY- and NBD-labeled lipids (11-BODIPY-PS corresponds to TopFluor-PS)

group is less hydrophilic than other probes (like NBD) and have no net charge (Fig. 1). BODIPY probes also exhibit excellent brightness with high extinction coefficients and high quantum yields (≈ 1) and photostability [22]. In order to examine the incorporation of the BODIPY fluorophore within the lipid bilayer, several studies have examined its insertion at incremental depths along the acyl tail [23, 24] even though the BODIPY moiety does not fully penetrate the lipid bilayer as a looping back towards the polar head also occurs, as for NBD [25, 26]. To overcome this drawback, a more hydrophobic derivative of BODIPY (dipyrromethenboron difluoride or TopFluor) has been recently introduced [27]. This new derivative was grafted at the end of a saturated 11-carbon acyl chain of a PS phospholipid, referred to hereafter as TopFluor-PS to examine PS distribution and dynamics inside live cells. For TopFluor-PS, the fluorophore sits 8.2 Å from the center of the bilayer, implying that it remains buried within the hydrophobic location. Thus, TopFluor-PS appears as a more suitable mimic for PS in cells than previously available probes.

2.3 Dye Derivatives of Smart Design

The third category of dye derivatives is constituted by smart specially designed molecules in which, by chemical substitutions in the fluorophore part, an anisotropic incorporation into the membrane with high affinity binding to lipids is provided. This binding determines the depth and location of the fluorophore in the membrane. Probably, the first example of this category of dyes is provided by the family of membrane probes derived from 2-(alkylamino)-6-acyl-naphtalene fluorophore. Their utility for membrane studies is based on the strong solvatochromic shift of

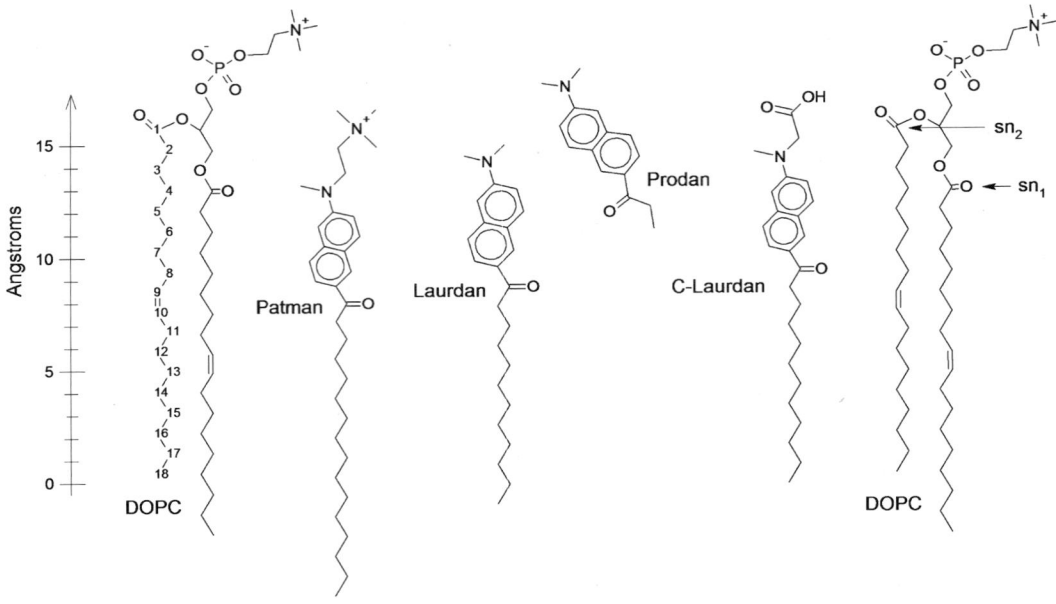

Fig. 2 Chemical structures of fluorescent probes from the Prodan family and their estimated location in lipid membranes (only one leaflet is shown). Adapted from refs. [29] and [38]

their emission spectra (~60 nm) occurring during membrane transitions either from gel or liquid-ordered phases to liquid-disordered phase [28]. Subsequently to the initial development of the smallest probe Prodan, which contains only a 3-carbon acyl chain, Laurdan and Patman were introduced with longer tails to better anchor them to a specific location in the membrane bilayer (Fig. 2) [29]. Patman's tail corresponds to a 16-carbon palmitic acid, but in addition it exhibits a trimethylammonium group at the opposite side of the fluorophore [30]. Such structural modifications confine the probe location within the membrane to a narrower vertical range compared to the shorter Prodan and Laurdan [31]. A specific advantage of Patman is that it can be used for studying plasma membrane of living cells in real time [32–34]. Obviously, the positive charge at the head of the probe restricts its subcellular distribution to the plasma membrane. Moreover, Laurdan, the corresponding 12-carbon dodecanoic acid analog of Patman, was used in order to visualize lipid rafts in living cells by the generalized polarization approach by using two-photon microscopy [35]. However, though Laurdan behaves ideally in giant unilamellar vesicles (GUVs) [36], it appeared not ideal in cells, probably due to its very low solubility in water. Therefore, the C-Laurdan molecule was designed by substituting one Laurdan's methyl group with a carboxymethyl group to improve water solubility while keeping the other functional groups intact [37, 38]. This probe demonstrated the existence of lipid rafts in mammalian cells by two-photon microscopy [37]. Moreover, C-Laurdan presents a greater sensitivity to the membrane polarity and brighter two-photon fluorescence images. It has also been advantageously used to measure membrane order in secretory vesicles [39]. The expected location of this family of probes in a lipid layer is shown in Fig. 2

It thus appears that an efficient and satisfactory probing cannot be obtained without providing the fluorophore the ability of occupying a definite location and orientation in the lipid bilayer. The most established approach to localize the fluorophore at a desirable site in the bilayer is to attach a positive charged group. This allows its location at the bilayer interface due to interaction with the negatively charged phosphate groups. In addition to Patman, another early example of a probe with defined location is TMA-DPH, the cationic derivative of DPH that was introduced three decades ago to selectively measure the fluidity of plasma membranes in live cells [40]. Other remarkable examples are the probes based on the 7-hydroxycoumarin fluorophore coupled to a myristoyl tail by a spacer group of varying length that were designed for monitoring the electrostatic potential or pH profiles at the external interface of biomembranes [41]. A positively charged group was included in these probes between the myristoyl and spacer groups to favor a float-like alignment in the membrane head-group region. This provided a ruler-type positioning of the fluorophores with distances of 2, 6, 10, and 13 Å from the surface. This predicted positioning was further supported by the binding properties of the probes and their fluorescence properties, including the accessibility of different quenchers.

A similar anchoring at the membrane interface was quite naturally obtained with styrylpyridinium voltage-sensitive probes like RH421 [42] and di-8-ANEPPS [43]. Both probes were designed in such a way that they are anchored to the lipid head region through their zwitterionic pyridinium-sulfonate group, while their rod-like fluorophore is oriented across the bilayer by attachment of hydrocarbon tails at the opposite side of the molecule. This alignment of the fluorophore appears to be a prerequisite for the probe to obtain a good sensitivity to the vectorial electric fields originated from the dipole and transmembrane potentials of the bilayer. Therefore, styrylpyridinium dyes, and di-8-ANEPPS in particular, are well established as reference probes for membrane electrostatics.

With these considerations in mind, a similar dedicated tailoring of fluorescent probes was applied to 3-hydroxyflavone (3-HF) dyes. 3-HF derivatives are very attractive fluorescent probes due to their ability to respond to small change in their microenvironment via a dramatic alteration of the relative intensities of their two well-separated emission bands resulting from an excited state intramolecular proton transfer (ESIPT) reaction and belonging to normal excited (N^*) and photo-tautomer (T^*) states [44].

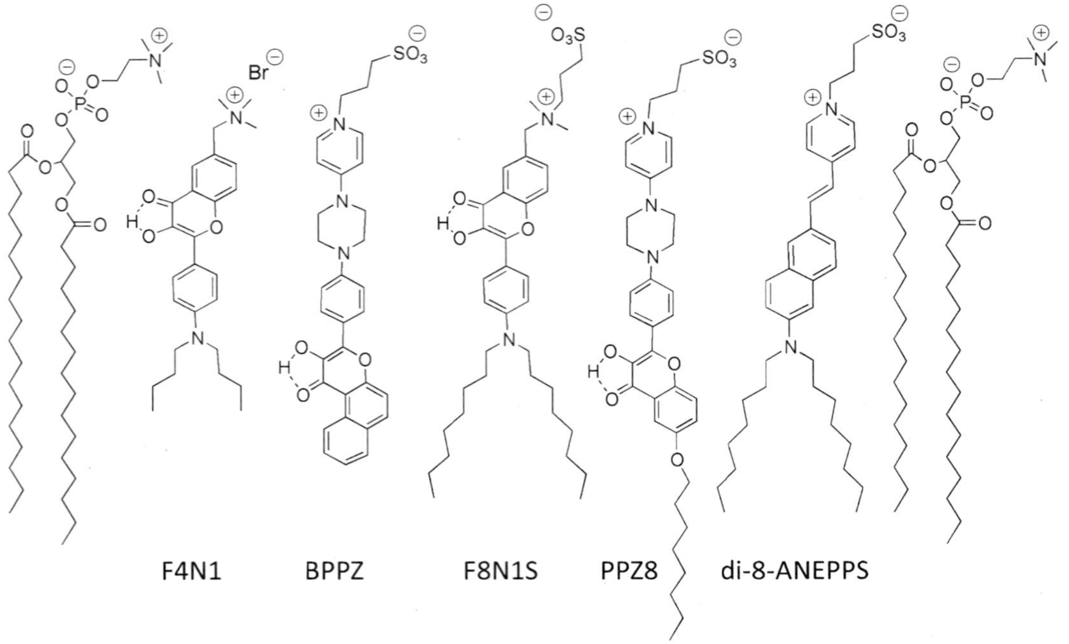

Fig. 3 Di-8-ANEPPS and examples of 3-HF probes featuring precise location and orientation in lipid membranes

Since these two emission bands originate from the two excited states N* and T* with different magnitudes and orientations of their dipole moments, the sensitivity of these states to polarity and electric field perturbation of their microenvironment is different. For example, the ESIPT reaction shows extreme sensitivity to intermolecular H-bond perturbations [45] that should have some directionality in space. Moreover, because of the asymmetric nature of the fluorophore and unidirectional nature of the ESIPT reaction, the fluorescence spectra of 3-HF probes are sensitive to polarity [46] and anisotropic properties of the environment and particularly to electric fields [47], with a particular reference in the study of biomembranes.

Tailoring processes were taken up with 3-HF dyes with strong concern toward maintaining the propensity of 3-HFs to undergo the ESIPT reaction [48]. In this pioneering work, not only the 3-HF moiety located at different depths but also two reverse orientations of the fluorophore were obtained, as shown in Fig. 3. In all cases, these ratiometric fluorescent probes provide multiparametric information on the properties of lipid membranes at different depths [1]. The pair of probes with a reverse orientation of the same chromophore (F4N1 and BPPZ) was used successfully to measure the dipole potential in phospholipid membranes (vide infra).

3 Location

The gradients of all physical parameters across the bilayer are very steep even in comparison with the fluorophore sizes. Therefore, the information on the order, dynamics, polarity, and hydration depend strongly on the fluorophore location. As a consequence, the knowledge of the precise probe location becomes crucial for evaluating this information. The most trivial approach using the simplest probes is to apply polar fluorescent probes to study the polar membrane interface, and apolar probes to study its hydrophobic interior. However, this approach often fails because the probe molecule may be distributed between several locations and orientations in the bilayer. Moreover, ground-state hydration of a probe capable of forming H-bonds is an additional feature that governs this distribution. Steady-state and time-resolved fluorescence data on Nile Red [49], coumarins [50], and Prodan [29, 50, 51] demonstrate that these probes can be located simultaneously in both polar and apolar regions of the membrane. Similarly, two ground-state forms of the uncharged and unsubstituted 3-HC dye 4-(dimethylamino)-3-hydroxyflavone (probe F)—one corresponding to the H-bonded form with a surface location, and the other to the non-H-bonded form located deeper—are simultaneously present in lipid bilayers [52]. Both forms show different emission profiles, reporting on different environments of the two locations.

Intramembrane location of a fluorophore needs to be determined with the highest possible precision. The most precise location and orientation of a specific fluorophore at the desired site can be achieved only from a special probe design based on uncharged and relatively nonpolar dye molecules that allow chemical modifications at their two opposite sides. The dyes of the 3-HF family satisfy these requirements [48, 53]. Addition of hydrophobic tails determines the membrane affinity and, if necessary, vertical orientation in the membrane, whereas the attached charged and polar groups fix the fluorophore location relative to the bilayer surface.

Structural changes in membranes may cause relocation and reorientation of the probe. Recent data on giant unilamellar vesicles (GUVs) directly show that Prodan and Laurdan do not present any preferential orientation in fluid phase membranes, whereas in liquid-ordered and gel phase membranes they present a constrained vertical orientation (non-H-bonded form) [54]. Moreover, hydrostatic pressure [51] and cholesterol [55] can cause Prodan relocation. Recently, an elegant combination of high-level quantum mechanical electronic structure calculations with a molecular field theory has been performed to get the positional–orientational–conformational distribution of Prodan and Laurdan probes in their ground and excited states inside of the lipid bilayer, taking into account at both levels the nonuniformity and anisotropy of the environment [56].

Thus, the features of the fluorescence spectra of Prodan and Laurdan were interpreted in relation to the position and orientation of their chromophores in the bilayer. It appears that the environment polarity is not sufficient to explain the large red shifts experimentally observed and that specific effects due to hydrogen bonding must be considered. The orientation of the probe was confirmed to be an important feature in determining the accessibility to water of the H-bond-acceptor group. Also, molecular dynamics investigations of Prodan lead to the conclusion that the orientation of this molecule inside a lipid bilayer is well defined, but its position appears very sensitive to the effect of the medium polarization [57].

Fluorophore relocation on modification of membrane properties was also proposed for NBD derivatives [58] and 3HC probe F [52]. Relocation of a probe along the polarity gradient may also occur during the time of emission, when the excited state is much more polar than the ground state [59]. The polarity of the fluorophore itself is also an important factor in its ground-state location. For example, as previously noted, covalent attachment of the relatively polar NBD moiety to flexible acyl chains does not allow its localization close to the bilayer center, because of its looping back to the interface [60, 61]. This and other unsuccessful attempts suggest that instead of modifying the lipids, artificial constructions that can fix the fluorophore at a desired orientation and depth should be preferable.

4   Perturbation of Structure

The fluorescence membrane probes are foreign molecules inserted in a host lipid matrix. Thus, perturbation of membrane structure, dynamics of its components and the thermotropic behavior of membrane domains cannot be excluded, even for low probe concentrations. The effect of six commonly used fluorescent membrane-anchored probes (Laurdan, TR-DPPE, Rh-DPPE, DiIC18, Bodipy-PC, and NBD-PC) on membrane mechanical properties has recently been studied through bending elasticity measurements [62]. Bending elasticity reflects any change in the organization of the main lipid component induced by the addition of an externally added component such as a fluorescent probe. The authors found that the effect of these probes on membrane mechanical properties is negligible.

Laurdan was confirmed as an environmentally sensitive probe because it can sense changes in the microenvironment (polarity, viscosity, or temperature) when inserted into the membrane. On the contrary, other data obtained by scanning force microscopy with NBD-PC [63] suggested that caution should be taken when studying the structure of model lipid membranes in the presence of extrinsic probes, especially at the nanometer scale. But, as the experiments were performed on gel phase DPPC membranes, it is not sure that this conclusion can be extended to biomembranes.

5 Information Extracted from Biomembrane Probes

High structural anisotropy combined with steep (on molecular level) gradients of hydration, polarity, and electrostatic potentials makes the analysis of molecular probing in biomembranes very challenging. In biological membranes (and in phospholipid bilayers as their models), the gradients of polarity extend on the same or even shorter length scales as the fluorophore sizes [64]. Correspondingly, the microenvironment of a solute varies from being highly hydrophobic inside the alkyl chains of the bilayer to being very polar at the interface with aqueous medium. Phospholipid segmental mobility in the membrane exhibits strong gradients increasing toward the chain ends, so that the diffusional mobility of incorporated small molecules depends on their location. Strong structural anisotropy of densely packed lipids results in anisotropic orientation of incorporated dyes. Depth-dependent and anisotropic binding of hydration water contributes additionally to this complexity.

Such complexity can lead to misinterpretation of many data in view of the fact that the fluorescence probes may exhibit heterogeneity of location sites, distribution between H-bonded and non-H-bonded forms and, even more, relocation during the excited-state lifetime. Moreover, the ordered charges and dipoles in bilayer structure together with adsorbed ions generate strong local electric fields leading to electrochromic effects in spectra. Therefore, deriving the "polarity" and "fluidity" values from fluorescence dyes having unclear location and preferential orientation in the membranes is not informative.

5.1 Hydration and Relaxation Dynamics

Hydration, which describes the highly depth-dependent distribution and dynamics of water molecules in the lipid membranes [65], is connected with membrane electrostatics since the oriented water dipoles solvating both phosphate and carbonyl groups contribute to the electric fields in the bilayer [66]. Because water is the most important H-bonding contributor, membrane hydration is generally evaluated with fluorescent probes presenting wavelength-shifts [67] or quenching [68] resulting from dipolar and H-bonding interactions with water. These fluorescent probes can detect the rate and extent of the structural relaxations around their excited-sate dipole, which describe mainly water dynamics [29].

Two aspects of membrane hydration in lipid bilayers should be considered, namely the distribution of water molecules across the bilayer and their dynamics. Both of these factors induce spectral shifts as well as time-resolved effects [29, 69]. Indeed, the relaxation dynamics of fluorescent probes in membranes, which includes the relaxation of water, is slower by several orders of magnitude than in neat solvent and is comparable to fluorescence lifetimes, thus leading to a red-edge effect on fluorescence spectra [70, 71].

Moreover, slower dynamics are observed at the level of the glycerol backbone than at the polar interface [59, 72]. Therefore, it is required to study relaxation dynamics in membranes taking into account the particular membrane site. For instance, a relatively polar chromophore such as NBD attached to the phospholipid polar head group will be used for studies at the level of glycerol and head groups [58, 71]. Laurdan [54] and Patman [50] allow the researcher to probe hydration and relaxation dynamics at the glycerol region only.

Hydration in the glycerol region is related to many membrane parameters. It increases with membrane curvature as revealed by relaxation dynamics of Prodan and Patman [50] and spectral response of F2N8, a 3-HF probe anchored at the glycerol region [73]. Mechanical stress [74] as well as external pressure [75] also affects membrane hydration. Moreover, a strong increase in hydration occurs when the temperature is increased within a fluid phase [29, 58, 73]. Cholesterol and other sterols, which decrease the water permeability of membranes [76], expel water from the fluid phase [67, 73], thus decreasing the water relaxation dynamics in this phase.

Due to the anisotropic location of the lipid polar groups, the relaxation of the anisotropically distributed water molecules in the membranes that determine the spectral shifts are more complicated than in isotropic media. Molecular ordering and generation of molecular-scale electric fields determine the pattern of dielectric relaxations that is usually derived from the time-dependent shifts of the spectra. Their signatures change dramatically across the bilayer influencing both the solvation energies and relaxation rates.

5.2 Hydration and Lipid-Order in Bilayers: Lipid Domains

Monitoring lipid order in membranes presents a particular importance due to the key role of lipid domains in biomembrane function [77]. As previously noted, a fluid phase, characterized by a low lipid order and high mobility of lipids, exhibits fast relaxation dynamics and a high level of hydration. In contrast, a gel phase, characterized by high lipid order and slow lipid mobility, exhibits much lower hydration. Viscosity-sensitive probes, such as DPH derivatives [2, 78], clearly show the differences between these two phases. Moreover, the hydration evaluated from TMA-DPH lifetimes also correlates well with phase states, the phase with a higher lipid order presenting a lower level of hydration [68]. A particularly important emission red-shift is observed with Prodan and Laurdan during the gel to fluid phases transition [28], confirming the strong increase in the hydration of the probes environment, probably together with some relocation of the probes. Indeed, while in a fluid phase, these probes show no preferential orientation in the bilayer, they adopt a vertical orientation with a concomitant poor accessibility to water in a gel phase [79]. Therefore, when analyzing the response of environment sensitive probes to phase transition, it is necessary to take into account the influence of lipid packing (order) on the probe location/orientation in the bilayer.

In many aspects (but not all), the same considerations are valid for the comparison between the so-called liquid-disordered (Ld) and liquid-ordered (Lo) phases, which are more relevant when biomembranes are considered. An Ld phase can be assimilated to a fluid phase composed of unsaturated phospholipids and cholesterol. A Lo phase consists of saturated lipids, either phosphatidylcholine but more often sphingomyelin (SM), and cholesterol. These Lo domains or "rafts" float in a pool of Ld domains [80, 81]. In cells, these rafts are believed to play a crucial role in the regulation of a number of functions involving membrane proteins and membrane transport [77, 82]. Environment-sensitive probes able to distinguish fluid from gel phases by changes in their emission color can of course do the same to distinguish Ld from Lo phases. Laurdan and Prodan, which exhibit clear-cut differences in their spectra between these two phases, are a perfect example [54, 81]. In the Lo phase, these probes reveal a less polar environment being close to that of a gel phase, with a blue-shifted emission compared to the Ld phase.

Later on, this approach was repeated with the 4-styrylpyridinium probe di-4-ANEPPDHQ that shows red-shifted absorption and emission spectra and a deeper insertion into the membrane [83–85]. Alternatively, since molecular rotors bound to the Lo phase show significant enhancement of their fluorescence, these phases were visualized using a molecular rotor based on a Prodan analog [86]. Very recently, Laurdan fluorescence lifetimes have been shown to be able to discriminate cholesterol content from changes in fluidity in living cells membranes [87].

Due to their high sensitivity to membrane properties, such as polarity and hydration, 3-HF probes, and particularly the most characterized F2N8 and F2N12S probes, are of particular interest to study lipid domains. In model membrane vesicles, mimicking different membrane phases, it was found that the hydration parameter increases in the following order: $Lo \ll gel \approx Ld$ (with cholesterol) < fluid (without cholesterol). Thus, the Lo phase is characterized by an exceptionally low hydration, a feature that could be used for detection and imaging of this phase by this probe. The data with F2N8 were compared with the changes of membrane fluidity as monitored by TMA-DPH fluorescence anisotropy [88]. Remarkably, the membrane fluidity shows a somewhat different trend, namely, $Lo \approx gel < Ld < fluid$. Thus, gel and Lo phases exhibit similar fluidity, while Lo phase is significantly less hydrated than gel phase. It was hypothesized that cholesterol, due to its specific H-bonding interactions with lipids and its ability to fill the voids in lipid bilayers, efficiently expels water molecules from the highly ordered gel phase to form the Lo phase. This strong difference in hydration leads to a clear distinction between these two phases.

The strong spectroscopic response of 3-HF probes to Lo phase can be applied to visualize lipid domains. The best model for these domains is constituted by giant unilamellar vesicles (GUVs), which due to their size (5–100 μm) can be directly used in fluorescence microscopy. Both F2N8 and F2N12S probes were used to label GUVs displaying Ld, Lo or both phases. F2N12S differs from F2N8 by the presence of a zwitterionic N-(3-sulfopropyl)-1-dodecylammonium instead of a cationic N-1-octylammonium anchor group. F2N8 was found to preferentially bind the Ld phase, so that it is nearly impossible to label the Lo phase when it coexists with the Ld phase. This is not the case with F2N12S which binds both phases, probably due to the presence of the longer dodecanoyl tail. The influence of long alkyl chains in favoring the partitioning in ordered phases (gel or Lo) is well known and has been evidenced for probes such as LcTMA-DPH [78, 89] or diI-C20 [90].

Steady-state fluorescence studies further showed that similarly to F2N8, the dual emission of F2N12S is strongly modified depending of the lipid phase and can be correlated with changes in hydration [91]. Using two-photon microscopy, F2N12S probe allows to individually visualize both Lo and Ld phases on GUVs, through the ratio of its two emission bands [92] as well as through the fluorescence lifetime of the T* band. Moreover, as shown in Fig. 4, by using linearly polarized excitation laser light, a strong photoselection was observed for F2N12S in the Lo phase, indicating that its chromophore is nearly parallel to the lipid chains of the bilayer. Thus, the Lo phase imposes a vertical orientation to the probe, a situation already observed with Laurdan [54], di-4-ANEPPDHQ [84, 85], and LcTMA-DPH [89]. In contrast, the marginal photoselection observed with Ld phase indicated in this case no preferential orientation of the probe. Interestingly, additional probes derived from the parent F2N12S have been recently designed to orient the dye vertically in the membrane [93] and thus obtain more sensitive response on apoptosis. In line with expectations, these new probes were found to be more sensitive than F2N12S to the lipid phase in lipid vesicles and to loss of lipid order in cell plasma membranes after cholesterol extraction or apoptosis induction.

Another interesting feature of biomembranes is the asymmetric distribution of lipids between their two leaflets. Indeed, while the outer leaflet contains a large amount of SM (up to 40 %), its fraction in the inner leaflet is marginal [94]. Therefore, a correct evaluation of Lo phases requires probes presenting with high specificity to this phase together with a high selectivity to the outer membrane leaflet and negligible flip-flop between the two leaflets. The challenge to develop an environmentally sensitive fluorescent probe staining quite exclusively the outer leaflet of cell membranes has been only recently solved. It was shown that the 3-HF probe F2N12S exhibits a selective localization at the plasma membrane

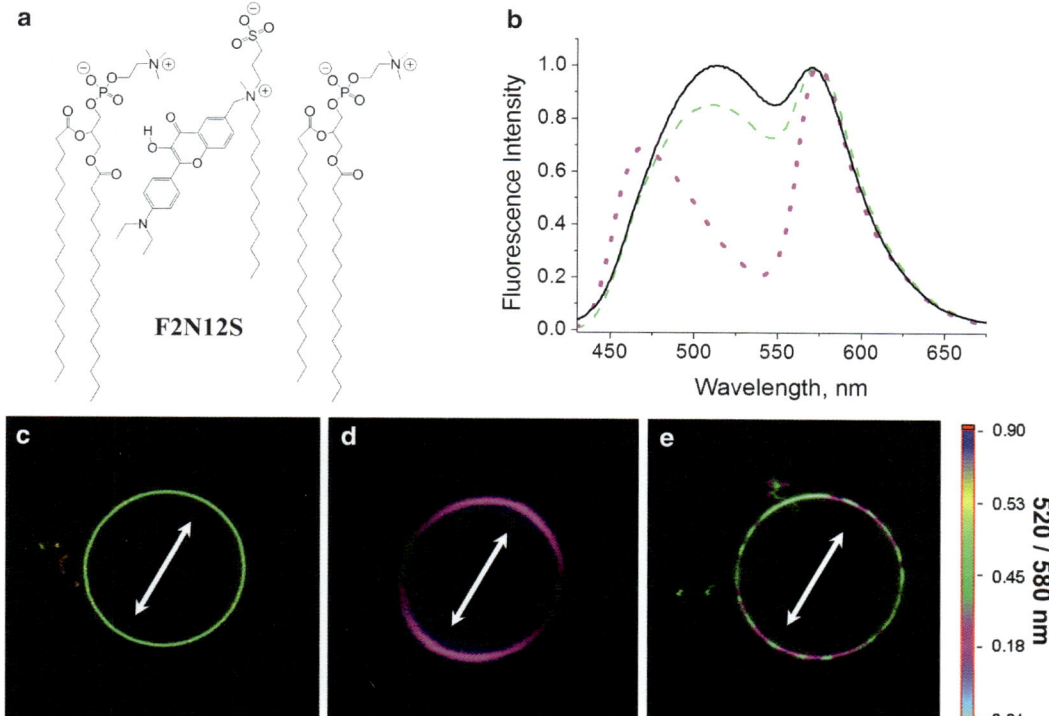

Fig. 4 Probe F2N12S (**a**) and its use for probing liquid ordered domains (**b–e**): (**b**) Fluorescence spectra of F2N12S in lipid vesicles composed of DOPC (*black solid curve*), DOPC + 35 % Cholesterol (*green dashed curve*), and SM + 35 % Cholesterol (*magenta dotted curve*). Fluorescence ratiometric images (520/580 nm) of GUVs composed of DOPC (**c**) and SM + 35 % Cholesterol (**d**) and the ternary mixture (DOPC/SM/Cholesterol in the molar ratios 1/1/0.7) (**e**). Two-photon excitation (830 nm) was used. *Arrows* indicate the orientation of light polarization. Sizes of the images were 50 × 50 μm (**b**) 15 × 15 μm (**c**), and 45 × 45 μm (**d**). Data adapted from ref. [91]

due to its zwitterionic head and long hydrophobic chain [95]. Though F2N12S was hypothesized to bind to the outer leaflet without fast flip-flop, no direct evidence supported this assumption. Indirect evidence was based on the strong response of the probe to apoptosis, the phenomenon that abolishes the asymmetry of cell membrane [95].

In fact, such asymmetric incorporation was proven for the NR12S probe that has recently been obtained by modifying the solvatochromic fluorescent dye Nile Red with the same amphiphilic anchor group as for F2N12S [96]. To evaluate its flip-flop rate, a methodology of reversible switching of its fluorescence at one leaflet using sodium dithionite was developed [96]. This method has shown that NR12S binds exclusively to the outer membrane leaflet of both lipid vesicles and plasma membranes of living cells, with negligible flip-flop within the time-scale of hours. Moreover, the emission of NR12S in model vesicles exhibits a significant blue

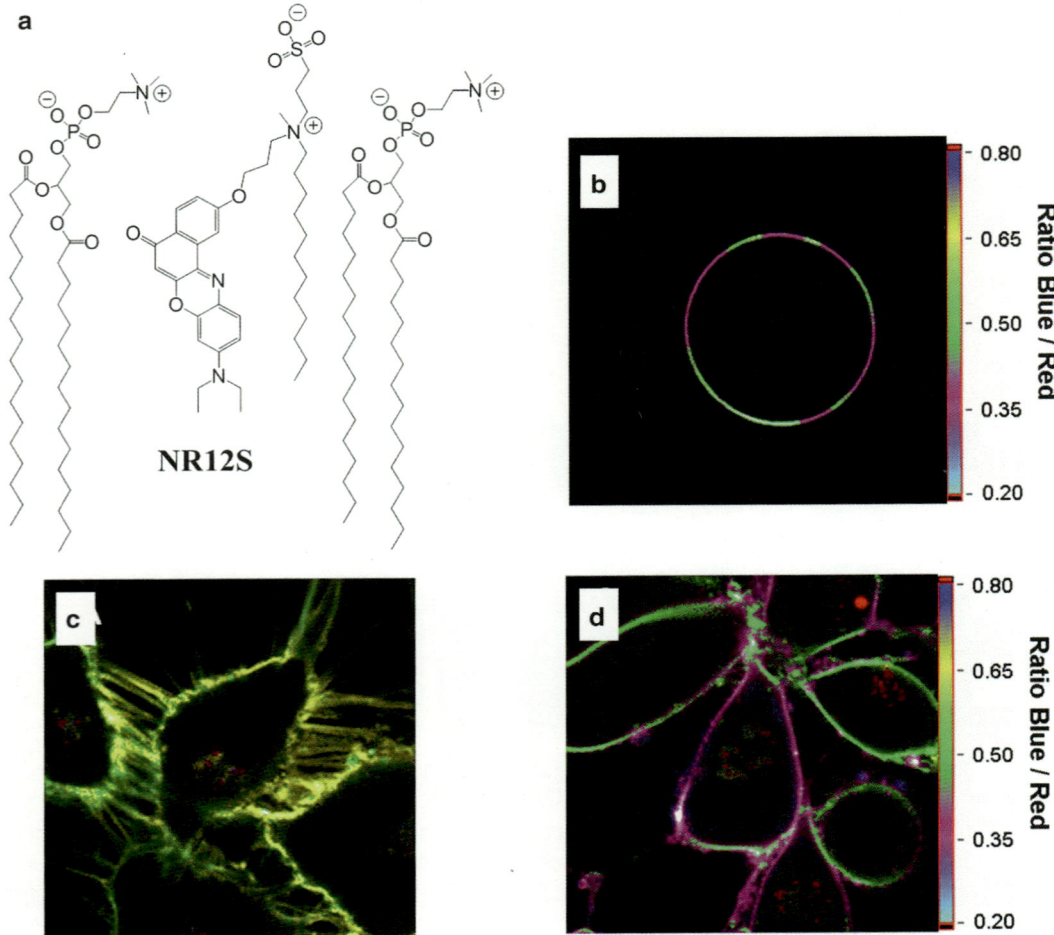

Fig. 5 Chemical structure of NR12S probe (**a**) and its application for fluorescence ratiometric imaging of giant lipid vesicles (**b**) and living astrocytoma cells. Giant vesicles were composed of Lo and Ld domains as visualized by NR12S. Intact cells (**c**) and cells after cholesterol extraction with methyl-β-CD (**d**) show different color in the ratiometric images. Sizes of the images were 50 × 50 μm. Data adapted from ref. [96]

shift in Lo phases as compared to Ld phases. Accordingly, these two phases could be clearly distinguished in NR12S-labeled GUVs by fluorescence microscopy imaging of intensity ratio between the blue and red parts of its emission spectrum. In living cells, NR12S binds predominantly, if not exclusively, to their plasma membranes, showing an emission spectrum intermediate between those for Lo and Ld phases of model membranes. Importantly, NR12S emission characteristics correlate well with the cholesterol content in cell membranes, which allows monitoring the cholesterol depletion process using methyl-β-cyclodextrin by fluorescence spectroscopy and microscopy, as shown in Fig. 5. Cholesterol extraction, oxidation, and SM hydrolysis were found to red-shift the emission spectrum, indicating a decrease in the lipid order at the outer plasma membrane leaflet. Remarkably, apoptosis produced very similar effects [97], suggesting that apoptosis also decreases the lipid order at this leaflet. This new concept in apoptosis sensing was further validated by fluorescence microscopy and flow cytometry [95, 97].

5.3 Electrostatic Membrane Potentials

The global electrostatic potential of a biomembrane, noted Ψ, does not change linearly across a lipid or cell membrane. In fact, Ψ follows a complex profile with the contribution of three different electrostatic potentials defined as the "trinity" of membrane potentials [98]: (1) $\Delta\Psi$, the transmembrane potential referred to the difference in ionic concentrations between the two aqueous membrane phases separated by the membrane; (2) Ψ_S, the surface potential which arises from charged residues at the membrane interfaces; and (3) Ψ_D, the dipole potential which results from the alignment of phospholipid dipolar residues and from associated water molecules within the membrane.

Among these membrane potentials, the dipole potential Ψ_D is the more difficult to control and measure. Indeed, it can be only theoretically calculated or inferred from indirect experimental methods. One of these indirect methods is the use of fluorescent voltage-sensitive probes with a location of their fluorophore in the region where Ψ_D is effective, namely between the apolar bilayer center and the membrane surface. Styrylpyridinium probes such as RH 421 or di-8-ANEPPS were the first popular dyes for the determination of Ψ_D [43, 99]. For example, di-8-ANEPPS allows a ratiometric recording of Ψ_D through the shift of its excitation spectra, provided that the sample could be excited with two different wavelengths. Despite consequent improvement brought last years to these styrylpyridinium probes [84, 100–102], a critical analysis revealed that further progresses require the development of probes that (a) do not possess a large ground-state dipole, (b) allow quantitative ratiometric recording in fluorescence emission and (c) allow desired location and orientation of the fluorophore. Indeed, it is much more convenient to dispose of dyes adapted for multicolor imaging microscopy, with a ratiometric response in emission rather than in excitation. Moreover, ratiometric measurements in emission eliminate any distortion of data caused by photobleaching, variation of probe loading and instrumental factors (such as light source stability). 3-HC dyes, and particularly 4′-(dialkylamino)-3-hydroxyflavones fit well with all these requirements.

By using a similar design as for the above-mentioned styrylpyridinium dyes, two probes (F4N1 and BPPZ, vide supra) were developed in which the fluorescent moieties and consequently the dipole moments are oriented vertically in lipid bilayers but inversely with respect of each other. In both cases, the probes strongly responded to variation of Ψ_D generated by Ψ_D-modifiers

6-ketocholestanol and phloretin [53]. This response was observed as a change in the intensity ratio of their two emission bands in an opposite way for the two probes, in accordance with their inverse orientation. Moreover, a multiparametric analysis based on band deconvolution showed that these changes are independent from the hydration parameter [103].

However, these probes were not applicable for cellular studies due to their fast internalization and/or poor staining of plasma membranes. Therefore, a second generation of probes was introduced, F8N1S and PPZ8 (Fig. 3) that are analogs of the previous pair, bearing zwitterionic groups and long hydrophobic tails at the opposite sides of the fluorophore. This second generation of probes demonstrates a fair selectivity for cell plasma membranes and exhibits strong ratiometric variations of their two emission bands versus Ψ_D in living cells [104].

Transmembrane potential ($\Delta\Psi$) is the most dynamic electrostatic component of cell plasma membranes and plays essential roles in a variety of cellular functions related to bioenergetics, ion transport, motility and cell communication. Intracellular microelectrode (or patch-clamp) recording and optical microscopy are the two major methods to monitor $\Delta\Psi$ in cells. Microelectrodes allow the recording of $\Delta\Psi$ with extremely high precision and temporal resolution. However, the observation of its spatial distribution within a single cell or cell assemblies with microelectrodes is not easy and requires multi-electrode techniques. This crucial limitation can be partly overcome using optical methods and fluorescent probes allowing the recording of $\Delta\Psi$ in single cells and multicellular assemblies with both high temporal and spatial resolution.

Better voltage-sensitive dyes of the RH series were developed for this aim [105]. These dyes bind to the external surface of cell membranes and act as molecular transducers transforming changes in $\Delta\Psi$ into changes in emitted fluorescence that occur on the microsecond time scale. These improvements facilitated in vivo imaging and lead to the possibility of Voltage Sensitive Dye Imaging [105]. However, some drawbacks are remaining such as responses in emission intensity, which are difficult to calibrate, being dependent on the probe concentration.

Another methodology of $\Delta\Psi$ measurement is based on fluorescence resonance energy transfer (FRET), using the hydrophobic oxonol dye as a FRET partner [106–108]. This dye undergoes a $\Delta\Psi$ driven translocation between the two leaflets of the cell membrane, allowing both fast and strong responses to $\Delta\Psi$ variations. This method has been used in a number of cellular applications [106–108] but requires double labeling of the cell plasma membrane with precise relative concentrations of the FRET components.

Styryl dyes introduced about 20 years ago [109, 110] have a fast and ratiometric response (recorded from the excitation spectrum) to $\Delta\Psi$ variations, but suffer from their low sensitivity

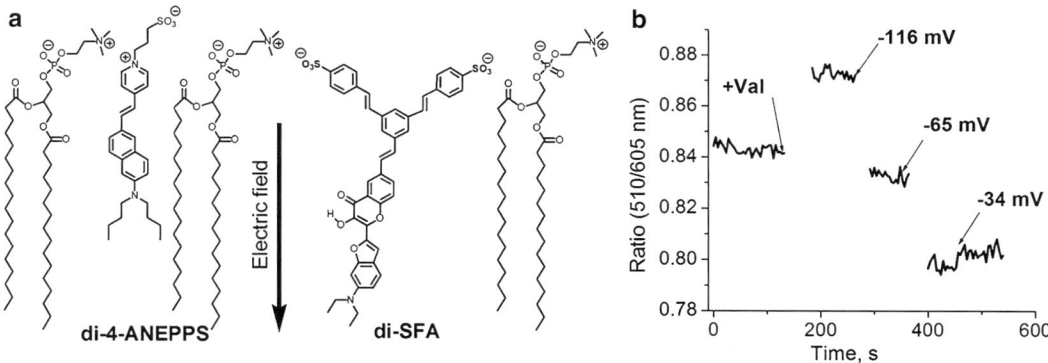

Fig. 6 New transmembrane potential probe di-SFA and the commercial probe di-4-ANEPPS. (**a**) Structure and location of the probes. (**b**) Ratiometric response of probe di-SFA to variation of transmembrane potential in suspensions of CEM cells. Data adapted from ref. [115]

($\approx 7\,\%/100$ mV). Only recently, their sensitivity has been improved twofold by increasing the fluorophore length and rigidity, reaching the physical limits of the electrochromic response [111, 112]. A last problem, namely the very low solubility in water of the zwitterionic dye ANNINE-6 containing six annelated benzene rings [113], thus rendering cellular staining impossible, was overcome by the synthesis of ANNINE-6+ , a double positively charged analog [114]. However, for fluorescence ratiometric imaging as previously underlined, probes with a ratiometric response in emission are more convenient. Again, a solution to this problem was found with the 3-HF family. Indeed, a 3-HF derivative (probe di-SFA) was designed, so that the fluorophore was coupled with two rigid arms bearing terminal anionic groups to impose a vertical orientation and a deep insertion of the fluorophore within the hydrophobic region in the middle of the bilayer [115]. Di-SFA shows a fluorescence spectrum corresponding to a very low polar environment in model and cellular membranes in line with a deep insertion. Cellular studies showed that the response of the probe to $\Delta\Psi$ is faster than 1 ms, with a sensitivity around $12\,\%/100$ mV (Fig. 6). Though this sensitivity is somewhat less than for the last representatives of the ANNINE family [113, 114], the key advantage of di-SFA resides in its ratiometric response in emission, which enables easier application in fluorescence microscopy by using a single excitation source and a two-color fluorescence detection set-up.

6 Conclusion

Investigation of the structure and the properties of biological membranes is a major challenge for many researchers. This is not only because the membranes are formed of many different components but also and mainly because of the strong structural anisotropy of the membrane components integrated into the bilayers. In addition to sub-nanoscale gradients of polarity and mobility, these systems demonstrate anisotropic molecular fields acting on fluorescence reporter dyes and influencing their location and orientation in the membrane and in this way, the fluorescence response. In addition to sub-nanoscale gradients of polarity and mobility, a strong structural anisotropy in the arrangement of charged and dipolar molecules generate strong electric fields sensed by the probing dyes.

Therefore, environment-sensitive fluorescent dyes constitute major tools for getting information about the structure and dynamics in biomembranes. Major trends in their future development will be focused on improving the probes' selectivity for a particular membrane property, increasing the information content of the probe response, and addressing the biophysical properties at a particular membrane location. In this respect, the development of multiparametric environment-sensitive probes capable of distinguishing between different physicochemical properties through additional spectroscopic parameters constitutes a substantial step forward. With the aid of these probes, essential cellular processes, such as apoptosis, or essential membrane features, such as lipid rafts, can be followed at the level of plasma membranes.

References

1. Demchenko AP, Mely Y, Duportail G, Klymchenko AS (2009) Monitoring biophysical properties of lipid membranes by environment-sensitive fluorescent probes. Biophys J 96:3461–3470
2. Lentz BR (1989) Membrane fluidity as detected by diphenylhexatriene probes. Chem Phys Lipids 50:171–190
3. Duportail G, Lianos P (1996) Probing of vesicles using pyrene and pyrene derivatives. In: Rosoff M (ed) Vesicles. Marcel Dekker, New York, pp 295–372
4. Gimpl G, Gehrig-Burger K (2007) Cholesterol reporter molecules. Biosci Rep 27:335–358
5. Ariola FS, Li ZG, Cornejo C, Bittman R, Heikal AA (2009) Membrane fluidity and lipid order in ternary giant unilamellar vesicles using a new bodipy-cholesterol derivative. Biophys J 96:2696–2708
6. Chattopadhyay A (1990) Chemistry and biology of N-(7-nitrobenz-2-oxa-1,3-diazol-4-yl)-labeled lipids: fluorescent probes of biological and model membranes. Chem Phys Lipids 53:1–15
7. Chattopadhyay A, Mukherjee S, Raghuraman H (2002) Reverse micellar organization and dynamics: a wavelength-selective fluorescence approach. J Phys Chem B 106:13002–13009
8. Rawat SS, Chattopadhyay A (1999) Structural transition in the micellar assembly: a fluorescence study. J Fluoresc 9:233–244
9. Lin S, Struve WS (1991) Time-resolved fluorescence of nitrobenzoxadiazole aminohexanoic acid—effect of intermolecular hydrogen-bonding on nonradiative decay. Photochem Photobiol 54:361–365
10. Sparrow CP, Patel S, Baffic J, Chao YS, Hernandez M, Lam MH, Montenegro J, Wright SD, Detmers PA (1999) A fluorescent cholesterol analog traces cholesterol absorption in hamsters and is esterified in vivo and in vitro. J Lipid Res 40:1747–1757
11. Pagano RE, Sleight RG (1985) Defining lipid transport pathways in animal cells. Science 229:1051–1057
12. Koval M, Pagano RE (1990) Sorting of an internalized plasma membrane lipid between recycling and degradative pathways in normal and Niemann-Pick, type A fibroblasts. J Cell Biol 111:429–442
13. Martin OC, Pagano RE (1987) Transbilayer movement of fluorescent analogs of phosphatidylserine and phosphatidylethanolamine at the

plasma-membrane of cultured-cells—evidence for a protein-mediated and Atp-dependent process(Es). J Biol Chem 262:5890–5898

14. Elvington SM, Nichols JW (2007) Spontaneous, intervesicular transfer rates of fluorescent, acyl chain-labeled phosphatidyl-choline analogs. Biochim Biophys Acta 1768:502–508

15. Bai JN, Pagano RE (1997) Measurement of spontaneous transfer and transbilayer move-ment of BODIPY-labeled lipids in lipid vesi-cles. Biochemistry 36:8840–8848

16. Chattopadhyay A, London E (1987) Parallax method for direct measurement of membrane penetration depth utilizing fluorescence quenching by spin-labeled phospholipids. Biochemistry 26:39–45

17. Abrams FS, London E (1993) Extension of the parallax analysis of membrane penetration depth to the polar-region of model mem-branes—use of fluorescence quenching by a spin-label attached to the phospholipid polar headgroup. Biochemistry 32:10826–10831

18. te Vruchte D, Lloyd-Evans E, Veldman RJ, Neville DCA, Dwek RA, Platt FM, van Blitterswijk WJ, Sillence DJ (2004) Accumulation of glycosphingolipids in Niemann-Pick C disease disrupts endosomal transport. J Biol Chem 279:26167–26175

19. Pagano RE, Martin OC, Kang HC, Haugland RP (1991) A novel fluorescent ceramide analog for studying membrane traffic in animal-cells—accumulation at the Golgi-apparatus results in altered spectral properties of the sphingolipid precursor. J Cell Biol 113:1267–1279

20. Martin OC, Pagano RE (1994) Internalization and sorting of a fluorescent analog of gluco-sylceramide to the Golgi-apparatus of human skin fibroblasts—utilization of endocytic and nonendocytic transport mechanisms. J Cell Biol 125:769–781

21. Golebiewska U, Nyako M, Woturski W, Zaitseva I, McLaughlin S (2008) Diffusion coefficient of fluorescent phosphatidylinositol 4,5-bisphosphate in the plasma membrane of cells. Mol Biol Cell 19:1663–1669

22. Boldyrev IA, Zhai X, Momsen MM, Brockman HL, Brown RE, Molotkovsky JG (2007) New BODIPY lipid probes for fluo-rescence studies of membranes. J Lipid Res 48:1518–1532

23. Sachl R, Boldyrev I, Johansson LBA (2010) Localisation of BODIPY-labelled phosphati-dylcholines in lipid bilayers. Phys Chem Chem Phys 12:6027–6034

24. Kaiser RD, London E (1998) Determination of the depth of BODIPY probes in model membranes by parallax analysis of fluores-cence quenching. Biochim Biophys Acta 1375:13–22

25. Menger FM, Keiper JS, Caran KL (2002) Depth-profiling with giant vesicle mem-branes. J Am Chem Soc 124:11842–11843

26. Armendariz KP, Huckabay HA, Livanec PW, Dunn RC (2012) Single molecule probes of membrane structure: orientation of BODIPY probes in DPPC as a function of probe struc-ture. Analyst 137:1402–1408

27. Kay JG, Koivusalo M, Ma XX, Wohland T, Grinstein S (2012) Phosphatidylserine dynamics in cellular membranes. Mol Biol Cell 23:2198–2212

28. Parasassi T, De Stasio G, Ravagnan G, Rusch RM, Gratton E (1991) Quantitation of lipid phases in phospholipid-vesicles by the gener-alized polarization of laurdan fluorescence. Biophys J 60:179–189

29. Jurkiewicz P, Olzynska A, Langner M, Hof M (2006) Headgroup hydration and mobility of DOTAP/DOPC bilayers: a fluorescence sol-vent relaxation study. Langmuir 22:8741–8749

30. Lakowicz JR, Bevan DR, Maliwal BP, Cherek H, Balter A (1983) Synthesis and character-ization of a fluorescence probe of the phase-transition and dynamic properties of membranes. Biochemistry 22:5714–5722

31. Olzynska A, Zan A, Jurkiewicz P, Sykora J, Grobner G, Langner M, Hof M (2007) Molecular interpretation of fluorescence sol-vent relaxation of Patman and H-2 NMR experiments in phosphatidylcholine bilayers. Chem Phys Lipids 147:69–77

32. Zipfel WR, Williams RM, Webb WW (2003) Nonlinear magic: multiphoton microscopy in the biosciences. Nat Biotechnol 21:1369–1377

33. Williams RM, Webb WW (2000) Single gran-ule pH cycling in antigen-induced mast cell secretion. J Cell Sci 113(Pt 21):3839–3850

34. Gibbons E, Pickett KR, Streeter MC, Warcup AO, Nelson J, Judd AM, Bell JD (2013) Molecular details of membrane fluidity changes during apoptosis and relationship to phospholipase A(2) activity. Biochim Biophys Acta 1828:887–895

35. Gaus K, Gratton E, Kable EPW, Jones AS, Gelissen I, Kritharides L, Jessup W (2003) Visualizing lipid structure and raft domains in living cells with two-photon microscopy. Proc Natl Acad Sci U S A 100:15554–15559

36. Bagatolli LA, Gratton E (1999) Two-photon fluorescence microscopy observation of shape changes at the phase transition in phospholipid giant unilamellar vesicles. Biophys J 77:2090–2101

37. Kim HM, Choo HJ, Jung SY, Ko YG, Park WH, Jeon SJ, Kim CH, Joo TH, Cho BR (2007) A two-photon fluorescent probe for lipid raft imaging: C-laurdan. Chembiochem 8:553–559

38. Barucha-Kraszewska J, Kraszewski S, Ramseyer C (2013) Will C-Laurdan Dethrone Laurdan in fluorescent solvent relaxation techniques for lipid membrane studies? Langmuir 29:1174–1182

39. Klemm RW, Ejsing CS, Surma MA, Kaiser HJ, Gerl MJ, Sampaio JL, de Robillard Q, Ferguson C, Proszynski TJ, Shevchenko A, Simons K (2009) Segregation of sphingolipids and sterols during formation of secretory vesicles at the trans-Golgi network. J Cell Biol 185:601–612

40. Kuhry JG, Fonteneau P, Duportail G, Maechling C, Laustriat G (1983) Tma-Dph—a suitable fluorescence polarization probe for specific plasma-membrane fluidity studies in intact living cells. Cell Biophys 5:129–140

41. Kraayenhof R, Sterk GJ, Sang HWWF (1993) Probing biomembrane interfacial potential and ph profiles with a new-type of float-like fluorophores positioned at varying distance from the membrane-surface. Biochemistry 32:10057–10066

42. Clarke RJ, Zouni A, Holzwarth JF (1995) Voltage sensitivity of the fluorescent-probe Rh421 in a model membrane system. Biophys J 68:1406–1415

43. Gross E, Bedlack RS, Loew LM (1994) Dual-wavelength ratiometric fluorescence measurement of the membrane dipole potential. Biophys J 67:208–216

44. Sengupta PK, Kasha M (1979) Excited-state proton-transfer spectroscopy of 3-hydroxyflavone and quercetin. Chem Phys Lett 68:382–385

45. Mcmorrow D, Kasha M (1984) Intramolecular excited-state proton-transfer in 3-hydroxy-flavone—hydrogen-bonding solvent perturbations. J Phys Chem 88:2235–2243

46. Klymchenko AS, Demchenko AP (2003) Multiparametric probing of intermolecular interactions with fluorescent dye exhibiting excited state intramolecular proton transfer. Phys Chem Chem Phys 5:461–468

47. Klymchenko AS, Demchenko AP (2002) Electrochromic modulation of excited-state intramolecular proton transfer: the new principle in design of fluorescence sensors. J Am Chem Soc 124:12372–12379

48. Klymchenko AS, Duportail G, Ozturk T, Pivovarenko VG, Mely Y, Demchenko AP (2002) Novel two-band ratiometric fluorescence probes with different location and orientation in phospholipid membranes. Chem Biol 9:1199–1208

49. Krishna MMG (1999) Excited-state kinetics of the hydrophobic probe nile red in membranes and micelles. J Phys Chem A 103:3589–3595

50. Sykora J, Jurkiewicz P, Epand RM, Kraayenhof R, Langner M, Hof M (2005) Influence of the curvature on the water structure in the headgroup region of phospholipid bilayer studied by the solvent relaxation technique. Chem Phys Lipids 135:213–221

51. Chong PLG (1988) Effects of hydrostatic-pressure on the location of Prodan in lipid bilayers and cellular membranes. Biochemistry 27:399–404

52. Klymchenko AS, Duportail G, Demchenko AP, Mely Y (2004) Bimodal distribution and fluorescence response of environment-sensitive probes in lipid bilayers. Biophys J 86:2929–2941

53. Klymchenko AS, Duportail G, Mely Y, Demchenko AP (2003) Ultrasensitive two-color fluorescence probes for dipole potential in phospholipid membranes. Proc Natl Acad Sci U S A 100:11219–11224

54. Bagatolli LA (2006) To see or not to see: lateral organization of biological membranes and fluorescence microscopy. Biochim Biophys Acta 1758:1541–1556

55. Bondar OP, Rowe ES (1999) Preferential interactions of fluorescent probe Prodan with cholesterol. Biophys J 76:956–962

56. Parisio G, Marini A, Biancardi A, Ferrarini A, Mennucci B (2011) Polarity-sensitive fluorescent probes in lipid bilayers: bridging spectroscopic behavior and microenvironment properties. J Phys Chem B 115:9980–9989

57. Nitschke WK, Vequi-Suplicy CC, Coutinho K, Stassen H (2012) Molecular dynamics investigations of PRODAN in a DLPC bilayer. J Phys Chem B 116:2713–2721

58. Alakoskela JMI, Kinnunen PKJ (2001) Probing phospholipid main phase transition by fluorescence spectroscopy and a surface redox reaction. J Phys Chem B 105:11294–11301

59. Demchenko AP, Shcherbatska NV (1985) Nanosecond dynamics of charged fluorescent-probes at the polar interface of a membrane phospholipid-bilayer. Biophys Chem 22:131–143

60. Loura LM, Ramalho JP (2007) Location and dynamics of acyl chain NBD-labeled phosphatidylcholine (NBD-PC) in DPPC bilayers. A molecular dynamics and time-resolved

fluorescence anisotropy study. Biochim Biophys 1768:467–478

61. Huster D, Muller P, Arnold K, Herrmann A (2001) Dynamics of membrane penetration of the fluorescent 7-nitrobenz-2-oxa-1,3-diazol-4-yl (NBD) group attached to an acyl chain of phosphatidylcholine. Biophys J 80:822–831

62. Bouvrais H, Pott T, Bagatolli LA, Ipsen JH, Meleard P (2010) Impact of membrane-anchored fluorescent probes on the mechanical properties of lipid bilayers. Biochim Biophys Acta 1798:1333–1337

63. Cruz A, Vazquez L, Velez M, Perez-Gil J (2005) Influence of a fluorescent probe on the nanostructure of phospholipid membranes: dipalmitoylphosphatidylcholine interfacial monolayers. Langmuir 21:5349–5355

64. Epand RM, Kraayenhof R (1999) Fluorescent probes used to monitor membrane interfacial polarity. Chem Phys Lipids 101:57–64

65. Disalvo EA, Lairion F, Martini F, Tymczyszyn E, Frias M, Almaleck H, Gordillo GJ (2008) Structural and functional properties of hydration and confined water in membrane interfaces. Biochim Biophys Acta 1778:2655–2670

66. Gawrisch K, Ruston D, Zimmerberg J, Parsegian VA, Rand RP, Fuller N (1992) Membrane dipole potentials, hydration forces, and the ordering of water at membrane surfaces. Biophys J 61:1213–1223

67. Parasassi T, Di Stefano M, Loiero M, Ravagnan G, Gratton E (1994) Cholesterol modifies water concentration and dynamics in phospholipid bilayers: a fluorescence study using Laurdan probe. Biophys J 66:763–768

68. Ho C, Slater SJ, Stubbs CD (1995) Hydration and order in lipid bilayers. Biochemistry 34:6188–6195

69. Sykora J, Kapusta P, Fidler V, Hof M (2002) On what time scale does solvent relaxation in phospholipid bilayers happen? Langmuir 18:571–574

70. Demchenko AP (2002) The red-edge effects: 30 years of exploration. Luminescence 17:19–42

71. Chattopadhyay A, Mukherjee S (1999) Depth-dependent solvent relaxation in membranes: wavelength-selective fluorescence as a membrane dipstick. Langmuir 15:2142–2148

72. Sykora J, Slavicek P, Jungwirth P, Barucha J, Hof M (2007) Time-dependent stokes shifts of fluorescent dyes in the hydrophobic backbone region of a phospholipid bilayer: combination of fluorescence spectroscopy and ab

initio calculations. J Phys Chem B 111:5869–5877

73. Klymchenko AS, Mely Y, Demchenko AP, Duportail G (2004) Simultaneous probing of hydration and polarity of lipid bilayers with 3-hydroxyflavone fluorescent dyes. Biochim Biophys Acta 1665:6–19

74. Zhang YL, Frangos JA, Chachisvilis M (2006) Laurdan fluorescence senses mechanical strain in the lipid bilayer membrane. Biochem Biophys Res Commun 347:838–841

75. Nicolini C, Celli A, Gratton E, Winter R (2006) Pressure tuning of the morphology of heterogeneous lipid vesicles: a two-photon-excitation fluorescence microscopy study. Biophys J 91:2936–2942

76. Schuler I, Milon A, Nakatani Y, Ourisson G, Albrecht AM, Benveniste P, Hartman MA (1991) Differential effects of plant sterols on water permeability and on acyl chain ordering of soybean phosphatidylcholine bilayers. Proc Natl Acad Sci U S A 88:6926–6930

77. Simons K, Ikonen E (1997) Functional rafts in cell membranes. Nature 387:569–572

78. Beck A, Heissler D, Duportail G (1993) Influence of the length of the spacer on the partitioning properties of amphiphilic fluorescent membrane probes. Chem Phys Lipids 66:135–142

79. Bagatolli LA, Gratton E (2000) Two photon fluorescence microscopy of coexisting lipid domains in giant unilamellar vesicles of binary phospholipid mixtures. Biophys J 78:290–305

80. Scherfeld D, Kahya N, Schwille P (2003) Lipid dynamics and domain formation in model membranes composed of ternary mixtures of unsaturated and saturated phosphatidylcholines and cholesterol. Biophys J 85:3758–3768

81. Dietrich C, Bagatolli LA, Volovyk ZN, Thompson NL, Levi M, Jacobson K, Gratton E (2001) Lipid rafts reconstituted in model membranes. Biophys J 80:1417–1428

82. Brown DA, London E (2000) Structure and function of sphingolipid- and cholesterol-rich membrane rafts. J Biol Chem 275:17221–17224

83. Owen DM, Rentero C, Magenau A, Abu-Siniyeh A, Gaus K (2012) Quantitative imaging of membrane lipid order in cells and organisms. Nat Protoc 7:24–35

84. Jin L, Millard AC, Wuskell JP, Dong X, Wu D, Clark HA, Loew LM (2006) Characterization and application of a new optical probe for membrane lipid domains. Biophys J 90:2563–2575

85. Jin L, Millard AC, Wuskell JP, Clark HA, Loew LM (2005) Cholesterol-enriched lipid domains can be visualized by di-4-ANEPPDHQ with linear and nonlinear optics. Biophys J 89:L4–L6

86. Kim HM, Jeong BH, Hyon JY, An MJ, Seo MS, Hong JH, Lee KJ, Kim CH, Joo TH, Hong SC, Cho BR (2008) Two-photon fluorescent turn-on probe for lipid rafts in live cell and tissue. J Am Chem Soc 130(2008):4246–4247

87. Golfetto O, Hinde E, Gratton E (2013) Laurdan fluorescence lifetime discriminates cholesterol content from changes in fluidity in living cell membranes. Biophys J 104:1238–1247

88. M'Baye G, Mely Y, Duportail G, Klymchenko AS (2008) Liquid ordered and gel phases of lipid bilayers: fluorescent probes reveal close fluidity but different hydration. Biophys J 95:1217–1225

89. Haluska CK, Schroder AP, Didier P, Heissler D, Duportail G, Mely Y, Marques CM (2008) Combining fluorescence lifetime and polarization microscopy to discriminate phase separated domains in giant unilamellar vesicles. Biophys J 95:5737–5747

90. Korlach J, Schwille P, Webb WW, Feigenson GW (1999) Characterization of lipid bilayer phases by confocal microscopy and fluorescence correlation spectroscopy. Proc Natl Acad Sci U S A 96:8461–8466

91. Oncul S, Klymchenko AS, Kucherak OA, Demchenko AP, Martin S, Dontenwill M, Arntz Y, Didier P, Duportail G, Mely Y (2010) Liquid ordered phase in cell membranes evidenced by a hydration-sensitive probe: effects of cholesterol depletion and apoptosis. Biochim Biophys Acta 1798:1436–1443

92. Klymchenko AS, Oncul S, Didier P, Schaub E, Bagatolli L, Duportail G, Mely Y (2009) Visualization of lipid domains in giant unilamellar vesicles using an environment-sensitive membrane probe based on 3-hydroxyflavone. Biochim Biophys Acta 1788:495–499

93. Darwich Z, Kucherak QA, Kreder R, Richert L, Vauchelles R, Yves M, Klymchenko AS (2013) Rational design of fluorescence membrane probes for apoptosis based on 3-hydroxyflavone. Methods Appl Fluoresc 1:025002

94. Zwaal RFA, Schroit AJ (1997) Pathophysiologic implications of membrane phospholipid asymmetry in blood cells. Blood 89:1121–1132

95. Shynkar VV, Klymchenko AS, Kunzelmann C, Duportail G, Muller CD, Demchenko AP, Freyssinet JM, Mely Y (2007) Fluorescent bio-

membrane probe for ratiometric detection of apoptosis. J Am Chem Soc 129:2187–2193

96. Kucherak OA, Oncul S, Darwich Z, Yushchenko DA, Arntz Y, Didier P, Mely Y, Klymchenko AS (2010) Switchable nile red-based probe for cholesterol and lipid order at the outer leaflet of biomembranes. J Am Chem Soc 132:4907–4916

97. Darwich Z, Klymchenko AS, Kucherak OA, Richert L, Mely Y (2012) Detection of apoptosis through the lipid order of the outer plasma membrane leaflet. Biochim Biophys Acta 1818:3048–3054

98. O'Shea P (2003) Intermolecular interactions with/within cell membranes and the trinity of membrane potentials: kinetics and imaging. Biochem Soc Trans 31:990–996

99. Zouni A, Clarke RJ, Holzwarth JF (1994) Kinetics of the solubilization of styryl dye aggregates by lipid vesicles. J Phys Chem 98:1732–1738

100. Wuskell JP, Boudreau D, Wei MD, Jin L, Engl R, Chebolu R, Bullen A, Hoffacker KD, Kerimo J, Cohen LB, Zochowski MR, Loew LM (2006) Synthesis, spectra, delivery and potentiometric responses of new styryl dyes with extended spectral ranges. J Neurosci Methods 151:200–215

101. Vitha MF, Clarke RJ (2007) Comparison of excitation and emission ratiometric fluorescence methods for quantifying the membrane dipole potential. Biochim Biophys Acta 1768:107–114

102. Millard AC, Jin L, Wei MD, Wuskell JP, Lewis A, Loew LM (2004) Sensitivity of second harmonic generation from styryl dyes to transmembrane potential. Biophys J 86:1169–1176

103. M'Baye G, Shynkar VV, Klymchenko AS, Mely Y, Duportail G (2006) Membrane dipole potential as measured by ratiometric 3-hydroxyflavone fluorescence probes: accounting for hydration effects. J Fluoresc 16:35–42

104. Shynkar VV, Klymchenko AS, Duportail G, Demchenko AP, Mely Y (2005) Two-color fluorescent probes for imaging the dipole potential of cell plasma membranes. Biochim Biophys Acta 1712:128–136

105. Grinvald A, Hildesheim R (2004) VSDI: a new era in functional imaging of cortical dynamics. Nat Rev Neurosci 5:874–885

106. Kuznetsov A, Bindokas VP, Marks JD, Philipson LH (2005) FRET-based voltage probes for confocal imaging: membrane potential oscillations throughout pancreatic islets. Am J Physiol Cell Physiol 289:C224–C229

107. Cacciatore TW, Brodfuehrer PD, Gonzalez JE, Jiang T, Adams SR, Tsien RY, Kristan

WB, Kleinfeld D (1999) Identification of neural circuits by imaging coherent electrical activity with FRET-based dyes. Neuron 23:449–459

108. Briggman KL, Abarbanel HDI, Kristan WB (2005) Optical imaging of neuronal populations during decision-making. Science 307:896–901

109. Montana V, Farkas DL, Loew LM (1989) Dual-wavelength ratiometric fluorescence measurements of membrane-potential. Biochemistry 28:4536–4539

110. Gross D, Loew LM (1989) Fluorescent indicators of membrane-potential—microspectrofluorometry and imaging. Method Cell Biol 30:193–218

111. Kuhn B, Fromherz P, Denk W (2004) New voltage-sensitive dye shows large fractional fluorescence changes with spectral-edge one- or 2-photon excitation. Biophys J 86:333–334

112. Kuhn B, Fromherz P (2003) Anellated hemicyanine dyes in a neuron membrane: molecular Stark effect and optical voltage recording. J Phys Chem B 107:7903–7913

113. Hubener G, Lambacher A, Fromherz P (2003) Anellated hemicyanine dyes with large symmetrical solvatochromism of absorption and fluorescence. J Phys Chem B 107: 7896–7902

114. Fromherz P, Hubener G, Kuhn B, Hinner MJ (2008) ANNINE-6plus, a voltage-sensitive dye with good solubility, strong membrane binding and high sensitivity. Eur Biophy J EBJ 37:509–514

115. Klymchenko AS, Stoeckel H, Takeda K, Mely Y (2006) Fluorescent probe based on intramolecular proton transfer for fast ratiometric measurement of cellular transmembrane potential. J Phys Chem B 110:13624–13632

Chapter 4

The Assembly and Use of Tethered Bilayer Lipid Membranes (tBLMs)

Charles Cranfield, Sonia Carne, Boris Martinac, and Bruce Cornell

Abstract

Because they are firmly held in place, tethered bilayer lipid membranes (tBLMs) are considerably more robust than supported lipid bilayers such as black lipid membranes (BLMs) (Cornell et al. Nature 387(6633): 580–583, 1997). Here we describe the procedures required to assemble and test tethered lipid bilayers that can incorporate various lipid species, peptides, and ion channel proteins.

Key words Tethered bilayer lipid membranes, AC impedance spectroscopy, Ion channels, Lipid bilayers

1 Introduction

Tethered membranes consist of a metal electrode, typically gold, to which tethering moieties are anchored preferably via benzyl disulfide groups (*see* **Note 1**). The tethers incorporate into one leaflet of a subsequently self-assembled lipid bilayer, thus tethering the bilayer to the metal surface [1] (Fig. 1). Interspersed with the tethers are *spacer* molecules that provide a *scaffolding* for the tether molecules. When the amphiphilic tethering molecules are interspersed with polar short-chained *spacer* molecules a volume between the metal surface and the bilayer is formed that creates an aqueous reservoir for ions crossing the membrane. Because they can be firmly held in place with a chemical attachment to the metal, tethered lipid bilayers are far more robust than solvent based BLMs. The formation of tBLMs is also a more predictable and controllable process than forming untethered BLMs. tBLMs remain intact for months, unlike untethered BLMs which typically have lifetimes of the order of minutes to hours.

The process of depositing smooth (<1.5 nm) ultrapure (99.9995 %) gold onto polymeric substrates with no contaminating intermediate metal layers involves considerable process development.

Dylan M. Owen (ed.), *Methods in Membrane Lipids*, Methods in Molecular Biology, vol. 1232,
DOI 10.1007/978-1-4939-1752-5_4, © Springer Science+Business Media New York 2015

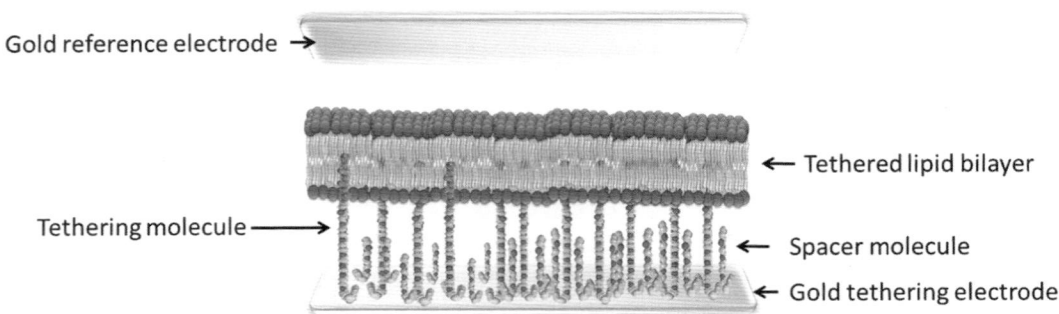

Gold reference electrode →

← Tethered lipid bilayer

Tethering molecule ——→

← Spacer molecule

← Gold tethering electrode

Fig. 1 Tethered bilayer lipid membrane (tBLM) schematic

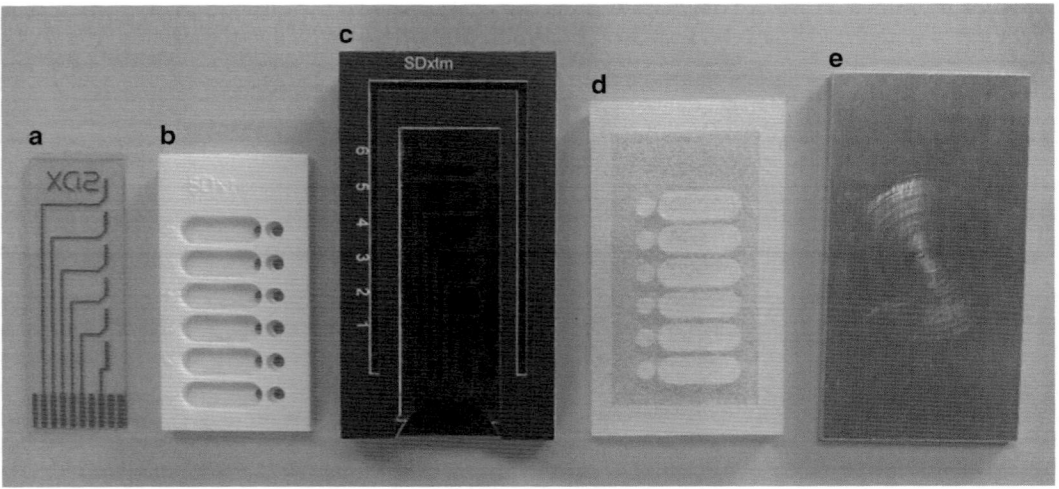

Fig. 2 Components required to form a tBLM flow cell. (**a**) Electrodes pre-coated with tethering chemistry. (**b**) A flow cell cartridge top. (**c**) Alignment jig for use when attaching the electrode to the flow cell cartridge (**d**) Silicon rubber pressure pad used when attaching the electrode to the flow cell cartridge (**e**) Aluminum pressure plate used when attaching the electrode to the flow cell cartridge

The resulting gold patterned electrodes are, however, ideally suited to the preparation of reproducible, well-sealed tethered membranes. The tethering chemistry is coated from an ethanol solution onto the gold surface immediately following gold deposition. This protocol describes the techniques required to use pre-prepared gold electrodes supplied from the company *SDx Tethered Membranes Pty Ltd* (SDx), the unique supplier in the world of such a tBLM platform. Gold patterned, chemically coated electrodes (as 25 mm × 75 mm slides (Fig. 2a)) are shipped to the user in hermitically sealed foil packages in ethanol solution. This format is chosen in order to provide the user the flexibility of forming tBLMs comprising their choice of lipid. The coated electrodes may also be stored at 4 °C for >1 year. Different tether:spacer densities are offered in order to accommodate protein or peptide components of molecular weights from 1 to 350 kDa within the membrane.

2   Materials

2.1   Electrode Selection

Ready-made SDx electrodes comprise a close-packed array of 2–4 nm strands of ethylene glycol that act as *spacers* and 4 nm strands of ethylene glycol terminated with a C20 phytanyl that act as membrane *tethers*. Typically electrodes are provided that have 10 % tether molecules and 90 % spacer molecules. Electrodes with this ratio of tethers to spacers have been designated as *T10 electrodes*. T10 electrodes create tBLMs that are very stable with little membrane leakage, and with the ability to incorporate up to 40 kDa of the membrane bound fraction of proteins and peptides. Although T1 electrodes provide greater capacity in the tBLM for molecular weights up to 350 kDa to penetrate the bilayer molecules, the resulting membrane is less stable. The length of the spacer molecules can also be altered. Hydrophilic spacers of hydroxyl-terminated lipid chains can stretch to the inner leaflet of the lipid bilayer membrane, or can extend only halfway to the inner leaflet (as depicted in Fig. 1) in order to create additional space between the gold tethering electrode and the tethered bilayer in order to accommodate protein loops extending beyond the membrane surface at the inner leaflet.

2.2   Cartridge Preparation Kit

The SDx six-channel measurement electrodes are assembled into a flow cell cartridge (Fig. 4a) which, in turn, plugs into the *SDx tethaPod*™ conductance and capacitance reader (Fig. 4b). A cartridge preparation kit is supplied by SDx which consists of:

- Individually packaged electrodes pre-coated with tethering chemistry (Fig. 2a)
- A flow cell cartridge top on which is coated the gold counter electrode (Fig. 2b)
- An alignment jig for use when attaching the electrode to the flow-cell cartridge (Fig. 2c)
- A silicon rubber pressure pad used when attaching the electrode to the flow cell cartridge (Fig. 2d)
- An aluminum pressure plate used when attaching the electrode to the flow cell cartridge (Fig. 2e)
- A pressure clamp also used when attaching the electrode to the flow cell cartridge (Fig. 3f)

2.3   Solutions

- Also supplied by SDx is a standard membrane forming lipid mixture that has been optimized to achieve the best electrical seal which comprises 3 mM ethanolic solution of a 70 %:30 %

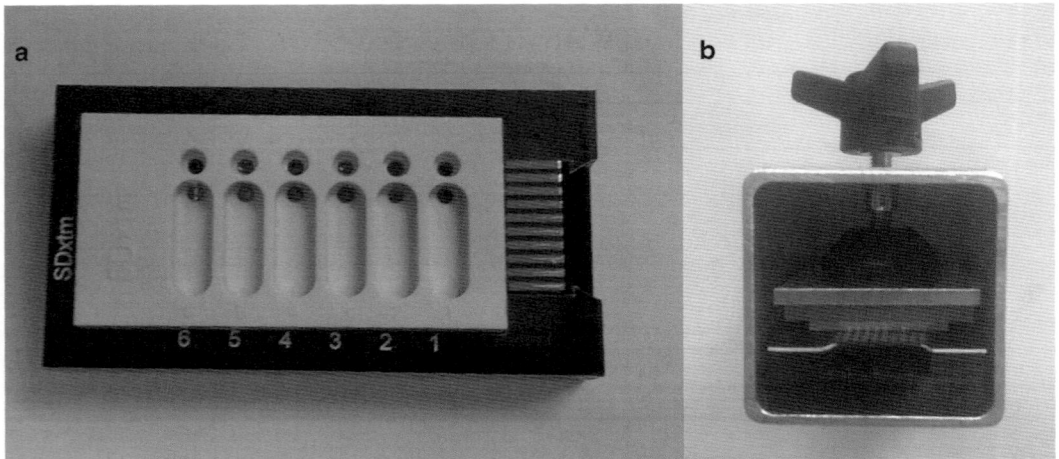

Fig. 3 (**a**) The flow cell cartridge attached to the electrode slide on the alignment jig. (**b**) The pressure clamp used when attaching the electrode to the flow cell cartridge

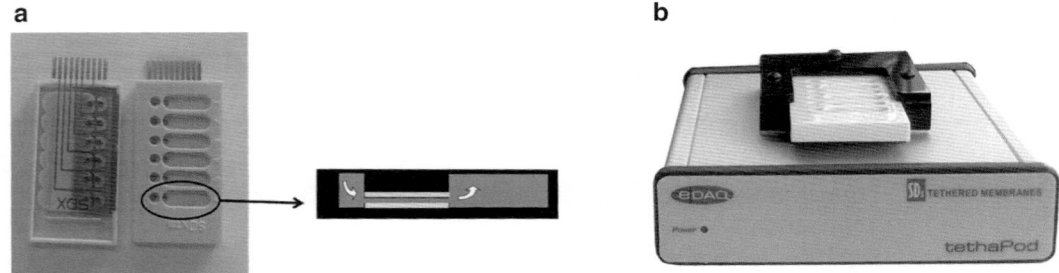

Fig. 4 (**a**) Assembled flow cell cartridge with electrodes. The flow cell cartridge provides the counter electrode which is overlayed onto the six tethering electrodes with a 0.1 mm gap for perfusion of reagents and buffer solutions. (**b**) Assembled flow cell cartridge fitted into a tethaPod™ AC impedance reader

mix of diether diphytanyl (C16) phosphatidyl choline:diether diphatanyl (C16) hydroxyl (mixture designated *AM199* by SDx). Alternative lipid combinations may be employed provided they are soluble in ethanol at 3 mM concentrations at room temperature (*see* **Note 2**).

- Preferred electrolyte solution such as Phosphate Buffered Saline (PBS).

3   Methods

3.1 Preparing Cartridges

1. Remove six-channel electrode from its sealed foil package using tweezers. Care should be taken not to touch the six gold regions that will form the gold tethered membrane. The side of the electrode where "SDX" appears inverted is the *up-side*

upon which the gold has been deposited (Fig. 2a). Allow 2–3 min for any residual ethanol to evaporate (*see* **Note 3**).

2. Place electrode into alignment jig so that the inverted "SDX" on the slide overlays the inverted "SDX" on the alignment jig (Fig. 2b). This will ensure the gold electrodes are correctly oriented.

3. Peel the thin plastic protective cover from the underside of the flow cell cartridge, *taking care to leave the 0.1 mm flow cell laminate and adhesive layer in place.*

4. Place the flow cell cartridge over the alignment jig with the adhesive laminate facing the electrode (Fig. 3a). The flow cell cartridge should be aligned such that the numbers 1–6 on the cartridge align with the 1–6 on the alignment jig. Insert the short end of the flow cell cartridge nearest to well 6 into the matching slot in the alignment jig and lower the cartridge onto the electrode.

5. Press the silicon rubber pressure pad into the flow cell cartridge top. Then position the aluminum pressure plate over the assembly and insert into the pressure clamp.

6. Compress by ¾ of a turn of the knob and leave for at least 1 min (Fig 3b).

7. Gently remove the assembly from the pressure clamp. Remove the aluminum pressure plate and silicon pressure pad taking care to prevent flow cell cartridge separating from the electrode.

8. By gently lifting from the underside of the electrode the electrode-flow cell cartridge assembly can be removed intact from the alignment jig. The remaining exposed gold surfaces are not critical to bilayer formation. The functional gold tethering electrodes are protected within the flow cell cartridge assembly (Fig 4a).

9. Once the cartridge is made it is important to attach a bilayer lipid membrane to the tethers as soon as practicable (within 1–2 min; *see* **Note 3**).

3.2 Creating a Tethered Bilayer Lipid Membrane Using Solvent Exchange

When forming a tBLM a solvent exchange method is employed as it permits the formation of tBLMs at low tethering ratios well beyond that possible employing liposomal fusion.

1. To well 1 add 8 µL AM199 or desired ethanolic lipid formulation via the circular opening in the cartridge top.

2. Wait 10 s and add 8 µL lipid solution to the second well, and wait a further 10 s before making additions to each of the subsequent four wells. A delay of 10 s provides a convenient

operational delay to permit each well to be incubated for same period of time.

3. Let each ethanolic solution of lipid incubate within the wells for exactly 2 min then rinse with 100 μL of PBS (or desired buffer solution, *see* **Note 4**) taking care not to introduce air bubbles into the flow cell chamber as this will damage membrane formation. Alternative lipid mixtures may require optimisation of their concentration and assembly times to achieve the highest possible membrane seal.

4. Rinse with at least 3×100 μL aliquots of the PBS. After the first wash with PBS time is no longer a critical factor. From the reservoir side of the flow cell cartridge remove 100 μL of the wash through solution and then repeat rinses to ensure the removal of any residual ethanol and lipids. Following formation, the assembled tethered bilayer must remain totally covered with aqueous solutions to prevent membrane disassociation (*see* **Note 5**).

3.3 Testing the Bilayer Using AC Impedance Spectroscopy

The conductance and capacitance of the tethered membrane may be measured by inserting the assembled electrode within the flow cell cartridge into a tethaPod™ reader. The reader simplifies the interpretation of the AC impedance spectrum and provides a measure of membrane conductance and capacitance. Typical conduction values for a freshly formed membrane using AM199 in PBS are 0.35 ± 0.15 μS and capacitance values of 18 ± 2 nF at room temperature. The conductance is proportional to the ion flux through the membrane and the capacitance is *inversely* proportional to the membrane thickness. A significant additional measure using a tethaPod™ is the *Goodness of Fit* (GOF). This indicates the quality of match between the experimental data and a model of the tethered membrane. GOF values of less than 0.1 indicate a good match of the data to this simple model, and suggest that the membrane is homogenous. The user also has the option to employ their own models to fit the raw impedance data.

3.4 Incorporation of Proteins and Peptides

3.4.1 Spontaneous Inserting Proteins and Peptides

Peptides and proteins such as *alamethicin*, *α-hemolysin*, *gramicidin A*, and *valinomycin* will spontaneously insert into pre-formed tBLMs [1–4]. The kinetics of their insertion can be measured by the time dependency of the increase in conduction. Although there are no established protocols for inserting ionophores into tBLMs the following guidelines should be considered:

1. Peptides dissolved in ethanol/buffer or methanol/buffer mixtures should be less than 10 % ethanol or methanol and be thoroughly rinsed to remove any residual solvent.

2. In order to obtain quantitative kinetics it is necessary to employ a controlled flow rate for the introduction of the peptide to the tBLM. This can be achieved using a syringe pump that couples to the flow cell cartridge (Fig. 5).

Fig. 5 An example of a syringe pump attachment to control flow across the tBLMs

3. In order to model the *kinetics* of the conduction increase following peptide or protein insertion, important considerations include:

    • The temperature of the solutions being added is equivalent to the temperature in the tBLM.

    • The aggregation state of the spontaneous inserting peptide or protein in the aqueous buffer. The form of the concentration versus conduction relationship will report on the aggregation state of the protein or peptide in the membrane, with a nonlinearity suggesting that a multimeric form of the peptide or protein is required for conduction.

4. Insertion of proteins and peptides may depend on the need for specific charged lipids being present in the tBLM.

5. The conduction of some peptides, such as *gramicidin A*, is selective for cations over anions and so applying a fixed negative potential bias to the tethering gold electrode will enhance conduction. The ability to apply such a bias is available on the TethaPod™ [1].

6. The insertion of peptides such as *alamethicin* is catalyzed by the application of a transmembrane potential which can be applied to the tBLM [3]

Table 1
Aggregation numbers as reported by Sigma-Aldrich

Detergent	Aggregation number	Concentration (μM)	Concentration (% w/v)
Brij 58	70	1	0.0001
TWEEN 20	Not reported	4	0.0005
Triton X-100	140	15.5	0.001
DDM	98	50	0.0025
CYMAL-5	47	400	0.02

3.4.2 Non-spontaneous Insertion of Proteins and Peptides

For membrane intrinsic proteins and peptides, their insertion occurs at the time of tBLM formation through their inclusion as a detergent micelle in the hydrating, membrane forming rinse step (*see* Subheading 3.2, **step 3** above). When selecting detergents to form micelles the general rule is to use *high aggregation number* detergents such as Brij 58, CYMAL 5, or DDM. Table 1 lists the upper tolerable concentration levels for AM199 T10 tBLMs for a list of common detergents.

4   Notes

1. The use of disulfides rather than thiols results in a greater stability of the coating solutions.

2. Lipids that are insoluble in ethanol may be dispersed in ethanol/methanol mixtures. Care should be taken not to use solvents, such as chloroform, which will degrade the polycarbonate cartridge.

3. Drying of the electrodes will result in adsorption of the tethering chemistry to the gold preventing its incorporation into the subsequently formed lipid bilayer. For this reason electrodes should not be left exposed to dry for longer than 2–3 min. Small droplets of residual ethanol on the electrodes will not affect the subsequent assembly process.

4. A series of impedance spectroscopy measurements have demonstrated that 2 min incubation in the lipid ethanolic solution of AM199 produces the best sealed tBLMs at room temperature.

5. Following formation of the membrane an adhesive seal can be applied across the cartridge opening to prevent evaporation and drying of the sample in the flow-cell chamber.

Acknowledgments

This work was supported by the Australian Research Council, the National Health and Medical Research Council of Australia (Grant 1047980). We declare that Bruce Cornell is a shareholder, and Sonia Carne is an employee, of *SDx Tethered Membranes Pty Ltd.*

References

1. Cornell BA, Braach-Maksvytis VLB, King LG, Osman PDJ, Raguse B, Wieczorek L, Pace RJ (1997) A biosensor that uses ion-channel switches. Nature 387(6633):580–583

2. Vockenroth IK, Atanasova PP, Jenkins ATA, Köper I (2008) Incorporation of α-hemolysin in different tethered bilayer lipid membrane architectures. Langmuir 24(2):496–502

3. Yin P, Burns CJ, Osman PD, Cornell BA (2003) A tethered bilayer sensor containing alamethicin channels and its detection of amiloride based inhibitors. Biosens Bioelectron 18(4):389–397

4. Raguse B, Braach-Maksvytis V, Cornell BA, King LG, Osman PD, Pace RJ, Wieczorek L (1998) Tethered lipid bilayer membranes: formation and ionic reservoir characterization. Langmuir 14(3):648–659

Chapter 5

Preparation of Detergent-Resistant Membranes (DRMs) from Cultured Mammalian Cells

Deborah A. Brown

Abstract

Detergent-resistant membranes (DRMs) isolated from cells are enriched in proteins and lipids with a high affinity for lipid rafts, or membrane microdomains in the liquid-ordered phase. Enrichment in DRMs provides a good indication that a protein is "raftophilic," and may be present in rafts in cell membranes before extraction. Here, I describe preparation of Triton X-100-insoluble DRMs from cultured mammalian cells on sucrose gradients. Methods for analyzing the distribution of particular proteins across the gradient, and for recovering DRMs for further use are presented.

Key words Lipid rafts, Liquid-ordered phase, Membrane microdomains, Triton X-100-insoluble membranes, DRMs

1 Introduction

Lipid rafts are membrane microdomains in the liquid-ordered (lo) phase [1]. Rafts form spontaneously in model membranes of appropriate lipid composition, and can coexist with microdomains in the familiar liquid-disordered (ld) phase. Lipids in both the lo and frozen gel phases are tightly packed and highly ordered, with extensive van der Waals interactions between acyl chains. At appropriate temperatures, order-preferring, saturated chain lipids form the gel phase in the absence of sterols, but form the sterol-dependent lo phase in the presence of sterols. Acyl chain interactions in both phases are strong enough to exclude many non-ionic detergents (including Triton X-100), which efficiently solubilize ld phase membranes. For this reason, lo- and gel-phase membranes are largely insoluble in these detergents. Gel or lo-phase domains that coexist with ld-phase domains in two-phase model membranes are similarly detergent-insoluble [2, 3]. When these two-phase model membranes are subjected to detergent extraction, ld-phase domains are largely solubilized and converted to detergent–lipid

Dylan M. Owen (ed.), *Methods in Membrane Lipids*, Methods in Molecular Biology, vol. 1232,
DOI 10.1007/978-1-4939-1752-5_5, © Springer Science+Business Media New York 2015

mixed micelles, while lo- or gel-phase domains can persist as detergent-resistant membranes (DRMs).

DRMs may not always correspond precisely to preexisting ordered-phase domains. Triton X-100 was reported to induce phase separation, allowing DRM formation [4, 5]. However, later work clarified this finding, showing that Triton X-100 does not actually trigger phase separation. Instead, Triton X-100 induces coalescence of rafts that were too small to be detected in the earlier studies [3].

lo phase DRMs can also be isolated from detergent-extracted cells [6, 7]. This finding initially created great excitement, as it suggested that rafts exist in cells. The cholesterol- and sphingolipid-rich plasma membrane is the most favorable for raft formation, and is the source of most DRMs isolated from cells. (Some intracellular membranes, especially the cholesterol-rich trans-Golgi network, may also support DRM formation [8].) "Raftophilic" proteins (those with a high affinity for ordered-phase membranes) that are recovered in DRMs may exist in rafts before detergent extraction, and this localization may be important in function.

Several different non-ionic detergents (particularly Brij 96, Brij 98, and Lubrol WX) have been used to isolate DRMs that can differ somewhat in protein and lipid composition from Triton X-100 DRMs [8–10]. These studies have been taken to suggest that distinct types of detergent-resistant membrane raft may coexist in cell membranes. However, caution in this interpretation is warranted. First, biophysical studies in model membranes do not support the existence of distinct types of sterol-dependent, ordered phase domains. Instead, several different order-preferring lipids all form lo phase domains when mixed with cholesterol [11, 12]. Thus, the physical basis underlying formation of distinct raft subtypes is not clear.

Second, as noted above, Triton X-100 solubilizes most ld phase lipid in model membranes containing co-existing lo and ld phase domains, leaving most lo phase lipid insoluble [2, 3]. Similar studies have not been performed using other detergents. It is possible that weak detergents like those of the Brij series do not fully solubilize ld phase membranes and their associated proteins. In sum, it is not known whether using different detergents to isolate DRMs uncovers preexisting microdomain heterogeneity in the membrane, or instead reflects differential interactions of different detergents with a single type of domain. For this reason, this chapter is limited to a description of Triton X-100 DRMs. Readers are referred to a previous description of DRMs made with different detergents [8].

In later years, rafts have been surrounded by controversy, and their nature and even their existence in cells have been questioned [13–16]. Rafts may form only transiently, in a regulated manner, from preexisting nano-clusters of raftophilic proteins and lipids [17]. As in model membranes, it is clear that DRMs do not correspond exactly to preexisting rafts in cells. First, cells must be extracted with

Triton X-100 at low temperatures to generate DRMs. Phase separation is highly temperature-dependent, and raft formation is greatly enhanced at low temperatures. Second, Triton X-100 probably induces raft coalescence in cell membranes, as it does in model membranes [3]. DRMs isolated from cells are large (micron-sized), while FRET between raftophilic glycosyl phosphatidylinositol (GPI)-anchored proteins shows that rafts—if they exist in resting cells—are nanoscale [17, 18].

In light of these many caveats, why isolate DRMs? One important reason is that enrichment of proteins in DRMs is a simple way of assessing their inherent affinity for rafts, or raftophilicity. DRM-enriched proteins were not necessarily present in rafts at the time of extraction, but are likely to partition efficiently into rafts if and when they form. Demonstrating raftophilicity by DRM association can be an important step in determining whether raft association is important in protein function. Furthermore, because DRMs have a relatively simple protein composition, DRM isolation can be an easy and inexpensive step in purification of DRM proteins. Finally, DRM formation can provide a clue as to whether a membrane has any tendency to form lo domains, and obtain a first estimate of their likely protein and lipid composition. For instance, DRM formation helped show that rafts exist in the outer membrane of the bacterium *Borrelia burgdorferi* [19].

This chapter describes extraction of whole adherent cultured mammalian cells with Triton X-100 and flotation of DRMs on sucrose gradients. It then outlines a method for analysis of protein distribution across the gradient, to determine the fraction of a protein in DRMs. Finally, a method for recovery of DRMs from gradients for further use is presented.

2   Materials

2.1   Buffers and Solutions

- TNE: 25 mM Tris–Cl, pH 7.4; 150 mM NaCl; 5 mM EDTA.

- TNE/TX-100: TNE + 1 % w/v Triton X-100.

- Na_2CO_3/TX-100; 0.1 M Na_2CO_3 + 1 % Triton X-100.

- TNE/sucrose; Dissolve sucrose to the desired concentration (w/v) in TNE. Warming TNE + 80 % sucrose on a heating stir plate will facilitate solution. Store all TNE/sucrose solutions at 4 °C.

- PI's: Just before use, add the following protease inhibitors (PI's) to buffers where indicated: 0.2 mM PMSF; 1 µg/ml leupeptin, 1 µg/ml pepstatin. All these are stored as 1,000× stocks; PMSF in ethanol at room temperature; pepstatin in methanol at 4 °C, and leupeptin in water, –20 °C.

- 4× Gel Loading Buffer: 40 mM Tris–Cl, pH 6.8; 8 % SDS; 40 % glycerol; 0.05 % bromophenol blue; 20 % beta mercaptoethanol.

2.2 Other Materials

- Teflon cell scrapers or rubber policemen.
- 5 ml syringes.
- 22 G needles.
- Beckman SW41 tubes; polyallomer (331372) or Ultraclear™ (344059).
- Beckman SW41 swinging bucket ultracentrifuge rotor (331362).

2.3 Equipment

- Ultracentrifuge, floor model, Beckman Coulter or other.
- Density gradient fractionator, ISCO Model 185 (optional).

3 Methods

3.1 Cell Lysate Preparation and Sucrose Gradient Formation

1. DRMs may be prepared from any mammalian cell type. This method is described for adherent cells, but may be easily adapted to pelleted suspension-grown cells. Grow adherent cells to confluence in a 37 °C, 5 % CO_2 incubator. Use two 10 cm dishes/SW41 gradient tube (*see* **Note 1**). Place cells on ice and rinse twice with ice-cold PBS.

2. For the standard method, add 1 ml of TNE/TX-100/PI's to each dish (*see* **Note 2**). Alternatively, add 1 ml of Na_2CO_3/TX-100/PI's to each dish (*see* **Note 3**).

3. Incubate on ice 20 min.

4. Scrape the lysate with a rubber policeman or Teflon scraper and transfer to a 10 ml glass tube. Pool the material from two dishes (*see* **Note 4**).

5. If cells were lysed with Na_2CO_3/TX-100/PI's, nuclei will lyse, releasing DNA and making the lysate viscous. In this case, push the lysate through a 22 G needle attached to a 5 ml syringe repeatedly until DNA is sheared and lysate is not viscous. (Omit this step if cells are lysed with TNE/TX-100/PI's, as the nuclei will not lyse and the solution will not be viscous. Nuclei will then pellet in the ultracentrifuge.)

6. Add 2.5 ml TNE/80 % sucrose/PI's, (or 1.25 ml for each 10 cm dish harvested) to the lysate. Mix well by pipetting but avoid bubbles.

7. Transfer lysate, carefully avoiding any bubbles, to a polyallomer or Ultraclear™ SW41 tube. (The floating DRM band is easier to see in Ultraclear™ than polyallomer tubes. However, Ultraclear™ tubes should be avoided if Na_2CO_3 is used, as they will crack.) Carefully overlay with 5 ml of TNE/38 % sucrose/PI's. Overlay this with 1.5–2 ml TNE/5 % sucrose/PI's, or enough to fill the tube to within about 2 mm of the top. For some applications, it may be desirable to include detergent in

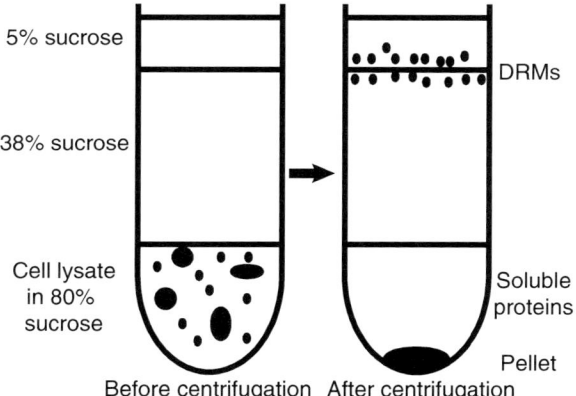

Fig. 1 Sucrose gradient before and after centrifugation. *Left*: gradient prepared as in the text. Whole cell Triton X-100 lysates, including detergent-insoluble cyto-skeleton, nuclear remnants, and DRMs (all represented as *filled ovals*) as well as detergent-solubilized proteins (not shown) adjusted to 80 % sucrose are placed in the bottom of the gradient and overlaid with 38 % and then 5 % sucrose solutions. *Right*: after centrifugation, DRMs (*small filled circles*) float to their equilibrium density position, at the 5 %–38 % sucrose interface. Cytoskeleton, nuclear remnants, and cytoskeleton-associated DRMs (often considerable) pellet. Soluble proteins sink slowly in the gradient but remain largely in the 80 % sucrose fractions at the bottom of the gradient. Solubilized detergent–lipid micelles (not shown) may float to the same position as the DRMs

the gradient solutions [20] (*see* **Note 5**). Caps are not required for the tubes. Balance carefully and place tubes in buckets for a SW41 swinging-bucket rotor. Place all six buckets (not only those containing samples) in the correct positions on the rotor and place the rotor in a prechilled ultracentrifuge. The vacuum should be on during prechilling, to avoid condensate in the ultracentrifuge.

8. Spin at 28–41K RPM ($100,000–288,000 \times g$) for at least 3 h, or as long as overnight, at 4 °C under vacuum (*see* **Note 6**). DRMs are visible as a cloudy band at the interface between the 38 and 5 % sucrose layers (Fig. 1).

9. There are two alternate methods from this step. An analytical method, used to determine how much of a protein is in DRMs by fractionating the gradient and Western blotting across the gradient, is described in Subheading 3.2. An alternate preparative method, used to recover DRMs, is described in Subheading 3.3.

3.2 Gradient Fractionation and DRM Protein Detection by Western Blotting

1. If available, insert the tube in a density gradient fractionator. This is equipped with a needle that slides up and down, positioned to puncture the bottom of the tube when in the up position. The needle is at the end of a piece of plastic tubing. Once the gradient tube is in place, push up the needle to

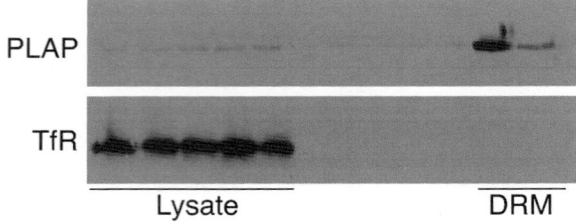

Fig. 2 Western blot across fractionated gradient showing positions of a raftophilic and a non-raftophilic protein. B16 melanoma cells transiently expressing the raftophilic GPI-anchored protein placental alkaline phosphatase (PLAP) were subjected to Triton X-100 extraction and gradient centrifugation as described in the text. The gradient was fractionated from the bottom. Proteins in aliquots of each fraction were analyzed by SDS-PAGE and Western blotting. (Gradient bottom is at the *left*). *Top blot*: PLAP. *Bottom blot*: endogenous transferrin receptor (TfR). Most PLAP floats to a low-density position in the gradient, while most TfR remains in the load fractions at the bottom of the gradient

puncture the tube. The liquid in the gradient will flow out through the tubing, starting from the gradient bottom. Set the fraction collector to collect 1 ml fractions in disposable 5 ml glass tubes or 1.5 ml microfuge tubes.

If a gradient harvester is not available, carefully remove 1-ml fractions from the top of the gradient with a pipette. If it is of interest to determine how much of a protein pellets in the gradient, after removing the gradient fluid, use 250 μl 2× Gel Loading Buffer (made by diluting 4× Gel Loading Buffer in water) to solubilize proteins from the gradient pellet. Transfer to a microfuge tube.

2. Add 330 μl 4× Gel Loading Buffer to each gradient fraction tube, mix well, and heat to 95 °C for 5 min. This will solubilize the DRMs and fully denature proteins for SDS-PAGE.

3. Load 25 μl of each gradient fraction (and 5 μl of the solubilized pellet fraction, if used) on a standard SDS-PAGE mini-gel (previously optimized for the protein(s) of interest), and run the gel as usual. Transfer proteins to nitrocellulose or PVDF. Probe the blot with antibodies to the protein of interest and detect by enhanced chemiluminescence or an equivalent method (*see* **Note 7**). Also probe for positive and negative control proteins (*see* **Note 8**) (Fig. 2).

3.3 DRM Recovery by Pelleting in the Ultracentrifuge

1. After Subheading 3.1, **step 7** (gradient centrifugation), follow the steps in this section to recover DRMs (as an alternate to Subheading 3.2). This procedure is useful in DRM protein purification, and for analysis of DRM proteins by methods other than SDS-PAGE. This method should also be used for DRM lipid analysis (*see* **Note 9**).

2. Use a 22 G needle on a 5 ml syringe to carefully harvest the visible band at the 5 %–38 % interface in the gradient. Drag the needle back and forth over the interface while drawing back on the syringe plunger until all visible membrane has been drawn into the syringe. You should have a total of 2–3 ml of fluid.

3. Dispense the contents of the syringe into an SW41 tube, and add enough TNE/PI's to fill the tube. Mix well by pipetting or inverting (after covering the tube with Parafilm). Dilution is important to reduce the density of the solution, allowing the DRMs to pellet. Balance the tube carefully and load into the SW41 rotor. Centrifuge at 28–41 K RPM ($100,000-288,000\times g$) for 1 h.

4. Remove and discard the supernatant from the tube. Dislodge the pellet with 100–200 µl of TNE/PI's. Transfer to a 1.5 ml microfuge tube. Rinse the ultracentrifuge tube twice with 100 µl of TNE/PI's and add to the DRMS in the microfuge tube. Spin the tube 10 min at top speed in a microfuge in the cold room to pellet the DRMs. Remove and discard supernatant. Pelleted DRMs are ready for further use (*see* **Note 10**).

4 Notes

1. DRM preparation can be scaled up or down. Scale the number of cells used proportionately to the volume of the gradient tube used.

2. A variety of other non-ionic detergents may be used instead of Triton X-100 to generate DRMs, which may have different properties [9, 10, 21]. Avoid octyl β-d-glucopyranoside (octyl glucoside) as this solubilizes DRMs [6]. DRMs are not very sensitive to buffer variation, and most common buffers could be substituted for Tris.

3. Lysing cells with Na_2CO_3/TX-100/PI's increases DRM yield. DRMs are often associated with cytoskeleton, increasing their density and causing them to pellet, instead of floating, in the sucrose gradient. Alkaline carbonate breaks most protein–protein interactions, but does not affect hydrophobic interactions of proteins with lipid, and is often used to strip peripheral proteins from membranes [22]. Preparing DRMs from cells treated with Na_2CO_3/TX-100/PI's releases associated cytoskeleton as well as peripheral membrane proteins from DRMs, allowing more DRMs with their associated integral membrane proteins to float in the gradient. Thus, this method is useful for determining whether a DRM-enriched protein associates directly with DRM lipids, or whether it binds to another DRM-associated protein. Both proteins will be enriched in DRMs prepared with TNE/TX100/PI's, but only the former will be present in DRMs prepared with Na2CO3/TX-100/PI's.

4. It is essential to keep the lysate and gradient cold throughout the procedure, to avoid solubilizing DRMs. Cells may be lysed and the gradient formed in the cold room.

5. Including detergent throughout the gradient can enhance full solubilization of weakly raftophilic proteins and minimize DRM rearrangement during centrifugation, especially when a low concentration of detergent is used in the initial extraction [20].

6. If scaling up or down, scale the volumes of each gradient component proportionately to the total volume of the tube. Spin for at least the equivalent of $100,000 \times g$ for 3 h, or up to overnight. DRMS will migrate to their equilibrium density position (at the 5 %–38 % sucrose interface), while solubilized detergent–protein micelles will remain in the load fraction at the bottom of the tube. Detergent–protein micelles will slowly migrate downward, but will still largely remain in the load fractions even after overnight centrifugation.

7. DRM proteins will be present in Fractions 10–12 (numbering from the bottom) in the gradient. Fully solubilized protein–detergent micelles will remain at the bottom of the gradient, in Fractions 1–5. Ideally, no protein should be present in the intervening gradient fractions (Fig. 2). The amount of the protein in DRMs can be determined by scanning the bands and quantitating the signal in each band by densitometry. The percent of the protein in DRMs is the sum of the signal in the DRM fractions, divided by the sum of the signal in all fractions, times 100.

8. It is important to examine positive and negative control plasma membrane proteins in the same gradient, to ensure that detergent solubilization and gradient conditions are sufficient to fully solubilize non-raftophilic proteins, while giving a good enrichment of raftophilic proteins in DRMS. Raftophilic glycosyl phosphatidylinositol (GPI)-anchored and caveolin proteins are good positive controls. Transferrin receptor is a good non-raftophilic negative control, as it is widely expressed in mammalian cells and sensitive antibodies with broad species reactivity that work well on blots are available (e.g., Life Technologies #13-6800). Proteins residing in intracellular membranes, such as the endoplasmic reticulum (ER), are not good negative controls. Because it is poor in sterol and sphingolipids, the ER may be fully solubilized under conditions that do not fully solubilize non-raftophilic plasma membrane proteins.

9. Solubilized lipids, in the form of detergent–lipid micelles, have a lower density than solubilized proteins, and may float in the gradient and may comigrate with DRMs. However, they will not pellet with DRMs after centrifugation. For this reason, the method described in **steps 12–14** should be used for DRM lipid analysis.

10. DRMs can be solubilized with TNE/1 % Triton X-100/PI's by incubating at 37 °C for 5 min, or with TNE/60 mM octyl glucoside/PI's in 20 min on ice. Proteins of interest can then be recovered by immunoprecipitation, or examined by SDS-PAGE and Western blotting. Alternatively, for Western blotting, the DRM pellet can be directly dissolved in 2× Gel Loading Buffer. In this case, the high amount of lipid in the sample may distort the gel near the dye front and make it difficult to detect very small proteins.

To do this, resuspend DRM pellet in 1 ml cold acetone, to dissolve the lipids and precipitate proteins. Vortex to disperse the precipitated protein. Incubate on ice 10 min, then spin 10 min at 4 °C. Remove the acetone and add 1 ml diethyl ether. (DRM lipids may not be fully solubilized by acetone, but are solubilized by diethyl ether.) Vortex and spin 10 min at 4 °C. Remove diethyl ether, air-dry the pelleted proteins, and solubilize in 1× Gel Loading Buffer.

Acknowledgement

This work was supported by NIH grant GM47897 to DAB.

References

1. Brown DA (2006) Lipid rafts, detergent-resistant membranes, and raft targeting signals. Physiology 21:430–439
2. Ahmed SN, Brown DA, London E (1997) On the origin of sphingolipid-cholesterol rich detergent-insoluble domains in cell membranes: physiological concentrations of cholesterol and sphingolipid induce formation of a detergent-insoluble, liquid-ordered lipid phase in model membranes. Biochemistry 36:10944–10953
3. Pathak P, London E (2011) Measurement of lipid nanodomain (raft) formation and size in sphingomyelin/POPC/cholesterol vesicles shows TX-100 and transmembrane helices increase domain size by coalescing preexisting nanodomains but do not induce domain formation. Biophys J 101:2417–2425
4. Heerklotz H (2002) Triton promotes domain formation in lipid raft mixtures. Biophys J 83:2693–2701
5. Lichtenberg D, Goni FM, Heerklotz H (2005) Detergent-resistant membranes should not be identified with membrane rafts. Trends Biochem Sci 30:430–436
6. Brown DA, Rose JK (1992) Sorting of GPI-anchored proteins to glycolipid-enriched membrane subdomains during transport to the apical cell surface. Cell 68:533–544
7. Ge M, Field KA, Aneja R et al (1999) ESR characterization of liquid ordered phase of detergent resistant membranes from RBL-2H3 cells. Biophys J 77:925–933
8. Waugh MG, Hsuan JJ (2009) Preparation of membrane rafts. Methods Mol Biol 462:403–414
9. Röper K, Corbeil D, Huttner WB (2000) Retention of prominin in microvilli reveals distinct cholesterol-based lipid microdomains in the apical plasma membrane. Nat Cell Biol 2:582–592
10. Drevot P, Langlet C, Guo X-J et al (2002) TCR signal initiation machinery is pre-assembled and activated in a subset of membrane rafts. EMBO J 21:1899–1908
11. Zhao J, Wu J, Heberle FA et al (2007) Phase studies of model biomembranes: complex behavior of DSPC/DOPC/cholesterol. Biochim Biophys Acta 1768:2764–2776
12. Konyakhina TM, Wu J, Mastroianni JD et al (2013) Phase diagram of a 4-component lipid mixture: DSPC/DOPC/POPC/chol. Biochim Biophys Acta 1828:2204–2214

13. Munro S (2003) Lipid rafts: elusive or illusive? Cell 115:377–388

14. Skwarek M (2004) Recent controversy surrounding lipid rafts. Arch Immunol Ther Exp (Warsz) 52:427–431

15. Shaw AS (2006) Lipid rafts: now you see them, now you don't. Nat Immunol 7:1139–1142

16. Kraft ML (2013) Plasma membrane organization and function: moving past lipid rafts. Mol Biol Cell 24:2765–2768

17. Kusumi A, Suzuki KG, Kasai RS et al (2011) Hierarchical mesoscale domain organization of the plasma membrane. Trends Biochem Sci 36:604–615

18. Sharma P, Varma R, Sarasij RC et al (2004) Nanoscale organization of multiple GPI-anchored proteins in living cell membranes. Cell 116:577–589

19. LaRocca TJ, Crowley JT, Cusack BJ et al (2010) Cholesterol lipids of Borrelia burgdorferi form lipid rafts and are required for the bactericidal activity of a complement-independent antibody. Cell Host Microbe 8:331–342

20. Korzeniowski M, Kwiatkowska K, Sobota A (2003) Insights into the association of Fc-gamma-RII and TCR with detergent-resistant membrane domains: isolation of the domains in detergent-free density gradients facilitates membrane fragment reconstitution. Biochemistry 42:5358–5367

21. Hooper NM, Turner AJ (1988) Ectoenzymes of the kidney microvillar membrane. Differential solubilization by detergents can predict a glycosyl-phosphatidylinositol membrane anchor. Biochem J 250:865–869

22. Fujiki Y, Hubbard AL, Fowler S et al (1982) Isolation of intracellular membranes by means of sodium carbonate treatment: application to endoplasmic reticulum. J Cell Biol 93:97–102

Chapter 6

Isolation of Giant Plasma Membrane Vesicles for Evaluation of Plasma Membrane Structure and Protein Partitioning

K.R. Levental and I. Levental

Abstract

Although investigation into the structure of eukaryotic cell membranes has been an intense focus of cell biology for the past two decades, definitive insights have been limited by the lack of coherent methods for the isolation of specific organelle membranes and the identification of membrane subdomains. Here we describe a method for the isolation of mammalian cell plasma membranes as Giant Plasma Membrane Vesicles (GPMVs) and strategies for imaging membrane lateral structure and quantification of protein partitioning between coexisting domains by fluorescence microscopy.

Key words Lipid rafts, Plasma membrane, GPMV, Phase separation

1 Introduction

The lipid raft hypothesis [1] posits nanoscale organization of mammalian cell membranes driven by preferential interactions between specific lipids and proteins that give rise to functional lateral membrane domains. The physicochemical basis for this organization is the cholesterol-dependent formation of a liquid-ordered membrane phase that can coexist with a liquid-disordered phase rich in unsaturated lipids and depleted of saturated lipids/cholesterol/glycosphingolipids [2]. Although these interactions and their resultant long-range organization have been extensively characterized in purified lipid model membranes [2, 3], the complexity of live cells, lack of robust experimental methods, and inability to directly image lipid rafts led to significant controversy regarding their functions [4] and even existence [5]. However, recent advances in plasma membrane isolation [6–9], nanoscale spectroscopy [10, 11], super-resolution [12] and electron microscopy [13], and lipidomics [14, 15] have led to definitive observations of lateral domains in biological membranes. Such structures

Dylan M. Owen (ed.), *Methods in Membrane Lipids*, Methods in Molecular Biology, vol. 1232,
DOI 10.1007/978-1-4939-1752-5_6, © Springer Science+Business Media New York 2015

can be generated either by lipid-driven phase separation (i.e., membrane rafts) or by cholesterol-independent protein–protein and protein–lipid interactions, likely with significant overlap and cooperation between these mechanisms [16]. In all cases, the result of domain formation is the functional segregation of membrane components.

In live cells, lipid-driven membrane rafts are rarely observable by light microscopy because of the length and time scales involved—domains are estimated to be tens of nanometers in size and are dynamic on millisecond timescales. Thus, advanced techniques are required for the measurement of domain compositions and properties. Giant plasma membrane vesicles (GPMVs) are one such method for efficient isolation of intact plasma membranes (PMs) maintaining the full diversity of native membrane components [17]. The coexistence of two liquid phases with distinct physical properties [18, 19] and compositions [20, 21] in these natural membranes provides compelling evidence for the central tenet of the lipid raft hypothesis. More importantly, GPMVs comprise an intermediate biological membrane model system, combining the compositional complexity and protein content of live cell membranes with the macroscopic phase separation and experimental malleability of synthetic vesicles. The most important advantage of this model system is that it allows quantitative measurement of protein partitioning [18, 20] and (potentially) function in large, stable, well-resolved domains. This capability has allowed investigation of the structural determinants of raft partitioning [20–22] and is likely to yield more insights about raft-dependent protein function in the coming years.

Here, we describe methods to isolate plasma membranes of mammalian cells as GPMVs, and observe their lateral compositional heterogeneity. The entirety of the experiment, comprising fluorescent labeling of live cell membranes, GPMV isolation, and microscopic visualization can be achieved in 2–3 h.

2   Materials

Prepare all solutions using ultrapure water (at least 18 MΩ) and analytical grade reagents. Reagent preparation can be done at room temperature, but buffers are stored at 4 °C. Some reagents used are toxic and proper regulations must be followed for their handling and disposal.

2.1   Reagents

1. Cell culture medium (varies depending on cell type).

2. Paraformaldehyde (PFA) *Caution: PFA is a suspected carcinogen. Use gloves and avoid contact.*

3. Dithiothreitol (DTT) *Caution: DTT is toxic. Use gloves and avoid contact.*

4. *N*-ethyl maleimide (NEM) *Caution: NEM is toxic. Use gloves and avoid contact.*

5. 3,3′-dilinoleyloxacarbocyanine perchlorate (FAST-DiO; Invitrogen, cat.no. D3898).

6. 1,1′-dilinoleyl-3,3,3′,3′-tetramethylindocarbocyanine, 4-chlorobenzenesulfonate (FAST-DiI; Invotrogen, cat. no. D7756).

7. Cholera Toxin B subunit with desired fluorescent tag (CTxB; Invitrogen, cat. no. C-34775, C-34776, C-34777, C-34778).

8. Sodium Chloride (NaCl).

9. Potassium Chloride (KCl).

10. Calcium Chloride ($CaCl_2$).

11. 4-(2-Hydroxyethyl)piperazine-1-ethanesulfonic acid (HEPES).

12. Potassium phosphate monobasic (KH_2PO_4).

13. Sodium phosphate dibasic (Na_2HPO_4).

14. Hydrochloric Acid (HCl).

15. Sodium Hydroxide (NaOH).

16. Bovine Serum Albumin (BSA).

17. Distilled water.

18. Vaseline® (Sigma-Aldrich, cat. no. 16415)—or other water-repellent sealant/lubricant.

19. PBS (*see* Reagent Setup).

20. GPMV buffer (*see* Reagent Setup).

2.2 Solutions to Prepare

1. *Phosphate-buffered Saline (PBS)*: 137 mM NaCl, 2.7 mM KCl, 10 mM Na_2HPO_4, 2 mM KH_2PO_4, pH 7.4. For 1 L solution, dissolve 8 g NaCl, 0.2 g KCl, 1.42 g Na_2HPO_4, and 0.27 g KH_2PO_4 in distilled water. Bring to 1 L. Adjust the pH with HCl or NaOH to 7.4.

2. *GPMV buffer*: 10 mM HEPES, 150 mM NaCl, 2 mM $CaCl_2$, pH 7.4. For 1 L solution, dissolve 8.75 g NaCl, 0.22 g $CaCl_2$ and 2.38 g HEPES in distilled water and bring to 1 L. Adjust the pH with HCl or NaOH to 7.4. Store at 4 °C.

3. *FAST DiO solution*: Dissolve dye in ethanol to make stock solutions of 0.5 mg/mL. Store at –20 °C in lightproof vials.

4. *4 % PFA*: Dissolve 4 g of PFA in 50 mL PBS. Heat to approximately 60 °C, and add a few drops of 1 M NaOH until the solution becomes clear. pH to 7.4 with HCl or NaOH, and bring final volume to 100 mL with PBS. Store in aliquots at –20 °C.

5. *1 M DTT*: Dissolve 1.54 g DTT in distilled H2O and bring to 10 mL. Aliquot and store at –20 °C.

6. *1 M NEM*: Dissolve 1.25 g NEM in 100 % ethanol and bring to 10 mL. Aliquot and store at –20 °C.

7. *1 mg/mL Bovine Serum Albumin (BSA)*: Dissolve 50 mg BSA in distilled water and bring to 50 mL. Store at 4 °C.

2.3 Equipment

1. 37 °C Incubator.
2. Tissue culture dishes.
3. Microcentrifuge tubes.
4. Microcentrifuge.
5. #1.5 coverslips.
6. Inverted epifluorescence microscope.
7. Temperature controller (Warner Instruments; cat. no. 64-0352).
8. Thermal insert (Warner Instruments; cat. no. 64-1636, 64-1646).
9. 40× air objective.
10. Circulating water bath with hoses.
11. Dehumidified air or nitrogen (N_2).
12. Tally Counter (with two positions).

3 Methods

3.1 Preparation of GPMVs

1. Seed the cells in a 35 mm dish (or other appropriate chamber—*see* **Note 1**).
2. Incubate the culture vessels in appropriate conditions until cells reach ~70 % confluency. Check American Type Culture Collection (ATCC) database for the appropriate conditions and media. Optional: Manipulate the cell culture (e.g., transfection, drug-treatment) depending on the experiment (*see* **Note 2**).
3. Label the cell membranes:
   (a) Wash cells 2× with 1 mL PBS.
   (b) Label cells with fluorescent dye. Add dye at appropriate concentration to 0.5 mL of PBS and carefully pipet onto to cells (*see* **Note 3**).
   (c) Incubate at 4 °C for 10 min to allow dye incorporation into membranes.
   (d) Aspirate dye solution, wash 5× with PBS to remove unincorporated dye, and proceed to GPMV isolation.
4. Wash cells 2× with 1 mL of GPMV buffer.
5. Add vesiculation agents to GPMV buffer and apply to cells (*see* **Note 4**):
   (a) For PFA/DTT (25 mM PFA/2 mM DTT): 18 µL of 4 % PFA solution and 2 µL of 1 M DTT solution to 1 mL of GPMV buffer.
   (b) For NEM (2 mM): 2 µL of 1 M NEM to 1 mL of GPMV buffer.

6. Incubate the cells at 37 °C for 1 h.

7. Remove chambers from incubator and check for presence of vesicles. These should be readily observable at 20× magnification as dark free-floating spheres just above the plane of the cells (Fig. 1d).

8. Transfer the GPMV-rich cellular supernatant into a microcentrifuge tube. After isolation, GPMVs can be stored at 4 °C for 1–2 days without visible degradation.

9. Concentrate GPMV suspension:

    (a) For biochemical or spectroscopic experiments where purity of plasma membrane is more important than vesicle morphology, centrifuge GPMV suspension at 100 rcf for 10 min to pellet cell debris. Then, centrifuge the supernatant at 20,000 rcf for 1 h to pellet GPMV membranes. Pellet can be stored at −20 °C, or resuspend in appropriate buffer for subsequent assays (protein biochemistry, lipid mass spectrometry, etc.).

    (b) For imaging experiments where purity is less important than GPMV integrity, leave GPMV suspension in microcentrifuge tube for 20–30 min—GPMVs will concentrate at the bottom of the tube.

3.2 Analysis of Isolated GPMVs: Imaging of GPMVs and Quantification of Partitioning

1. Concentrate vesicles without centrifugation as in **step 9b** in Subheading 3.1.

2. Prepare BSA-coated coverslips:

    (a) The bottom coverslip of the imaging chambers should be coated with BSA to avoid nonspecific sticking/bursting of the GPMVs on the glass. Incubate coverslips in 1 mg/mL BSA solution for at least 1 h. Prior to use, rinse with distilled H_2O and dry using a paper wipe (e.g., Kimwipes®). For LabTek chambers, add 250 μL of 1 mg/mL BSA solution into the wells and incubate at least 1 h. Wash 2× with PBS.

3. Prepare the cooling system:

    (a) Connect the temperature controlled stage and/or objective cooler to a circulating, temperature-controlled water bath with rubber hoses, sealed at the connection points. Set the water bath temperature to approximately the desired chamber/sample temperature. If using a Peltier-element temperature controller, water circulation should begin prior to temperature regulation, to remove the heat produced in the device. If the cooling setup in Fig. 1 is used with an air immersion objective, cooling the objective is unnecessary. However, condensation can form on the sample surface, hindering imaging. To avoid this, a stream of dehumidified air or N_2 can be blown directly onto the

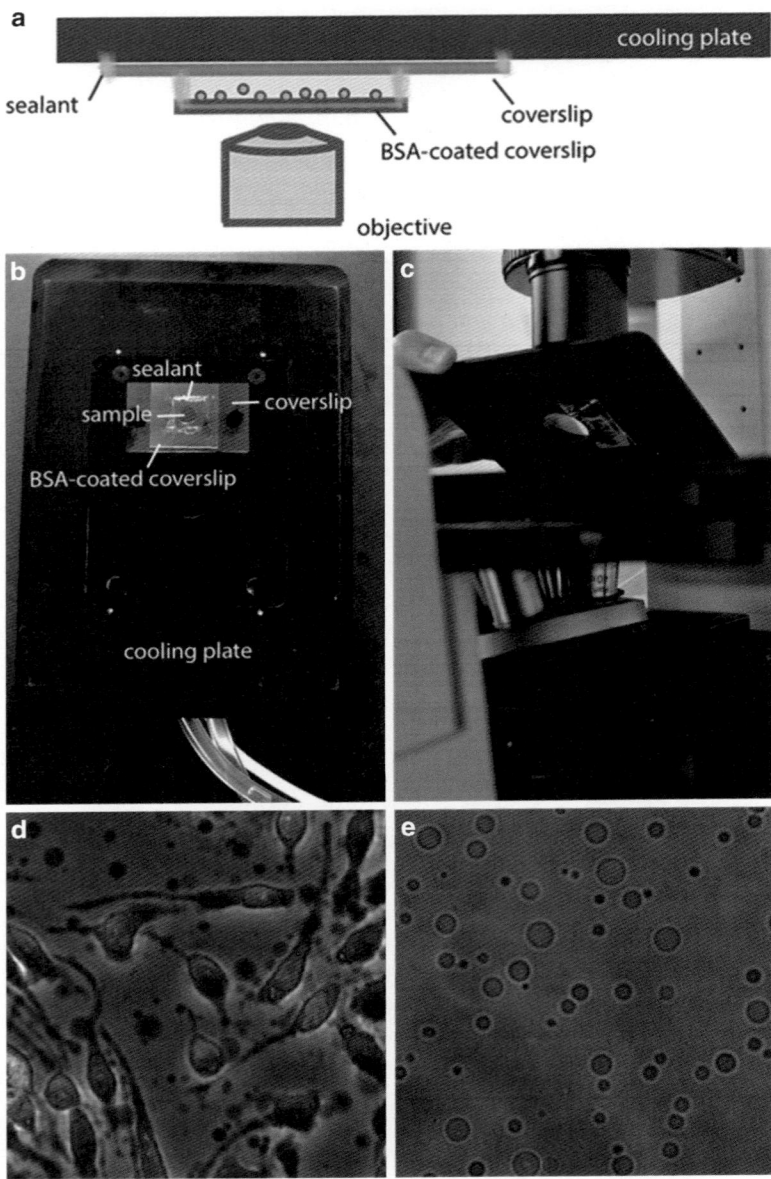

Fig. 1 Experimental setup and expected results. (**a**) Graphical representation of the microscopic setup comprising the GPMV suspension between two coverslips mounted on a temperature-controlled microscope stage. (**b**) Photograph of sample mounted on the cooling plate and (**c**) then on the microscope stage. (**d**) Free-floating and cell-attached GPMVs above adherent cells imaged at 20× using DIC microscopy after **step 7**. (**e**) Isolated GPMVs imaged at 40× after **step 9**

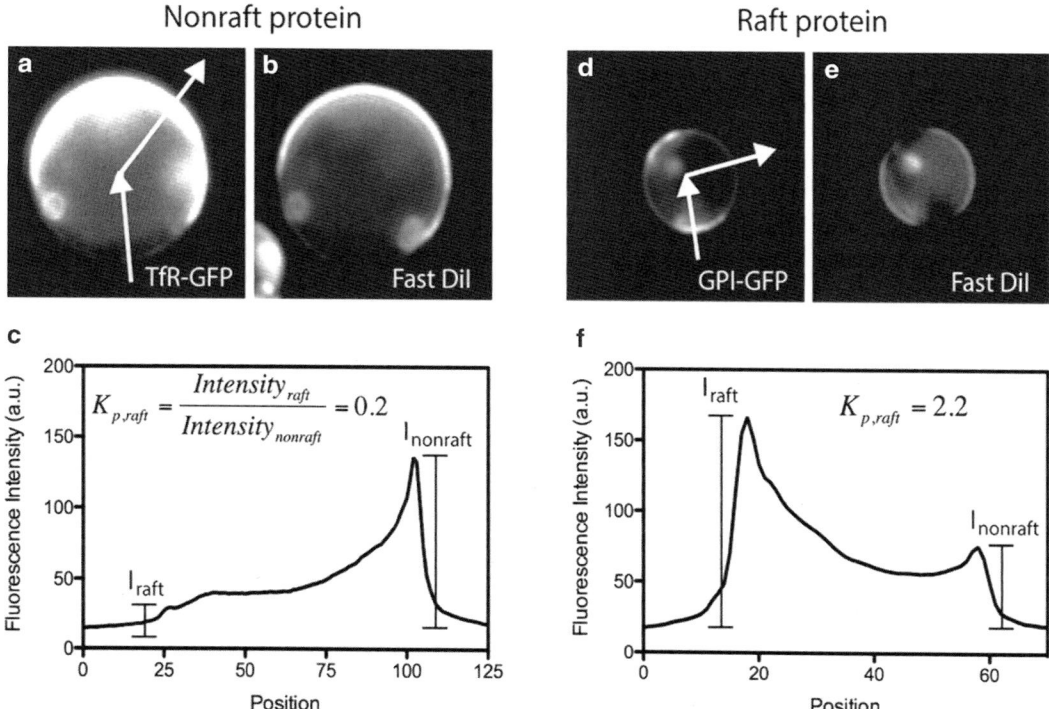

Fig. 2 Quantification of raft partitioning. (**a**) A non-raft, GFP-tagged protein (Transferrin Receptor—TfR-GFP) is imaged in the same vesicles with a non-raft marker lipid (FAST Dil, **b**). (**c**) $K_{p,raft}$ is quantified by the ratio of the background-subtracted fluorescence intensities in the two domains. (**d–f**) Analogous images and quantification for a raft-enriched protein (GPI-GFP)

sample or in the space between the objective and sample. If immersion objectives are used to image GPMVs, objective cooling may be necessary to avoid heat flow to the sample. In this case, care must be taken to thermally isolate the objective from the rest of the microscope to avoid condensation inside the microscope box.

4. Construct imaging chamber as shown in Fig. 1a–c and let temperature equilibrate for at least 10 min. This time will also allow the vesicles to sink to the bottom of the chamber.

5. Decrease the temperature below 15 °C for GPMVs derived with PFA/DTT and below 10 °C for those derived with NEM (*see* **Notes 4–7**).

6. One channel should be used to image a well-characterized marker of either the raft or non-raft phase (e.g., unsaturated lipids for non-raft; fluorescently labeled cholera toxin for raft) while imaging the component of interest in the other fluorescent channel (as in Fig. 2; *see* **Note 8**). Image each channel sequentially at the same focal position.

7. Draw a segmented line scan starting from outside of the raft phase to the middle of the GPMV, continuing through the middle of the nonraft phase. Raft partitioning can be quantified by calculating $K_{p,raft}$ from the line scans as shown in Fig. 2 (*see* **Note 9**).

3.3 Analysis of Isolated GPMVs: Imaging of GPMVs and Quantification of Miscibility Temperature

1. Concentrate vesicles without centrifugation as in **step 9b** in Subheading 3.1.

2. Construct imaging chamber as shown in Fig. 1a–c and let temperature equilibrate for at least 10 min. This time will also allow the vesicles to sink to the bottom of the chamber.

3. Increase the temperature to 25 °C.

4. Using a tally counter, count the number of vesicles that are and are not phase-separated (*see* Fig 3b, c for examples of each).

5. If the percentage of phase-separated vesicles is more than 10 %, increase the temperature to 30 °C and count again. If the percentage of phase-separated vesicles is less than 10 %, decrease the temperature by 4 °C, and count again. Repeat this process of decreasing the temperature and counting the percentage of phase-separated vesicles until the vesicles are 95–100 % phase-separated.

6. Plot the percent of phase-separated GPMVs versus the temperature and fit to a sigmoidal function (e.g., Eq. 1) as shown in Fig. 3d to derive the miscibility transition temperature at which 50 % of vesicles would be expected to be phase separated.

(a)

$$\% \text{ phase separated} = 100 \times \left(A - \frac{1}{1 - e^{-B \times \text{Temp}}} \right) \tag{1}$$

A and B are fit parameters.

4   Notes

1. Protocol is optimized for cells seeded in a 35-mm cell culture dish. For microscopy experiments, a 70 % confluent 35-mm dish should be sufficient for a number of individual samples. For biochemical assays, it is often necessary to start with a 10-cm dish. Seed the cells on #1.5 glass slide (or a glass-bottom chamber slide) if imaging or spectroscopy of cell-attached GPMVs is needed.

2. Some common transfection agents (e.g., Lipofectamine™) use cationic lipids to form complexes with DNA for transmembrane delivery. If possible, these reagents should be avoided

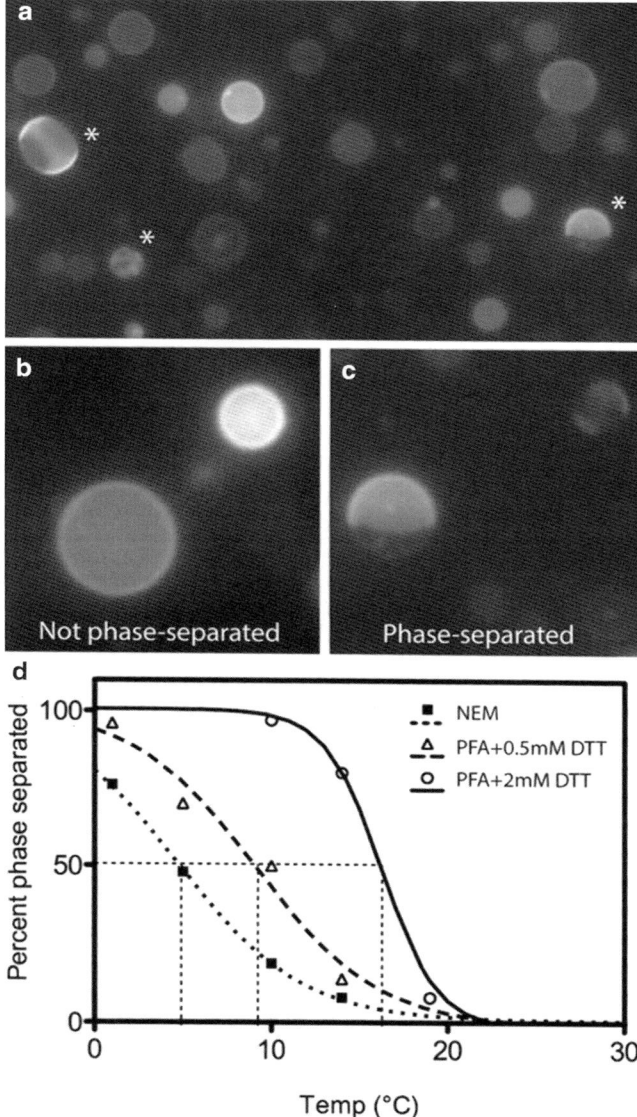

Fig. 3 Quantification of phase separation temperature. Exemplary images of GPMVs isolated from cells prestained with FAST DiO. (**a**) A mixed population of phase separated and non-phase separated GPMVs (*asterisks* mark phase separated GPMVs); (**b**) not phase separated GPMVs; and (**c**) phase separated GPMVs. Dozens of GPMVs can be quickly manually scored for phase separation and the fraction of phase-separated vesicles can be quantified for various temperatures. (**d**) Sigmoidal fits to this data yield the miscibility transition temperature, where 50 % of vesicles would be expected to show phase separation

Table 1
Fluorescent lipid dyes used in labeling GPMVs

Dye	Excitation (nm)	Emission (nm)	Phase partitioning	Working concentration (µg/mL)
FAST DiO	484	501	Nonraft/disordered	0.5–5
FAST DiI	549	565	Nonraft/disordered	0.5–5
Fluorescently labeled Cholera Toxin B	[a]	[a]	Raft/ordered	1–10
NBD-DSPE	470	530	Raft/ordered	5–20

[a]Excitation and emission is dependent on choice of fluorescent label for Cholera Toxin B

due to unknown effects of exogenous lipids on membrane composition and properties. This is particularly relevant for biophysical experiments on the phase behavior of isolated membranes where lipid contaminants could cause major artifacts. Nucleofection and calcium phosphate transfection are efficient alternatives.

3. Concentrations and incorporation conditions should be optimized for all dyes, although the conditions listed in Table 1 have been optimized for the dialkylcarbocyanine (i.e., FAST-DiO, FAST-DiI, etc.) dyes used here and can be used as a reasonable starting point for untested dyes. FAST DiO/DiI are recommended both because of their good fluorescence qualities and because their unsaturated acyl chains impart a high preference for the disordered/non-raft phase (compared to DiO/DiI which have saturated acyl chains and thus a low preference for either phase). High dye concentrations can affect the physical and biological properties of membranes, and these should always be kept to the minimal adequate concentrations (no more than 0.5 mol.%). For natural membranes (like GPMVs), mol fractions are not quantifiable, but can be estimated by comparing the brightness of incorporated dye to a synthetic membrane with well-defined dye concentration.

4. The most common preparation for GPMVs uses 25 mM formaldehyde and 2 mM DTT as the chemical agents that induce vesiculation, largely because it prevents detachment of adherent cells and therefore yields a cleaner and more efficient vesicle isolation. However, this preparation includes a number of undesired artifacts: (a) cross-linking of proteins by the aldehyde, which interferes with many biochemical analyses (e.g., PAGE); (b) cleavage of protein disulfides and thioesters [20]; (c) coupling of specific lipid headgroups (phosphatidylethanolamines) to proteins [18]. These effects (especially c) combine

to significantly affect phase behavior/properties of GPMVs, progressively increasing the phase coexistence temperature by 10–15 °C (*see* Fig. 3). The requirement for vesiculation appears to be treatment with a chemical that crosses the plasma membrane and covalently blocks sulfhydryls, e.g., *N*-ethyl maleimide (NEM) (for a non-exhaustive list, *see* ref. 23). Non-crosslinking, non-reducing vesiculants are suggested to avoid the artifacts listed above. To observe extensive, microscopic phase coexistence in these preparations, it is often necessary to cool these samples below 10 °C (Fig. 3; *see* **Note 10**).

5. GPMV isolation conditions need to be optimized for different cell types. Although the common preparations (25 mM PFA + 2 mM DTT or 2 mM NEM) will yield GPMVs for most common cell types, vesiculant concentrations, chemistries, and incubation times/temperatures can be varied to achieve optimal GPMV yields and purities. Temperatures as low as 4 °C can be used, though incubation times up to 24 h are necessary to produce efficient vesicle yields.

6. Although the GPMV isolation protocol described here yields vesicles large enough for most microscopic applications (radius ~1–5 μm), it is possible to increase vesicle size by reducing the osmolarity of the GPMV isolation buffer (*see* Subheading 2.2). Reducing [NaCl] down to 50 mM will yield larger vesicles at little cost to yield.

7. A significant experimental hurdle for observing phase separation is the requirement for cooling the microscopic sample. A simple way to increase the phase separation temperature is to treat isolated vesicles with sub-mM concentrations of deoxycholate [24]; ~0.5 mM will often allow observation of phase separation at room temperature. This is only recommended for preliminary experiments, to ensure that phase separation can be observed. It must be stressed that the effects of these biological detergents on PM structure, domain formation, and component partitioning are not characterized, and thus great care must be taken in generalizing results derived with such treatments.

8. For microscopic evaluation of component partitioning (as in Fig. 2), it is important to be aware of fluorescent signal bleeding between channels. This is particularly important if GFP-labeled proteins are being observed in the presence of bright, red fluorescent dyes (e.g., FAST-DiI). Using standard microscope/filter settings, the intensity of DiI in the green channel (excitation ~480–500 nm; emission ~510–550 nm) can be up to 20 % of that in the red channel (excitation ~520–550 nm ; emission ~560–600 nm).

9. The $K_{p,raft}$ calculation shown in Fig. 2 is a very reasonable approximation of the thermodynamic partition coefficient for

fluorescent chimeras of transmembrane proteins because the fluorophores are far removed from the membrane and are thus unlikely to be affected by the differing physical properties of the coexisting membrane phases. This is not true for fluorescent lipids, whose fluorophores are typically embedded in the membrane. Additionally, a more rigorous calculation for component partitioning than the simple $K_{p,raft}$ ratio shown in Fig. is the logarithm of $K_{p,raft}$ (i.e., $\log P$).

10. It must be emphasized that while the GPMV preparation is a useful model of the cell plasma membrane, and clearly more compositionally representative of natural membranes than synthetic membranes, it is certainly not completely representative of the PM of live cells. These are several PM changes that are known to occur during GPMV formation, including ATP depletion, PIP2 hydrolysis, detachment of the cytoskeleton, and exposure of phosphatidylserine coupled to the loss of strict transbilayer asymmetry. Additionally, GPMVs represent the PM at equilibrium, a state that is incompatible with life. On top of these are the myriad unknown (and likely unknowable) changes that cells likely undergo during the hour-long incubation in the presence of reactive chemicals. These artifacts must be considered when GPMV results are extrapolated to live cells physiology.

Acknowledgments

We thank the Cancer Prevention and Research Institute of Texas (R1215) for funding support.

References

1. Simons K, Ikonen E (1997) Functional rafts in cell membranes. Nature 387(6633):569–572
2. Simons K, Vaz WL (2004) Model systems, lipid rafts, and cell membranes. Annu Rev Biophys Biomol Struct 33:269–295
3. McConnell HM, Vrljic M (2003) Liquid-liquid immiscibility in membranes. Annu Rev Biophys Biomol Struct 32:469–492
4. Kenworthy AK (2008) Have we become overly reliant on lipid rafts? Talking Point on the involvement of lipid rafts in T-cell activation. EMBO Rep 9(6):531–535
5. Munro S (2003) Lipid rafts: elusive or illusive? Cell 115(4):377–388
6. Baumgart T et al (2007) Large-scale fluid/fluid phase separation of proteins and lipids in giant plasma membrane vesicles. Proc Natl Acad Sci U S A 104(9):3165–3170
7. Levental I et al (2009) Cholesterol-dependent phase separation in cell-derived giant plasma-membrane vesicles. Biochem J 424(2):163–167
8. Lingwood D, Ries J, Schwille P, Simons K (2008) Plasma membranes are poised for activation of raft phase coalescence at physiological temperature. Proc Natl Acad Sci U S A 105(29):10005–10010
9. Ayuyan AG, Cohen FS (2008) Raft composition at physiological temperature and pH in the absence of detergents. Biophys J 94:2654–2666
10. Sengupta P, Holowka D, Baird B (2007) Fluorescence resonance energy transfer between

lipid probes detects nanoscopic heterogeneity in the plasma membrane of live cells. Biophys J 92(10):3564–3574

11. Swamy MJ et al (2006) Coexisting domains in the plasma membranes of live cells characterized by spin-label ESR spectroscopy. Biophys J 90(12):4452

12. Eggeling C et al (2009) Direct observation of the nanoscale dynamics of membrane lipids in a living cell. Nature 457(7233):1159–1162

13. Prior IA, Muncke C, Parton RG, Hancock JF (2003) Direct visualization of Ras proteins in spatially distinct cell surface microdomains. J Cell Biol 160(2):165–170

14. Klemm RW et al (2009) Segregation of sphingolipids and sterols during formation of secretory vesicles at the trans-Golgi network. J Cell Biol 185(4):601–612

15. Gerl MJ et al (2012) Quantitative analysis of the lipidomes of the influenza virus envelope and MDCK cell apical membrane. J Cell Biol 196(2):213–221

16. Hancock JF (2006) Lipid rafts: contentious only from simplistic standpoints. Nat Rev Mol Cell Biol 7(6):456–462

17. Sezgin E et al (2012) Elucidating membrane structure and protein behavior using giant plasma membrane vesicles. Nat Protoc 7(6):1042–1051

18. Levental I, Grzybek M, Simons K (2011) Raft domains of variable properties and compositions in plasma membrane vesicles. Proc Natl Acad Sci U S A 108(28):11411–11416

19. Kaiser HJ et al (2009) Order of lipid phases in model and plasma membranes. Proc Natl Acad Sci U S A 106(39):16645–16650

20. Levental I, Lingwood D, Grzybek M, Coskun U, Simons K (2010) Palmitoylation regulates raft affinity for the majority of integral raft proteins. Proc Natl Acad Sci U S A 107(51):22050–22054

21. Sengupta P, Hammond A, Holowka D, Baird B (2008) Structural determinants for partitioning of lipids and proteins between coexisting fluid phases in giant plasma membrane vesicles. Biochim Biophys Acta 1778(1):20–32

22. Diaz-Rohrer BB, Levental KR, Simons K, Levental I (2014) Membrane raft association is a determinant of plasma membrane localization. Proc Natl Acad Sci 111(23):8500–8505

23. Scott RE (1976) Plasma membrane vesiculation: a new technique for isolation of plasma membranes. Science 194(4266):743–745

24. Zhou Y et al (2013) Bile acids modulate signaling by functional perturbation of plasma membrane domains. J Biol Chem. 288(50):35660–35670

Chapter 7

Asymmetric Giant Lipid Vesicle Fabrication

Peichi C. Hu and Noah Malmstadt

Abstract

Synthetic lipid bilayers have long been used as models of cell membranes. The compositional asymmetry in the eukaryotic plasma membrane is a key chemical characteristic of this membrane that has traditionally been difficult to reproduce in synthetic systems. In this chapter, we describe recent technologies for fabricating compositionally asymmetric giant unilamellar lipid vesicles (GUVs) and provide detailed protocols for a microfluidic-based fabrication technique.

Key words Giant unilamellar vesicles, Microfluidic droplets, Asymmetric giant vesicles, Cell membrane, Lipid bilayer

1 Introduction

Artificial or synthetic cell membranes are important tools used for studying membrane biophysics, the biochemistry of membrane proteins, and electrophysiology. Most biological cells have a size of 1–100 μm, and due to their similar length scale and morphology, giant unilamellar lipid vesicles (GUVs) are important models of the cell membrane for in vitro biomembrane studies. Electroformation [1] and gentle hydration [2] are the most widely used techniques for giant vesicle formation, although these techniques produce samples with significant size polydispersity and they are limited in the lipid compositions and buffer conditions they can accommodate. Microfluidic methods reduce polydispersity, and recently reported methods include microfluidic jetting [3], microfluidic capillary double emulsion [4], microfluidic droplet emulsion transfer [5], transient membrane ejection [6], ice droplet hydration [7], solvent extraction-based vesicles [8], and microfluidic assembly line formation [9]. Droplet emulsion transfer [10, 11], which relies on transferring a water-in-oil emulsion across another lipid-coated oil–water interface, is another, non-microfluidic, option to fabricate GUVs.

Dylan M. Owen (ed.), *Methods in Membrane Lipids*, Methods in Molecular Biology, vol. 1232, DOI 10.1007/978-1-4939-1752-5_7, © Springer Science+Business Media New York 2015

Eukaryotic plasma membranes are compositionally asymmetric, with different lipid species on the interior and exterior leaflets of the membrane. Notably, sphingomyelin (SM) lipids tend to be concentrated on the exterior leaflet, whereas phosphatidylserine (PS) and phosphatidylethanolamine (PE) lipids tend to be concentrated on the interior leaflet. In cells, compositional asymmetry is actively maintained, with transport proteins such as flippases, floppases, and scramblases responsible for aiding movement of phospholipids between the two leaflets. Mimicking compositional asymmetry in a GUV system requires setting up the bilayer with an initial asymmetric composition and relying on the high flip-flop energy barrier to maintain this asymmetry for an experimentally relevant timescale. Several approaches to setting up asymmetric GUV bilayers have emerged in recent years.

Fabrication of nanoscale asymmetric liposomes was pioneered by Zhang et al. [12], Xiao et al. [13, 14], and Pautot et al. [15]. These researchers all transferred water-in-oil emulsions across a water–oil interface to form bilayer-bounded liposomes. In this technique, vesicles are assembled in two independent steps during which each of the bilayer leaflets is generated separately. The first leaflet is formed at the interface of nanoscale water-in-oil droplets and bulk oil. The second lipid leaflet is formed at the interface between bulk oil and bulk water. The nanoscale water droplets, coated by lipid monolayers, pick up a second lipid monolayer when they cross the bulk oil–water interface. The key to adapting this technique to GUVs is generating monolayer-coated droplets of the appropriate size. Droplets necessary for GUVs have been made by tapping [16], vortexing [17, 18] or pipetting [19, 20]. The monodispersity of the water emulsion, and therefore the uniformity of the vesicles formed, can be improved by using a microfluidic device [21, 22] to generate the droplets. The first monolayer is formed inside of the microfluidic device with a multiphase droplet flow configuration consisting of a continuous oil stream in which water droplets are formed. These droplets are dispensed into a vessel containing a layer of oil over a layer of water. The lipid-stabilized oil–water interface contributes the second lipid monolayer when the droplets pass through, driven by centrifugal force. The asymmetric GUVs are formed when the droplets are wrapped with the second layer of lipid molecules and transferred into the lower water phase. The details of this method are described in this chapter.

Asymmetric GUVs have also been formed by jetting picoliter volumes of liquid through an asymmetric membrane [23]. By encapsulating two kinds of small unilamellar vesicles (SUVs) in two water-in-oil droplets and bringing these droplets in contact, an asymmetric lipid membrane is formed at where two droplets attach [24]. Ejecting the piezoelectric-driven liquid across the thin lipid membrane results in a single asymmetric giant unilamellar vesicle.

2   Materials

Chloroform and SU-8 developer are volatile solvents and potential inhalation health risks. Use of a fume hood is recommended. Care should be taken in the use of sharp blades.

2.1   Lipids

Asolectin from soybean (Sigma-Aldrich)

1. 1,2-dipalmitoyl-*s,n*-glycero-2-phosphoethanolamine-N-(biotinyl) (biotin-DPPE) (Avanti Polar Lipids).
2. 1,2-dipalmitoyl-*sn*-glycero-3-phosphocholine (DPPC) (Avanti).
3. 1,2-dioleoyl-*sn*-glycero-3-phosphocholine (DOPC) (Avanti).
4. Plant cholesterol (Avanti).
5. Porcine brain L-α-phosphatidylserine (PS) (Avanti).

2.2   Buffers

1. Tris buffer: 5 mM Tris, pH 7.4. Weigh out 605.7 mg Tris base, transfer to volumetric flask, and add about 900 mL Milli-Q water. Dissolve and adjust pH with HCl (*see* **Note 1**). Fill up to 1 L with Milli-Q water. Store at room temperature.
2. Sucrose buffer: 0.5 M sucrose, 5 mM Tris, pH 7.4. Weigh out 17.12 g sucrose, transfer to a volumetric flask, and add about 80 mL Tris buffer. Dissolve and fill up to 100 mL with Tris buffer. Check the pH after the solution is made and adjust if necessary with HCl or NaOH.
3. Glucose buffer: 0.5 M glucose, 5 mM Tris, pH 7.4. Weigh out 9.00 g glucose and prepare 100 mL solution as in previous step (*see* **Note 2**).

2.3   Lipid Solutions for Biotin Asymmetric Vesicles

1. Lipids for inner leaflet: weigh out 0.5 mg DOPC and transfer to a glass vial. Dissolve in 1 mL chloroform. Dry the lipid solution with an argon stream in a fume hood until a dry lipid film forms (*see* **Note 3**). Add 1 mL decane into the glass vial and vortex to dissolve lipid in solvent completely.
2. Lipid for outer leaflet: weigh out 0.5 mg DOPC. Transfer to a glass vial and add 12.5 of a 1 mg/mL biotin-DPPE in chloroform solution. Dissolve to a total volume of 1 mL chloroform. Dry the lipid solution with argon stream in a fume hood until a dry lipid film forms. Add 1 mL decane and squalene mixture (1:1 volume ratio) into the glass vial and vortex to dissolve lipid in solvent completely (*see* **Note 4**).

2.4   Lipid Solutions for PS Asymmetric Vesicles

Vesicles were formed with PS either on the outer or inner leaflet.

1. Inner lipid solution for PS at inner leaflet: weigh out 1.4 mg DOPC, 1.4 mg DPPC, and 0.7 mg cholesterol and transfer to a glass vial. Add 30 wt% PS (1.5 mg) into the same glass vial

and dissolve in 1 mL chloroform. Dry the lipid solution with an argon stream in a fume hood until a dry lipid film forms. Add 1 mL decane into glass vial and vortex to dissolve lipid in solvent completely.

2. Outer lipid solution for PS at inner leaflet: weigh out 2 mg DOPC, 2 mg DPPC, and 1 mg cholesterol and transfer to a glass vial. Dissolve in 1 mL chloroform. Dry the lipid solution with an argon stream in a fume hood until a dry lipid film forms. Add 1 mL decane and squalene mixture (1:1 volume ratio) into glass vial and vortex to dissolve lipid in solvent completely.

3. Inner lipid solution for PS at outer leaflet: weigh out 2 mg DOPC, 2 mg DPPC, and 1 mg cholesterol and transfer to a glass vial. Dissolve in 1 mL chloroform. Dry the lipid solution with an argon stream in a fume hood until a dry lipid film forms. Add 1 mL decane into glass vial and vortex to dissolve lipid in solvent completely.

4. Outer lipid solution for PS at outer leaflet: weight 1.4 mg DOPC, 1.4 mg DPPC, and 0.7 mg cholesterol and transfer to a glass vial. Add 30 wt% PS (1.5 mg) into the same glass vial and dissolve in 1 mL chloroform. Dry the lipid solution with an argon stream in a fume hood until a dry lipid film forms. Add 1 mL decane and squalene mixture (1:1 volume ratio) into glass vial and vortex to dissolve lipid in solvent completely.

2.5 Microfluidic Devices

SU-8 50 negative photoresist (MicroChem Corporation).

1. Silicon wafers (3 in. (111) mech grade, University Wafer).

2. Poly(dimethylsiloxane) (PDMS) prepolymer and curing agent kits (Sylgard 184, Dow Corning).

2.6 Microfluidic System

1. Two glass syringes; 50 and 500 µL (Syringes 80962 and 81262, Hamilton Company).

2. Syringe adaptors (P-702, P-200X, and P-235, IDEX, Inc.).

3. Tubing (PEEK Grn 1/16 × 0.03, IDEX, Inc.).

4. Two syringe pumps (11 plus, Harvard Apparatus Inc.).

2.7 Observation

1. Sykes-Moore chamber (Bellco Glass, Inc.).

2. 25 mm round #1 cover glass slide (Bellco).

3. Fluorescent labels can be included in any of the lipid mixtures described above. We typically included Texas Red-labeled DPPE (Life Technologies) at 0.5 wt% lipid. We observed asymmetric GUVs with spinning disk confocal microscopy performed using a Yokagawa CSUX confocal head on a Nikon TI-E inverted microscope with a 60× objective (Apo TIRF 1.49 oil immersion). Illumination was provided by 50 mW

solid-state lasers at 491, 561, or 640 nm. Bright field microscopy was performed using the same microscope with transmitted light illumination.

3 Methods

3.1 Microfluidic Device Fabrication: Silicon Master Wafer Fabrication

1. Silicon wafer master fabrication should be performed in a clean environment (*see* **Note 5**).

2. Design a negative photomask pattern of the microchannel (20,000 dpi, transparency mask film, CAD/Art Services). Device geometry is shown in Fig. 1.

3. Rinse silicon wafer with the following solvents: acetone, ethanol, and isopropyl alcohol. Finally, rinse the silicon wafer with DI water and air dry.

4. Bake silicon wafer on a hot plate for 5 min at 200 °C.

5. Place the wafer on a spin-coater (e.g., a Solitech, Inc. 5110) and turn on the stage vacuum.

6. Pour 1 mL SU-8 50 directly onto the center of wafer.

7. Start the spinning program at 500 rpm for 10 s. Increase speed to 2,000 rpm for 30 s.

8. After the wafer is coated, allow it to rest for 5 min to release the air trapped inside of photoresist.

9. Soft-bake the wafer for 5 min at 65 °C and 15 min at 95 °C on a contact hot plate.

10. Allow the wafer to cool to room temperature.

11. Expose the photomask pattern on the wafer for 400 mJ/cm² (using, for instance, a Karl-Suss MJB3 contact aligner with a 365 nm UV source) (*see* **Note 6**).

12. Spin another layer of photoresist on the wafer following the protocol described in **steps 6** and **7**. This double-spin protocol gives a photoresist pattern height of ~50 μm.

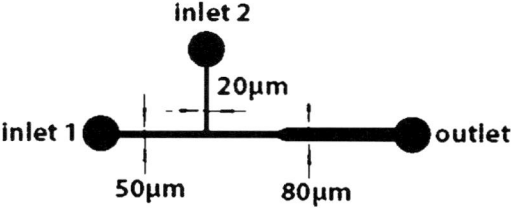

Fig. 1 Schematic of the microfluidic channel used for vesicle fabrication. At the water inlet, the channels are 20 μm wide; at the oil inlet, the channels are 50 μm wide; after the channel intersection, channels are 80 μm wide. The channels height is 30–40 μm

13. Repeat **steps 8–11**, taking care to align the exposed areas of the photomask with the exposed areas of the first layer of photoresist. The pattern on the first layer of photoresist should be readily visible through the second, unexposed layer.

14. Post-exposure bake the wafer for 5 min at 65 °C and 10 min at 95 °C on a contact hot plate.

15. Develop in SU-8 developer for 3–5 min. Development is complete when the unexposed area has completely dissolved.

16. Rinse with IPA and dry with a nitrogen stream.

17. Store the master wafer in a 3-in. petri dish.

3.2 Microfluidic Device Fabrication: Silanizing the Master Wafer to Facilitate Pull-Off of Molded PDMS

1. Place an open bottle containing 1 mL of trichloro(1,1,2,2-perfluorooctyl)silane in a vacuum desiccator.

2. Place the patterned master wafer in the desiccator as well.

3. Connect the dessicator to vacuum for 2 min. Then, close the valve to the vacuum line, isolating the dessicator with the open silane container inside from vacuum. Allow the silane vapor to deposit on the master surface for 20 min.

4. Gently vent the desiccator.

5. Remove the master wafer from the desiccator.

6. Pull vacuum on the dessicator and purge with argon prior to resealing the silane bottle. Seal the silane bottle with Parafilm to minimize exposure to atmospheric water.

3.3 Microfluidic Device Fabrication: Preparation of Molded PDMS Channels

1. Measure 20 g PDMS base and 2 g curing agent in plastic weighing dishes.

2. Mix the PDMS base and curing agent completely (*see* **Note 7**).

3. Degas the PDMS mixture by placing it in a vacuum desiccator under vacuum until all of the air bubbles disappear (about 20 min).

4. Wrap the bottom of master wafer with a 4-in. by 4-in. square of aluminum foil (*see* Fig. 2a–c). (*See* **Note 8**).

5. Pour degassed PDMS on the master wafer gently to avoid trapping air bubbles. Pour PDMS to a height of 2–5 mm (*see* Fig. 2d).

6. Degas the poured PDMS as in **step 3**.

7. Place the mold in an oven at 65 °C for 2 h to cure the PDMS.

8. Cut away the aluminum foil, leaving the molded and cured PDMS on the master wafer.

9. Peel off the PDMS replica from master wafer gently and cut out the channels, leaving a clearance of 5 mm from the edge of each replicated feature to allow for device bonding. Using the device design shown in Fig. 1, this produces a $23 \times 12 \times 2$ mm slab of PDMS with the channels replicated on the bottom.

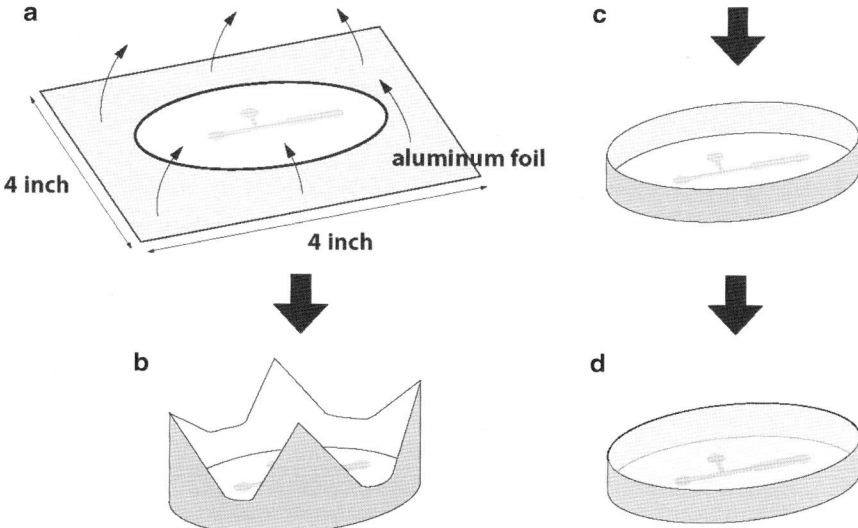

Fig. 2 Wrap the silicon master with an aluminum foil. (**a**) Place the patterned silicon master on top of a 4 in. by 4 in. aluminum foil square. (**b**) Wrap the master by pulling foil corners upwards. (**c**) Cut the excess foil with scissors and make sure the foil is in a good contact with the wafer without air trapped between. (**d**) Pour PDMS into the wrapped master to a height of 2–5 mm

10. Punch a hole with a Harris Uni-Core tissue punch (tip ID 1.5 mm, OD 1.91 mm) at each device inlet (as indicated in Fig. 1).

11. Clean the molded side of the PDMS slab by repeatedly adhering and removing a strip of clear adhesive tape (Scotch transparent film 605 clear tape, 3 M).

3.4 Microfluidic Device Fabrication: Preparing Glass Parts for Sealing Devices

1. Prepare a microscope glass slide (1 mm thick). Drill a hole (1/32 in. in diameter) through the glass slide using an air spindle (Neiko Professional Grade Micro Air Die Grinder) with a diamond bit (220 G, Starlite, Inc.); the hole will be used as the outlet of the device (*see* **Note 9**).

2. Clean the glass slide by soaking in 1 M NaOH for 2 h and rinse thoroughly with water before drying in an oven at 60 °C until dried completely.

3.5 Microfluidic Device Fabrication: Binding the PDMS Channel Replica to the Glass Slide

Clean the PDMS replica and glass with an air stream to remove any dust.

1. Oxidize both surfaces to be bonded by corona treatment (BD-20AC, Elecro-Technic Products); treat the surfaces with plasma evenly for 10 s per cm^2 (*see* **Note 10**).

2. Align the hole on the glass slide and the outlet of the microfluidic channel (as labeled on Fig. 1).

3. Bring the treated surfaces together. As bonding takes place, you can see the bonded area spreading at the interface between the PDMS slab and the glass slide.

4. Place the bonded device in oven at 65 °C for 2 h.

3.6 Microfluidic Device Fabrication: Device Outlet Preparation

Prepare a Nanoport assembly (N-333, IDEX, Inc.). Attach the Nanoport base to the outlet hole drilled in the glass slide with epoxy (Devcon Dev-Pak Adhesive Cartridges 5 min epoxy) and wait for 3 h to cure completely.

1. Insert a 1-in. length of tubing into the Nanoport fitting (*see* **Note 11**).

2. Screw the fitting into the Nanoport base.

3.7 Biotin Asymmetric Vesicle Formation

Biotin-avidin binding is used to immobilize the GUV at the observation area.

Asymmetric vesicles are formed in two steps: water in oil emulsion preparation and outer lipid leaflet preparation.

Water in oil emulsion preparation

1. Draw the inner lipid solution into a 500 µL glass syringe and connect to inlet 1 of the microfluidic device (Fig. 3a) (*see* **Note 12**).

2. Draw the sucrose buffer solution into a 50 µL glass syringe and connect to inlet 2 of the microfluidic device (Fig. 3a).

3. Pump lipid solution into the device at 100 µL/h and the sucrose solution at 20 µL/h.

4. Collect the water-in-oil emulsion in a 1 mL microcentrifuge tube.

5. Allow the water-in-oil emulsion to stabilize by resting for 1 h at room temperature.

Outer lipid leaflet preparation

1. Add 400 µL glucose buffer solution to a 1.5 mL polypropylene-coated microcentrifuge tube (E&K scientific products, Inc.).

2. Layer 200 µL outer lipid oil solution on top of the glucose buffer solution.

3. Allow the tube to rest at room temperature for 30 min in order to stabilize the outer lipid monolayer (Fig. 3b).

4. Gently dispense 50 µL of the water-in-oil emulsion into the same centrifuge tube (Fig. 3c).

5. Transfer the water droplets through the outer lipid monolayer by centrifugation ($200 \times g$) for 2 min.

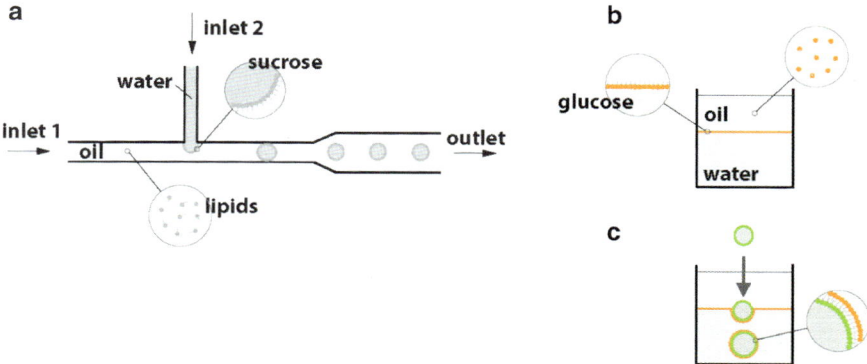

Fig. 3 Fabrication scheme for asymmetric GUVs. (**a**) Droplets of an aqueous solution were formed in an immiscible oil stream at a channel junction; lipids in the oil form a monolayer at the phase interface. (**b**) The second monolayer was formed by allowing lipids at an oil–water interface to stabilize for at least half hour. (**c**) The water droplets were poured into a vessel with a lipid monolayer at an oil–water interface. Centrifugation was used to force the vesicles through the interface, forming the outer leaflet of the bilayer

3.8 PS Asymmetric Vesicle Formation	Follow the procedure described in Subheading 3.7 using the lipid solution prepared as described **item 1** of Subheading 2.3
3.9 Liposome Suspension Preparation (for Surface Coating of the Observation Chamber)	1. Weight 2 mg asolectin and 0.2 mg biotin-DPPE; transfer to glass vial. 2. Dissolve in 1 mL chloroform. Dry the lipid solution with an argon stream in a fume hood until a lipid film forms. 3. Dry the lipid film in a vacuum oven at room temperature for 90 min. 4. Add Tris buffer into the glass vial and vortex until the lipid film is detached from the vial walls and suspended in the buffer. 5. Sonicate the lipid suspension at 35 °C for 30 min (Using, for instance, an Elma Elmasonic E15H bath sonicator) (*see* **Note 13**). 6. Filter the resulting liposome suspension through a syringe filter with 0.2 μm pores. 7. Store the liposome suspension at 4 °C. It should be used within 2 weeks.
3.10 Observation Chamber Preparation	1. Clean the metal parts of the Sykes-Moore chamber by soaking in 1 M NaOH solution for 2 h and rinsing with Milli-Q water. Then sonicate for 15 min in methanol at room temperature and dry at 60 °C before every use.

2. Sonicate cover glass slides in methanol for 15 min at room temperature and soak in 1 M NaOH for 2 h. Rinse glass slides with Milli-Q water and then rinse with methanol before drying in an oven at 60 °C.

3.11 Supported Lipid Bilayer Preparation

A supported bilayer at the surface of the observation chamber provides a gentle surface for GUVs to rest on and allows for biotin-streptavidin binding and GUV immobilization.

1. Seal a clean cover slip into Sykes-Moore chamber and remove any dust with an air stream.

2. Add 150 μL liposome suspension into the chamber and allow it to incubate at room temperature for 15 min (*see* **Note 14**).

3. Wash the chamber with Tris buffer five times.

4. Add 150 μL avidin solution in Tris buffer (0.1 mg/mL) into the chamber and allow it to incubate at room temperature for 15 min.

5. Wash the chamber with Tris buffer five times.

6. Add 300 μL glucose buffer to the chamber.

3.12 Transfer Sample to Chamber

1. Use a needle syringe to withdraw the outer oil in the sample tube prepared in Subheadings 3.7 and 3.8.

2. Withdraw approximately 10 μL of glucose buffer from the top of the sample tube before withdrawing the vesicles. This helps prevent contaminating the GUV sample with oil.

3. Transfer 200 μL of the vesicle solution from the bottom of the sample tube and into the observation chamber. Incubate for 30 min at room temperature to allow biotin in the GUV membranes to bind the avidin-coated surface (*see* **Note 15**).

4 Notes

1. The HCl volume should be included in the total buffer volume to get an accurate concentration. Add small amounts (~2 μL at a time) of HCl (1 M) to avoid a sudden drop in pH.

2. Adding 10 mL buffer into the volumetric flask before pouring glucose prevents glucose from sticking to the flask bottom. Heating the solution up to 40 °C helps to dissolve glucose in water.

3. Applying the argon stream evenly to the surface of the lipid solution helps to evaporate chloroform more quickly.

4. Purge with argon and cap immediately after every usage to prevent lipid oxidation. Store at −20 °C.

5. We performed photolithography in a Class 10 clean room.

6. The required exposure time may differ for different alignment/exposure systems.

7. Make sure the PDMS is well mixed otherwise the mold will not be solidify evenly.

8. Pull the aluminum foil up and fold the extra foil along the wafer edge. Cut the extra foil with scissors. After wrapping the silicon wafer with aluminum foil, make sure there is no air trapped between the wafer and the foil by pressing down on the wafer while it is resting on a table.

9. Drill in a water bath to cool the glass during drilling. Our air spindle required a pressure of at least 40 psi to generate sufficient tool speeds.

10. The required plasma treating time may vary. Over-treated surfaces may fail to bond.

11. Shaping the inner surface of the outlet tubing into a sloped geometry helps droplets to move toward the center of the tubing bore hole. Do not push the tubing firmly against the glass surface; allowing a small gap also facilitates droplet transfer (Fig. 4).

12. Injecting the oil phase into the microfluidic device before injecting the water phase helps to make a hydrophobic surface that will facilitate water droplet formation.

13. The liposome suspension goes from milky to translucent following sonication, indicating that multilamellar large vesicles have been converted to small unilamellar vesicles.

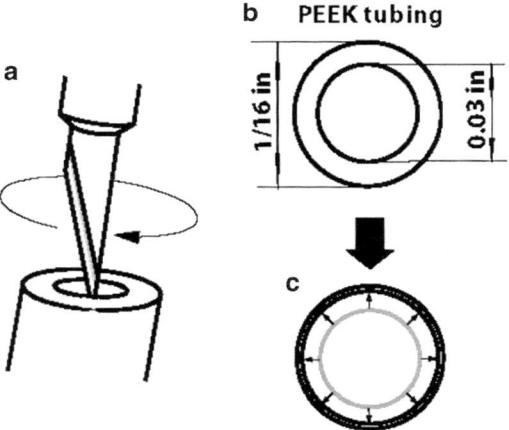

Fig. 4 Sloping the PEEK tubing inner surface with a blade. (**a**) Rotate the blade to reduce the thickness of the tubing (**b**) The PEEK tubing is 1/16 in. outer diameter and 0.03 in. inner diameter. (**c**) Decreasing the tubing thickness to about 1/5 of the original thickness

14. A supported bilayer formed by the binding and fusion of the biotinylated liposomes to the glass surface acts as a cushion to prevent GUVs from bursting.

15. Biotin-avidin immobilization facilitates observation of the GUVs.

Acknowledgement

This work was supported by the National Institutes of Health under awards 1R21AG033890 and 1R01GM093279.

References

1. Angelova MI, Dimitrov DS (1986) Liposome electroformation. Faraday Discuss Chem Soc 81:303

2. Reeves JP, Dowben RM (1969) Formation and properties of thin-walled phospholipid vesicles. J Cell Physiol 73:49–60

3. Stachowiak JC et al (2008) Unilamellar vesicle formation and encapsulation by microfluidic jetting. Proc Natl Acad Sci U S A 105:4697–4702

4. Shum HC et al (2008) Double emulsion templated monodisperse phospholipid vesicles. Langmuir 24:7651–7653

5. Tan YC et al (2006) Controlled microfluidic encapsulation of cells, proteins, and microbeads in lipid vesicles. J Am Chem Soc 128:5656–5658

6. Ota S et al (2009) Microfluidic formation of monodisperse, cell-sized, and unilamellar vesicles. Angew Chem Int Ed Engl 48:6533–6537

7. Sugiura S et al (2008) Novel method for obtaining homogeneous giant vesicles from a monodisperse water-in-oil emulsion prepared with a microfluidic device. Langmuir 24:4581–4588

8. Teh SY et al (2011) Stable, biocompatible lipid vesicle generation by solvent extraction-based droplet microfluidics. Biomicrofluidics 5:44113–4411312

9. Matosevic S, Paegel BM (2011) Stepwise synthesis of giant unilamellar vesicles on a microfluidic assembly line. J Am Chem Soc 133:2798–2800

10. Pautot S et al (2003) Production of unilamellar vesicles using an inverted emulsion. Langmuir 19:2870–2879

11. Yamada A et al (2007) Spontaneous generation of giant liposomes from an oil/water interface. Chembiochem 8:2215–2218

12. Zhang LG et al (1997) Preparation of liposomes with a controlled assembly procedure. J Colloid Interface Sci 190:76–80

13. Xiao ZD et al (1998) Novel preparation of asymmetric liposomes with inner and outer layer of different materials. Chem Lett 225–226

14. Xiao ZD et al (1998) Preparation of asymmetric bilayer-vesicles with inner and outer monolayers composed of different amphiphilic molecules. Supramol Sci 5:619–622

15. Pautot S et al (2003) Engineering asymmetric vesicles. Proc Natl Acad Sci U S A 100:10718–10721

16. Hamada T (2008) Construction of asymmetric cell-sized lipid vesicles from lipid-coated water-in-oil Microdroplets. Phys Chem B 112(47):14678–14681

17. Noireaux V, Libchaber A (2004) A vesicle bioreactor as a step toward an artificial cell assembly. Proc Natl Acad Sci U S A 101:17669–17674

18. Yamada A (2006) Spontaneous transfer of phospholipid-coated oil-in-oil and water-in-oil micro-droplets through an oil/water interface. Langmuir 22(24):9824–9828

19. Takiguchi K et al (2008) Entrapping desired amounts of actin filaments and molecular motor proteins in giant liposomes. Langmuir 24:11323–11326

20. Yanagisawa M et al (2011) Oriented reconstitution of a membrane protein in a giant unilamellar vesicle: experimental verification with the potassium channel KcsA. J Am Chem Soc 133:11774–11779

21. Hu PC et al (2011) Microfluidic fabrication of asymmetric giant lipid vesicles. ACS Appl Mater Interfaces 3:1434–1440

22. Nishimura K et al (2012) Size control of giant unilamellar vesicles prepared from inverted emulsion droplets. J Colloid Interface Sci 376:119–125

23. Richmond DL et al (2011) Forming giant vesicles with controlled membrane composition, asymmetry, and contents. Proc Natl Acad Sci U S A 108:9431–9436

24. Bayley H et al (2008) Droplet interface bilayers. Mol Biosyst 4:1191–1208

Chapter 8

Cholesterol Depletion Using Methyl-β-cyclodextrin

Saleemulla Mahammad and Ingela Parmryd

Abstract

Cholesterol is an essential component of mammalian cells. It is the major lipid constituent of the plasma membrane and is also abundant in most other organelle membranes. In the plasma membrane cholesterol plays critical physical roles in the maintenance of membrane fluidity and membrane permeability. It is also important for membrane trafficking, cell signalling, and lipid as well as protein sorting. Cholesterol is essential for the formation of liquid ordered domains in model membranes, which in cells are known as lipid nanodomains or lipid rafts. Cholesterol depletion is widely used to study the role of cholesterol in cellular processes and can be performed over days using inhibitors of its synthesis or acutely over minutes using chemical reagents. Acute cholesterol depletion by methyl-β-cyclodextrin (MBCD) is the most widely used method and here we describe how it should be performed to avoid the common side-effect cell death.

Key words Cholesterol, Methyl-β-cyclodextrin

1 Introduction

Cholesterol is essential for mammalian cells and is the major component of their plasma membrane [1]. Cholesterol has a small polar headgroup and large hydrophobic moiety which gives it a conical shape (Fig. 1a, c). In membranes, cholesterol associates with phospholipids in order to avoid unfavorable exposure to water [2]. Depletion or manipulation of cholesterol content is often used to study its role of various processes such as cell signalling, membrane fluidity, lipid sorting, membrane trafficking, and membrane permeability. Cholesterol depletion is also frequently used in lipid raft studies, but since it is not specific for a particular cholesterol pool this use should be discouraged [3].

Cellular cholesterol depletion can be achieved in several ways; treatment with metabolic inhibitors, growing cells in lipoprotein deficient medium or using chemicals to immobilize or to extract cholesterol. Examples of these methods are inhibition of cholesterol biosynthesis by statins [4], the culture of cells in the absence of exogenous cholesterol [5], the use of cholesterol binding agents

Dylan M. Owen (ed.), *Methods in Membrane Lipids*, Methods in Molecular Biology, vol. 1232,
DOI 10.1007/978-1-4939-1752-5_8, © Springer Science+Business Media New York 2015

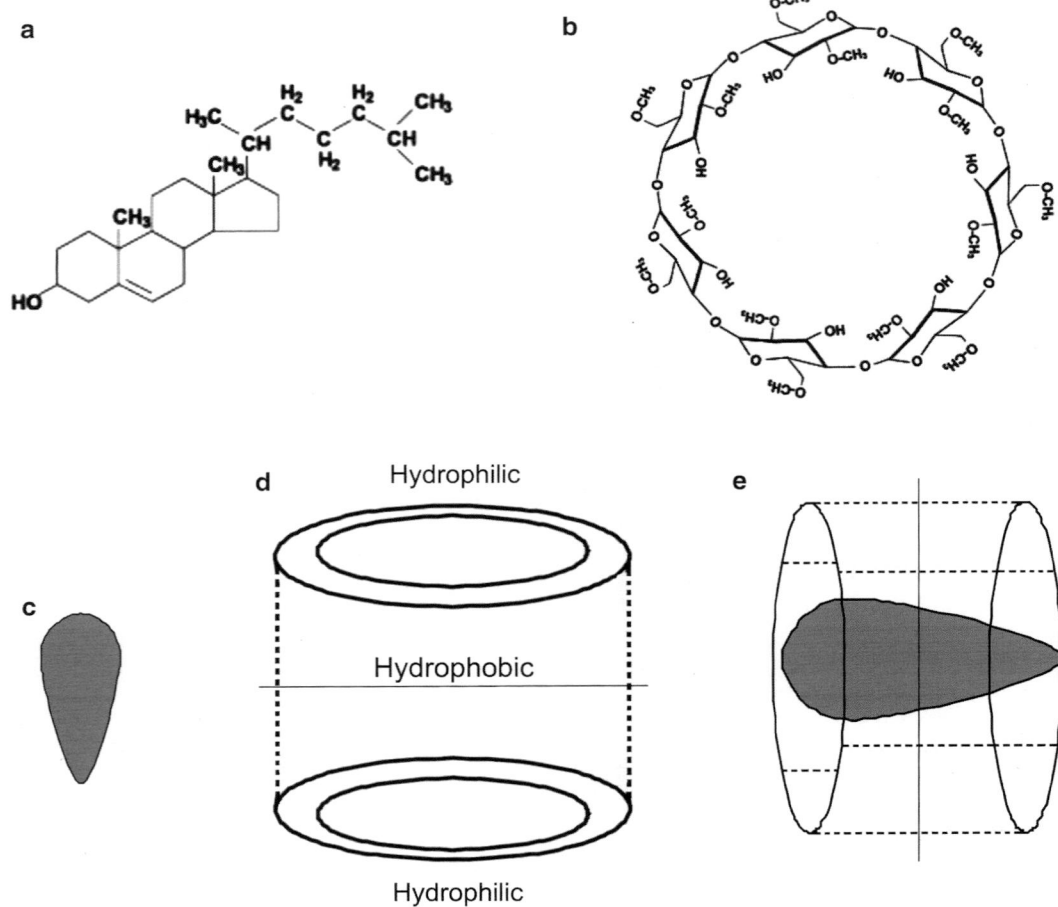

Fig. 1 The structures of cholesterol and MBCD and the formation of cholesterol–MBCD dimer complexes. (**a**) The structure of cholesterol. (**b**) The structure of MBCD. (**c**) Cholesterol represented as a *cone*. (**d**) A MBCD-dimer represented as a *hollow cylinder*. (**e**) A cholesterol–MBCD dimer complex

such as digitonin, filipin, and saponin [6, 7], and oxidation of cholesterol [8]. The most commonly used method is acute cholesterol depletion using methyl-β-cyclodextrin (MBCD). Unlike the cholesterol binding agents mentioned above that incorporate into membranes, MBCD has a central cavity able to form a 2:1 complex with cholesterol [9, 10]. Inhibition of de novo cholesterol synthesis or growing cells in lipoprotein deficient medium require treatment over days which is plenty of time to allow redistribution of the remaining cholesterol [3]. The cholesterol binding agents are not compatible with live cells; digitonin and saponin because they are detergents and filipin because it forms crystals with immobilized cholesterol resulting in leaky membranes [11]. Oxidation of cholesterol means the introduction of a new membrane component making it difficult to ascribe any changes to the depletion of one molecule or the gain of another. In comparison MBCD is

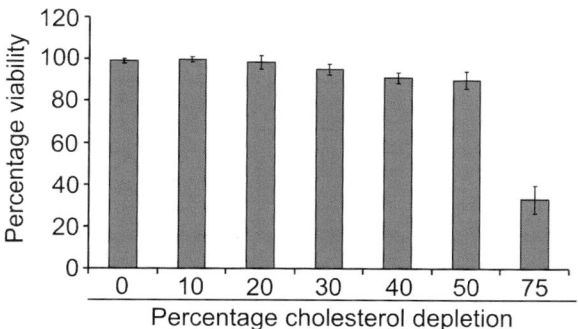

Fig. 2 Cell viability after MBCD treatment. Jurkat T cells were treated with MBCD at 37 °C to deplete progressive amounts of cholesterol and examined for viability by Trypan blue exclusion. Data shown are means ± SD from three experiments where 300 cells were counted for each condition (900 in total). The bulk of these data were originally published in BBA 1778, 1251–1258 [3]

faster, does not produce degradation products and, if used correctly, compatible with live cell studies. Moreover, MBCD has the advantage of acting strictly at the membrane surface [12]. A disadvantage with MBCD is that it is not specific for cholesterol [13], but this limitation can be addressed by performing appropriate controls.

Cyclodextrins are water-soluble oligosaccharides containing hydrophobic cavities [14, 15]. Cyclodextrins contain α-(1-4) linked D-glycopyranose units which can form hexamers (α-CD), heptamers (β-CD) (Fig. 1b), or octamers (γ-CD). The number of glycopyranose units in the ring structure and the degree of polymerization determines the size of the hydrophobic cavity and the affinity towards specific compounds. The water solubility of cyclodextrins can be improved by various modifications including methylation [14, 15]. Based on the dimensions of their cavities, α-CD can form inclusion complexes only with low-molecular-weight molecules or compounds with aliphatic side chains, β-CD can form complexes aromatic or heterocyclic compounds, while γ-CD can accommodate a wider variety of large organic compounds such as macrocycles and steroids [16]. MBCD has a high capacity to bind cholesterol and is relatively cheap [17]. Moreover, it is highly water soluble [18]; properties which makes MBCD the most widely used cyclodextrin for manipulating cholesterol levels in cells [17, 19]. In addition to cholesterol depletion, treatment of cells with MBCD complexed with cholesterol can be used to increase cholesterol levels or as equilibrium controls [20, 21].

Due to the requirement for cholesterol in creating the plasma membrane permeability barrier, cells cannot cope with loosing too much of their cholesterol. How much will differ between cell types and for instance Jurkat T cells remain viable until they loose more than 50 % of their total cholesterol (Fig. 2). Once a cell is dead, it is hardly surprising that a cellular process ceases to function and it is incorrect to ascribe this failure to the importance of cholesterol and/or lipid rafts. This was the case in the T cell signalling field [22] and care must be taken to make sure that cells remain viable.

2   Materials

To increase the reliability for the data, samples should be prepared in parallel. For instance if different levels of cholesterol depletion are desired, all samples should be prepared concurrently to minimize the risk that cell culture density/confluency or variations in protocols and reagents contribute to the results.

2.1   Cells

Grow cells of desired type under standard cell culture conditions. Make sure that cells that grow in suspension are not overgrown or that adherent cells are at no more than 80 % confluency (*see* **Note 1**).

2.2   Reagents for Cholesterol Depletion

1. Methyl-β-cyclodextrin (Sigma, St Louis, MO).
2. Cholesterol (Nu-Chek Prep, Inc., Elysian, MN).
3. Cell culture medium.
4. HEPES (if not included in the cell culture medium).

2.3   Reagents for Viability Assessment

1. Trypan blue (Sigma, St Louis, MO).
2. Phosphate buffered saline.

2.4   Reagents for Assessment of the Level of Cholesterol Depletion by Radioactivity Tracing

1. [^{3}H]-Cholesterol (Perkin Elmer, Waltham, MA).
2. Scintillation fluid (Perkin Elmer, Waltham, MA).

2.5   Reagents for Assessment of the Level of Cholesterol Depletion by Cholesterol Oxidation

1. Chloroform.
2. Methanol.
3. Water.
4. 10-acetyl-3,7-dihydroxyphenoxazine (Amplex Red) (Synchem OHG, Felsberg, Germany).
5. Potassium phosphate buffer, pH 7.4.
6. NaCl.
7. Cholic acid (Sigma, St Louis, MO).
8. Triton X-100 (Sigma, St Louis, MO).

9. Horse radish peroxidase (Sigma, St Louis, MO).

10. Cholesterol oxidase (Sigma, St Louis, MO).

11. Nitrogen gas.

2.6 Equipment for Preparing MBCD–Cholesterol Complexes

1. Glass tubes.
2. Microcaps.
3. Water bath.
4. Sonicator.
5. Vortex.

2.7 Equipment for Assessing Cell Viability

1. Hemocytometer.
2. Light microscope.

2.8 Equipment for Assessment of the Level of Cholesterol Depletion by Radioactivity Tracing

1. Glass tubes.
2. Scintillation counter.

2.9 Equipment for Assessment of the Level of Cholesterol Depletion by Cholesterol Oxidation

1. Glass tubes.
2. Fluoroscanner.
3. 96 well plates.
4. Microcaps.
5. Water bath.

3 Methods

3.1 Preparation of Cells for Cholesterol Depletion

1. Trypsinize the cells (*see* **Notes 2** and **3**).
2. Count the cells.
3. Transfer the desired number of cells to a tube.
4. Wash the cells in serum-free media. Repeat twice.
5. Suspend the cells in serum-free media at a concentration of 20×10^6 cells/ml.

3.2 Preparation of Methyl-β-cyclodextrin Solutions

A freshly made MBCD solution should always be used (*see* **Note 4**).

1. Dissolve the desired amount of MBCD in medium supplemented with 25 mM HEPES (*see* **Notes 5** and **6**).

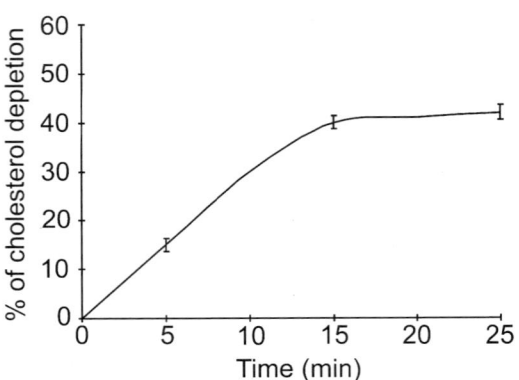

Fig. 3 Cholesterol extraction from cells by MBCD. Jurkat T cells (10×10^6/ml) labelled with [^{3}H]-cholesterol were treated with 2.5 mM MBCD in serum-free RPMI medium at 37 °C. At the indicated times, the cells were pelleted by a brief centrifugation and the MBCD-containing supernatant was transferred to a fresh tube. The cell pellet was resuspended in PBS and aliquots from both fractions were subjected to scintillation counting. Data shown are means ± SD, $n = 3$. These data were originally published in BBA 1778, 1251–1258 [3]

Table 1
Cholesterol extraction by MBCD at 37 °C

MBCD concentration (mM)	% of cholesterol extraction
0.5	10
1.0	20
1.5	30
2.5	40
3.0	50

Jurkat T cells at 10×10^6 cells/ml were treated with different concentrations of MBCD at 37 °C for 15 min to reach the maximal level of cholesterol extraction for each concentration. The bulk of these data were originally published in BBA 1778, 1251–1258 [3]

3.3 Cholesterol Depletion at Steady State

Cholesterol extraction is biphasic with an initial fast phase followed by a second slower phase that reaches a specific percentage extraction for each MBCD concentration [3] (Fig. 3 and *see* **Note 7**).

1. Prepare the cells as described under Subheading 3.1.

2. Mix the cell suspension with the MBCD solution at a 1:1 ratio. A summary of what percentage of cholesterol depletion different concentrations of MBCD can achieve is presented in Table 1.

3. Incubate at 37 °C for 15 min. Tap the tubes intermittently during the MBCD-treatment so that the cells remain suspended in the solution and do not settle at the bottom of the tube.

4. Proceed with the subsequent analysis immediately after the MBCD-treatment (*see* **Note 8**).

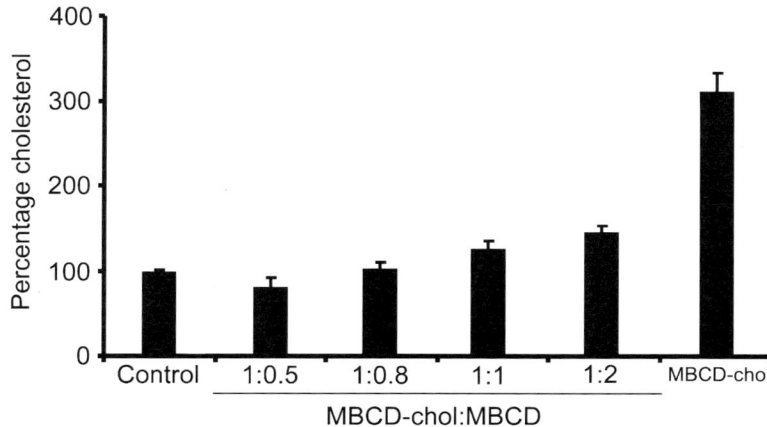

Fig. 4 Cholesterol equilibrium during MBCD-treatment. Jurkat T cells were treated for 15 min with 2.5 mM MBCD mixed with 2.5 mM MBCD–cholesterol complexes at 37 °C. The cells were then washed and lipids extracted. The cholesterol content was estimated using the Amplex red assay. Control cells were defined as containing 100 % cholesterol. Data shown are means ± s.e.m., $n = 4$. These data were originally published in BBA 1801, 625–634 [22]

3.4 Preparation of Methyl-β-cyclodextrin–cholesterol Complexes

A freshly made MBCD–cholesterol solution, starting from **step 2** below, should always be used (*see* **Note 4**).

1. Prepare a 25 mg/ml cholesterol solution in chloroform–methanol (1:1, v:v) in a glass tube (*see* **Note 9**).

2. Transfer 200 μl of the 25 mg/ml cholesterol stock solution to a glass tube using a glass microcap.

3. Heat the tube to 80 °C to evaporate the solvents.

4. Add 10.36 ml of 5 mM MBCD freshly made in serum-free medium.

5. Sonicate at 50–60 % amplitude for 5 min in cycles of 4 s with 2 s on time and 2 s off time. After the sonication, the solution should appear clear.

6. Vortex the tube vigorously for 3 min.

7. Incubate the tube at 37 °C overnight with constant stirring to avoid cholesterol crystallization.

3.5 Establishing Extraction Conditions Where No Net Cholesterol Extraction Takes Place

The hydrophobic pocket that forms when two MBCD molecules form a complex is of a suitable size to accommodate cholesterol. However, MBCD is by no means specific for cholesterol and other lipids may also be extracted [13]. Therefore control experiments where no net cholesterol extraction takes place need to be performed (*see* **Note 10**). Figure 4 illustrates how the level of cholesterol extraction varies in Jurkat T cells incubated with cholesterol-loaded MBCD and MBCD at different ratios.

1. Mix 5 mM MBCD with the 5 mM MBCD–cholesterol solution at different ratios.

2. Add the mixes at a 1:1 ratio to cells prepared as above, i.e., at a final cell density of 10×10^6 cells/ml and a final MBCD concentration of 2.5 mM.

3. Incubate at 37 °C for 15 min.

4. Assess the level of cholesterol depletion using either the radioactivity tracing or the cholesterol oxidation method.

3.6 Assessment of Cell Viability

As with any treatment, it is important to check whether the cells remain viable after MBCD-treatment. There are many ways of doing this of which one method that is easy to perform is the Trypan blue exclusion method (*see* **Note 11**). Viable cells exclude the Trypan blue dye while dead cells are stained and appear blue.

1. Trypsinize the cells (if adherent cells).

2. Prepare a cell suspension (*see* **Note 12**).

3. Dilute the cell suspension 1:1 with 0.4 % (w:v) Trypan blue in PBS. Mix well.

4. Fill the counting chambers of a hemocytometer or slides of an automatic counter with the cell suspension.

5. Count the number of both stained and unstained cells.

6. Calculate the percentage of unstained cells, which represents the percentage of viable cells (*see* **Note 13**).

3.7 Assessment of Cholesterol Depletion

It is recommended that the extent of cholesterol extraction is always assessed. For many experiments, it may also be informative to vary the level of extraction to study its effect.

3.7.1 By Radioactivity Tracing

1. Prepare cells as described under Subheading 3.1, **steps 1–4**.

2. Culture 2×10^6 cells in RPMI containing 2.5 % FCS and 10 μCi [^{3}H]-cholesterol for 40 h.

3. Wash the labelled cells with serum-free medium and mix them with non-labelled cells at a 2:3 ratio.

4. Resuspend the cells at 20×10^6 cells/ml in serum-free medium.

5. Perform cholesterol depletion with MBCD as outlined above.

6. Quickly pellet the cells by centrifugation.

7. Transfer the MBCD-containing supernatants to fresh tubes.

8. Resuspend the cell pellets in PBS.

9. Mix aliquots from the pellet and supernatant fractions with scintillation fluid.

10. Perform scintillation counting (*see* **Note 14**).

3.7.2 By Assessing Cholesterol Oxidation

1. Lyse the control and cholesterol depleted cells in 2.3 ml chloroform–methanol–water (1:1:0.3, v:v:v). Use glass tubes.

2. Add 2 ml chloroform, 1 ml methanol and 0.7 ml water to change the chloroform–methanol–water ratios to 3:2:1 (v:v:v). This composition will separate into two phases.

3. Transfer the lower organic phases to glass tubes.

4. Extract the water-phase with chloroform–methanol (3:1, v:v).

5. Pool the two organic phases.

6. Place the tubes in a heated (60 °C) water bath in a fume hood and evaporate the solvents under a stream of nitrogen.

7. Dissolve the residue in 200 μl of Amplex red assay buffer (0.1 M potassium phosphate pH 7.4, 50 mM NaCl, 5 mM cholic acid, and 0.1 % (w/v) Triton X-100 (TX)).

8. Incubate at 37 °C for 2 h.

9. Transfer 10 μl sample aliquots using microcaps to wells of a 96 well plate containing 90 μl Amplex red, horse radish peroxidase and cholesterol oxidase in Amplex red assay buffer. The final concentration should be 300 μM Amplex red and 2 U/ml each of horse radish peroxidase and cholesterol oxidase.

10. Incubate the 96 well plate at 37 °C for 150 min (*see* **Note 15**).

11. Read the fluorescence by excitation at 544 nm and emission at 590 nm in a Fluoroscanner (*see* **Notes 16** and **17**).

12. The level of cholesterol depletion is calculated by comparing the values of control and MBCD treated samples.

4   Notes

1. In overgrown cells cholesterol may become a limiting factor for continued growth, which could effect its distribution in membranes and lipid bodies/droplets.

2. If may be tempting to perform the cholesterol depletion of adherent cells in plates, dishes, or flasks, but it is crucial that the cell to MBCD ratio is kept constant. Assessing confluence simply is not as accurate as counting cells.

3. For cells that grow in suspension, trypsinization is not required. Using a system capable of correctly counting adherent cells without the need to first deattach them is a way to avoid the trypsinization step also for adherent cells.

4. Stored MBCD solutions are not as efficient at cholesterol extraction as are freshly made MBCD solutions. Aliquots of MBCD can be weighed concurrently and stored in capped tubes until the start of an experiment.

5. When dissolved in water MBCD is acidic and in order not to affect the cells a substantial amount of buffer, for instance 25 mM HEPES, is required.

6. The molecular mass of MBCD is rarely stated, but its average level of substitution is. The molecular mass ranges between 1,310.00 and 1,331.36 depending on the level of substitution.

7. The extractions stop at a steady level when equilibrium between free and cholesterol-bound MBCD in water solution is reached. Once this happens, cholesterol still moves between the cell and MBCD-complexes but no net cholesterol extraction takes place.

8. The cholesterol will inevitably be extracted from the plasma membrane, but there is a continuous traffic of lipids in cells. Given the importance of cholesterol in maintaining the cell's permeability barrier and thus keeping it alive, cholesterol from intracellular compartments will be relocated to replenish the cholesterol that has been extracted from the plasma membrane [3].

9. Cholesterol, like any lipid, can adhere to plastic surfaces. To allow quantification of cholesterol it is therefore crucial to avoid plastic, including pipette tips.

10. An alternative to control experiments where no net cholesterol extraction takes place is to add cholesterol at the end of a depletion experiment using cholesterol loaded MBCD to see if the effect of cholesterol extraction is reversed. However, we do not favor this method since the cholesterol added may not distribute like the cholesterol removed and cholesterol addition can cause alterations in for instance cell signalling originating at the plasma membrane [23, 24].

11. The Trypan blue exclusion method is simple and provides a non-graded answer to the question whether a cell is alive or not. It will, however, not distinguish cells committed to dying from live cells, which require more elaborate methods like Annexin V staining of phosphatidyl serine [25].

12. If a hemocytometer is used cell densities of around $0.5–2 \times 10^6$ are appropriate. For automatic cell counters, follow the protocol of the manufacturer.

13. It is crucial to account also for cells that are lost. If a short time series is performed, the assumption is that the cell density remains constant and that any lowering of the total cell count is attributable to lysed cells no longer present in the solution. Thorough suspension of the cells before cell aliquots are removed for viability assessment is of utmost importance.

14. Control cells should have 100 % of scintillation counts in the cell pellet and no counts in the supernatant. For the MBCD-treated

cells, the scintillation counts should decrease in cell pellet samples and increase in the supernatant as the concentration of MBCD is increased.

15. When many samples are run in parallel, it saves time and increases accuracy if components are first mixed and then added to each sample in a single pipetting step. Prepare a mix for the number of samples plus two since small drops of liquid tend to be lost both at the outside of the pipette tip and in the tube where the mix is prepared. If the mix is added before the sample aliquots, there is no need to change pipette tips between wells.

16. The levels of cholesterol are estimated using 10-acetyl-3,7-dihydroxyphenoxazine (Amplex Red), which produces the fluorophore resorufin when combined with H_2O_2. Hydrogen peroxide is formed when cholesterol is subjected to cholesterol oxidase.

17. For comparison of samples, it is important that their cholesterol concentrations are within the response range of the assay. To test this, the control sample should be run at two different dilutions for instance 1× and 2×. If the response of the undiluted sample is not twice that of the 2× diluted sample, the samples need to be further diluted until a doubling in the response is achieved.

Acknowledgements

This work was supported by grants from Magnus Bergvall's Foundation, Signhild Engkvist's Foundation, and O.E. and Edla Johansson's Foundation.

References

1. Maxfield FR, van Meer G (2010) Cholesterol, the central lipid of mammalian cells. Curr Opin Cell Biol 22:422–429

2. Huang J, Feigenson GW (1999) A microscopic interaction model of maximum solubility of cholesterol in lipid bilayers. Biophys J 76: 2142–2157

3. Mahammad S, Parmryd I (2008) Cholesterol homeostasis in T cells. Methyl-beta-cyclodextrin treatment results in equal loss of cholesterol from Triton X-100 soluble and insoluble fractions. Biochim Biophys Acta 1778:1251–1258

4. Taraboulos A, Scott M, Semenov A, Avrahami D, Laszlo L et al (1995) Cholesterol depletion and modification of COOH-terminal targeting sequence of the prion protein inhibit formation of the scrapie isoform. J Cell Biol 129:121–132

5. Esfahani M, Bigler RD, Alfieri JL, Lund-Katz S, Baum JD et al (1993) Cholesterol regulates the cell surface expression of glycophospholipid-anchored CD14 antigen on human monocytes. Biochim Biophys Acta 1149:217–223

6. Nazih-Sanderson F, Pinchon G, Nion S, Fruchart JC, Delbart C (1997) HDL3-signalling in HepG2 cells involves glycosyl-phosphatidylinositol-anchored proteins. Biochim Biophys Acta 1346:45–60

7. Cerneus DP, Ueffing E, Posthuma G, Strous GJ, van der Ende A (1993) Detergent insolubility of alkaline phosphatase during biosynthetic transport and endocytosis. Role of cholesterol. J Biol Chem 268:3150–3155

8. Nguyen DH, Taub DD (2003) Inhibition of chemokine receptor function by membrane cholesterol oxidation. Exp Cell Res 291:36–45

9. Nishijo J, Moriyama S, Shiota S (2003) Interactions of cholesterol with cyclodextrins in aqueous solution. Chem Pharm Bull (Tokyo) 51:1253–1257

10. Loftsson T, Magnusdottir A, Masson M, Sigurjonsdottir JF (2002) Self-association and cyclodextrin solubilization of drugs. J Pharm Sci 91:2307–2316

11. Behnke O, Tranum-Jensen J, van Deurs B (1984) Filipin as a cholesterol probe. II. Filipin-cholesterol interaction in red blood cell membranes. Eur J Cell Biol 35:200–215

12. Lopez CA, de Vries AH, Marrink SJ (2011) Molecular mechanism of cyclodextrin mediated cholesterol extraction. PLoS Comput Biol 7:e1002020

13. Leventis R, Silvius JR (2001) Use of cyclodextrins to monitor transbilayer movement and differential lipid affinities of cholesterol. Biophys J 81:2257–2267

14. Davis ME, Brewster ME (2004) Cyclodextrin-based pharmaceutics: past, present and future. Nat Rev Drug Discov 3:1023–1035

15. van de Manakker F, Vermonden T, van Nostrum CF, Hennink WE (2009) Cyclodextrin-based polymeric materials: synthesis, properties, and pharmaceutical/biomedical applications. Biomacromolecules 10:3157–3175

16. Li Z, Wang M, Wang F, Gu Z, Du G et al (2007) Gamma-cyclodextrin: a review on enzymatic production and applications. Appl Microbiol Biotechnol 77:245–255

17. Ohtani Y, Irie T, Uekama K, Fukunaga K, Pitha J (1989) Differential effects of alpha-, beta- and gamma-cyclodextrins on human erythrocytes. Eur J Biochem 186:17–22

18. Puglisi G, Ventura CA, Spadaro A, Campana G, Spampinato S (1995) Differential effects of modified beta-cyclodextrins on pharmacological activity and bioavailability of 4-biphenylacetic acid in rats after oral administration. J Pharm Pharmacol 47:120–123

19. Ohvo H, Slotte JP (1996) Cyclodextrin-mediated removal of sterols from monolayers: effects of sterol structure and phospholipids on desorption rate. Biochemistry 35:8018–8024

20. Levitan I, Christian AE, Tulenko TN, Rothblat GH (2000) Membrane cholesterol content modulates activation of volume-regulated anion current in bovine endothelial cells. J Gen Physiol 115:405–416

21. Christian AE, Haynes MP, Phillips MC, Rothblat GH (1997) Use of cyclodextrins for manipulating cellular cholesterol content. J Lipid Res 38:2264–2272

22. Mahammad S, Dinic J, Adler J, Parmryd I (2010) Limited cholesterol depletion causes aggregation of plasma membrane lipid rafts inducing T cell activation. Biochim Biophys Acta 1801:625–634

23. Cox BE, Griffin EE, Ullery JC, Jerome WG (2007) Effects of cellular cholesterol loading on macrophage foam cell lysosome acidification. J Lipid Res 48:1012–1021

24. Nguyen DH, Espinoza JC, Taub DD (2004) Cellular cholesterol enrichment impairs T cell activation and chemotaxis. Mech Ageing Dev 125:641–650

25. Koopman G, Reutelingsperger CP, Kuijten GA, Keehnen RM, Pals ST et al (1994) Annexin V for flow cytometric detection of phosphatidylserine expression on B cells undergoing apoptosis. Blood 84:1415–1420

Chapter 9

A Comparative LC-MS Based Profiling Approach to Analyze Lipid Composition in Tissue Culture Systems

G. Ekin Atilla-Gokcumen and Ulrike S. Eggert

Abstract

Although lipids participate in many cellular processes both as signaling and structural molecules, our understanding of the roles of individual lipids as well as global changes in lipid composition are limited. Here we describe an LC-MS based method to identify lipids that change in a biological process. This method describes the isolation of lipids from tissue culture cells, sample preparation for LC-MS, the LC-MS run, and the subsequent data processing steps to compare the global lipid profiles and identify species that are enhanced or depleted. Identifying lipids that change is the first step towards functional studies to unravel their roles.

Key words Lipids, LC-MS based lipid profiling, Untargeted lipid profiling

1 Introduction

1.1 Lipids

Lipids are hydrophobic or amphiphilic compounds that make up the vast majority of metabolite species in cells. Based on their structure and function, lipids are classified in eight major families: fatty acids, glycerolipids, glycerophospholipids, sphingolipids, sterols, prenol lipids, saccharolipids, and polyketides [1]. This diversity in structure is further expanded by the diversity in size (with the incorporation of fatty acids with different length alkyl chains) and shape (different degrees of unsaturation) within each lipid family, which creates thousands of different possible lipid structures. The synthesis, turnover, and transport of lipids are regulated by hundreds of different proteins.

Our understanding of the involvement of lipids in cellular processes is limited. While it is clear that lipids have both structural and signaling roles [2], the molecular mechanisms underlying their functions are not known except for a few examples [3–7]. Identifying key lipid players is the first step towards functional characterization. One way to achieve this is to study the compositional changes in the cellular lipidome during different biological events. We have recently shown the active temporal and spatial regulation

Dylan M. Owen (ed.), *Methods in Membrane Lipids*, Methods in Molecular Biology, vol. 1232,
DOI 10.1007/978-1-4939-1752-5_9, © Springer Science+Business Media New York 2015

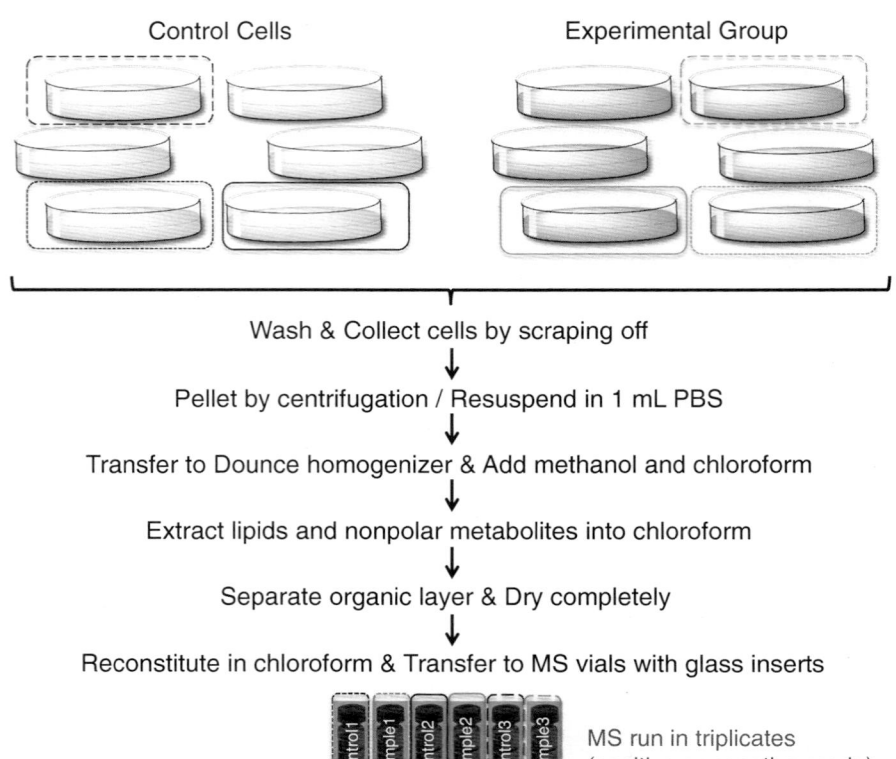

Fig. 1 Sample preparation. Cells are grown in tissue culture flasks, collected by scraping off and centrifugation. Lipids are extracted into chloroform, concentrated by rotary evaporation, and transferred to MS vials

of lipid composition in dividing cells [8]. Since cells actively regulate their lipid levels, it is likely that a particular lipid plays a role if its levels change during an event.

Here, we discuss how mass spectrometry can be used to determine the identity of lipids that change significantly throughout a given cellular event. Specifically, we introduce mass spectrometry based lipid profiling [9] and describe the experimental and analytical steps to study processes in cultured cells—from sample preparation [10] (Fig. 1) to data analysis (Fig. 2) and lipid identification (Fig. 3) [11]. This tag-free technique involves preparing tissue culture models to study the process of interest and the direct comparison of total lipid extracts in different samples and allows for the identification of lipids that change during the process.

1.2 Liquid Chromatography– Mass Spectrometry (LC-MS) Based Lipid Analysis

Recent developments in high resolution and high mass accuracy methods to identify and to distinguish large numbers of lipids allowed the emergence of comparison-based approaches to study changes in lipid composition in complex mixtures [12]. Mass spectrometry-based profiling provides a direct comparative and qualitative analysis of metabolites in different samples in a label-free format [11].

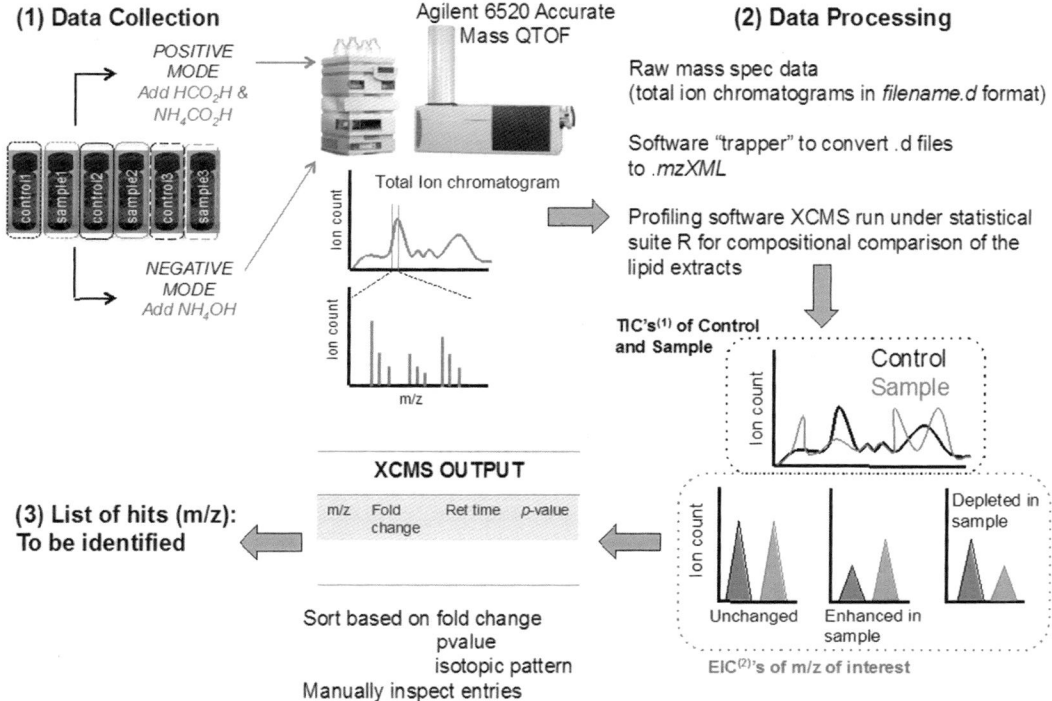

Fig. 2 Workflow for the data collection and analysis. LC-MS samples are run in positive and negative mode. Raw data is channeled through XCMS for profiling studies to identify depleted/enhanced species in sample. (*1*) TIC is the total ion chromatogram of the sample. (*2*) EIC is the extracted ion chromatogram for a specific *m/z*

Raw mass spectrometry data can be channeled through different software applications for profiling studies. These analyses allow us to fingerprint each data set and identify characteristic components (sample vs. control). Additional tools overlay these characteristic components and identify those that change, i.e., are either elevated or depleted in different samples. The sensitivity of mass spectrometry along with advances in the electrospray ionization (ESI) technique are making mass spectrometry-based approaches applicable to studying a broad range of phenotypes [13, 14].

1.3 Identification and Characterization of Lipids That Are Enhanced or Depleted

Mass spectrometry-based profiling approaches to identify lipids that change in a certain process require the establishment of two distinct experimental samples to be compared to each other. Once the raw LC-MS data of these distinct states are collected, they can be overlaid and the candidate differences between the samples are identified. For example, in order to evaluate changes in lipid composition caused by a decreased activity of a protein (either by a small molecule inhibitor or RNAi), LC-MS profiles of cells that are exposed to treatment (small molecule inhibition or RNAi knockdown) and control (empty vehicle) need to be obtained. Depending on their chemical composition and charge status, different lipids

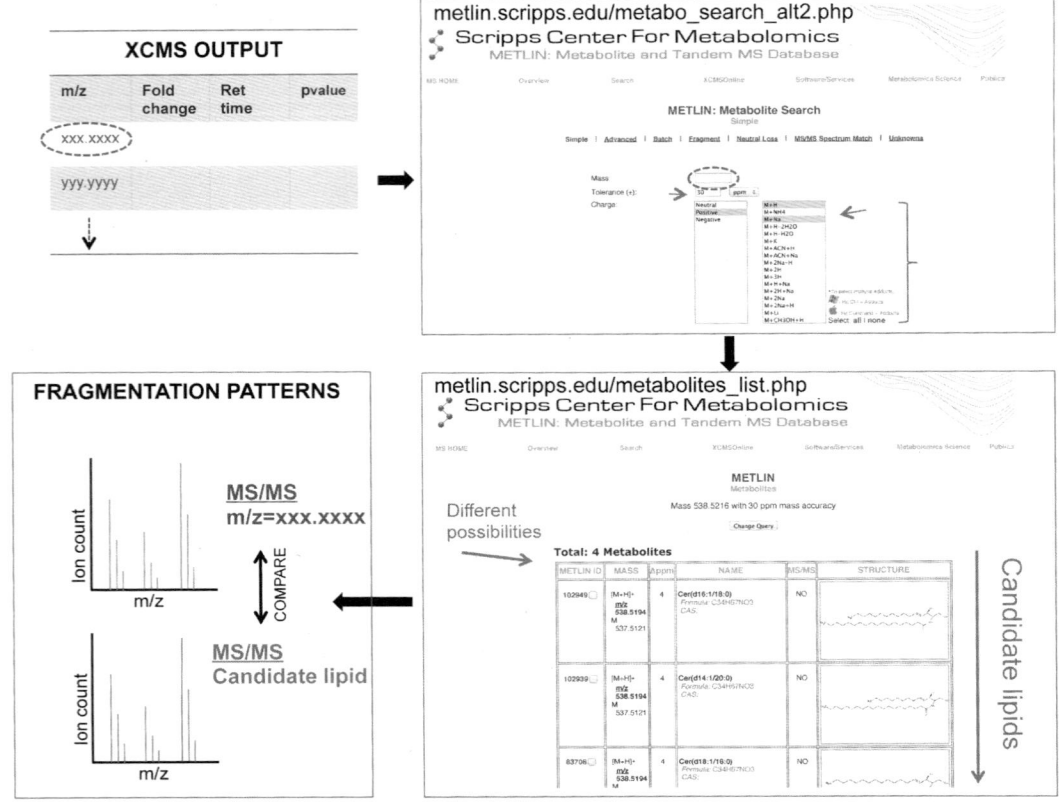

Fig. 3 Workflow for lipid identification. Candidate lipids for a specific *m/z* are identified based on high resolution mass in Metlin database. Screenshots from Metlin [17] are shown. Candidate lipids (or representatives for the lipid family with structural similarity) are purchased and the fragmentation pattern of the purchased lipid is compared to that of *m/z*

will be detected in negative and positive ionization modes. Therefore profiling experiments are carried out in both modes to detect and study as many different lipids as possible. Once ions are generated, corresponding mass to charge ratios (*m/z*) of individual ions are calculated by a high-resolution mass analyzer. The work flow described here is optimized for an electrospray ionization source instrument that is coupled to a time-of-flight (TOF) detection system.

The differences between the two samples (experimental vs. control) can be studied in an untargeted manner where the overall changes in the lipid composition are analyzed without focusing on a particular lipid group or in a targeted manner where the levels of specific lipids are compared. In untargeted lipid profiling, *m/z*'s that change in the experiment are identified in an unbiased manner whereas in targeted lipid profiling, the levels of specific lipids of interest are calculated and compared in control and experimental samples.

In untargeted profiling, once the species are identified, the next step is to characterize m/z's to determine which specific lipid species they correspond to [15]. To do this, tandem MS experiments need to be carried out to obtain the fragmentation patterns of these species, preferably by comparison to a known standard that is identical with or similar to the candidate lipid. There are several databases that are useful as they provide candidate lipids (and other metabolites) for a corresponding m/z value. This narrows down the different possibilities for a given m/z and allows analysis of the fragmentation patterns in a more manageable way, for example by comparisons to known standards. LIPIDMAPS (http://www.lipidmaps.org/) and METLIN (http://metlin.scripps.edu/index.php) provide mass information for lipids (and metabolites for METLIN) that are known as well as fragmentation information for a subgroup of metabolites (METLIN) [1, 16, 17]. Although ultimately the characterization process relies on tandem MS experiments, these databases provide clues on which lipid a certain m/z might correspond to.

2   Materials

2.1   Reagents

1. Cell culture medium (varies depending on cell type).
2. PBS (*see* Reagents).
3. Trypsin (optional).
4. Bradford Dye.
5. Bovine serum albumin (BSA) (as standard for Bradford assay).
6. LC-MS solvents: analytical grade methanol, isopropanol, water.
7. Analytical grade chloroform.
8. Ammonium hydroxide (28–30 %).
9. Formic acid.
10. Ammonium formate.
11. ^{13}C-oleic acid (1 mM in chloroform).

2.2   Equipment

1. Cell culture incubator.
2. Inverted microscope.
3. Hemocytometer.
4. Cell scraper.
5. Lysis buffer for protein concentration measurements. (M-PER Mammalian Protein Extraction Reagent from Thermo).
6. Dounce homogenizer (7 mL).
7. Glass vials (8, 4, and 2 mL).
8. Glass vial inserts for 2 mL vials.

9. Centrifuge.

10. Rotary evaporator with vial adaptors.

11. HPLC columns and guard columns.
    Column choice can vary depending on the HPLC system. Below is a column pair that can be used robustly and reproducibly:
    Gemini C18 reversed phase column (5 µm, 4.6 mm × 50 mm) from Phenomenex (for negative mode).
    Luna C5 reversed phase column (5 µm, 4.6 mm × 50 mm) from Phenomenex (for positive mode).
    Corresponding guard columns are available for Gemini and Luna systems from Phenomenex.

12. Mass spectrometer.
    The protocol described is optimized for an Agilent 6520 Accurate Mass ESI-QTOF.

2.3 Solutions to Prepare

1. *Phosphate-buffered Saline (PBS)*: 137 mM NaCl, 2.7 mM KCl, 10 mM Na_2HPO_4, 2 mM KH_2PO_4, pH 7.4.

2. *BSA standard solutions for Bradford assay*: Prepare 2 mg/mL BSA stock solution in PBS. By diluting into PBS, prepare 1.5, 1.0, 0.75, 0.5, 0.25, and 0.125 mg/mL BSA solutions.

3. *5 M ammonium formate*: Dissolve 78.75 g ammonium formate in 250 mL water.

4. *LC-MS stationary and mobile phase*: Prepare water–methanol, 95:5 for stationary phase (A), and 2-propanol–methanol–water, 60:35:5 for mobile phase (B). All solvents are analytical grade and should be filtered (0.2 µm Nylon membrane is recommended). The mixtures can be stored at 4 °C. Solvent modifiers (0.1 % formic acid and 5 mM ammonium formate for positive ionization mode, and 0.1 % ammonium hydroxide, for negative ionization mode) should be added just before the LC-MS run.

    (a) For negative ionization mode, add 1.42 mL 28–30 % ammonium hydroxide to 400 mL stationary and mobile phases.

    (b) For positive ionization mode, add 400 µL formic acid and 5 M ammonium formate to 400 mL stationary and mobile phases.

5. *Internal standard for ionization efficiency*: Prepare 1 mM ^{13}C-oleic acid in chloroform.

6. *Chloroform with internal standard to redissolve extracted lipids*: Prepare 10 µM ^{13}C-oleic acid in chloroform and use this solution to resuspend extracted lipids (*Solution C*).

3   Methods

3.1   Preparation of Samples from Tissue Culture

Cell growth conditions should be optimized based on the cell type. Once the biological conditions are established for the "experimental" and "control" samples, these should be prepared in biological replicates. All steps including the lipid extraction should be done on ice (Fig. 1).

1. Remove media before collecting the cell pellets and wash the cells gently with PBS (twice).

2. Collect cells by scraping off the flask. Avoid trypsinization to minimize the effect on proteins in the plasma membrane and the overall composition of the plasma membrane. Collect cell pellets by centrifugation at $200 \times g$ for 5 min.

3.2   Normalization and Lipid Extraction

3.2.1   Normalization Based on Protein Concentration Prior to Lipid Extraction

1. Resuspend the pellets in PBS. Use 1 mL PBS for 4–6×10^6 cells. This usually corresponds to one or two 10 cm dishes depending on the cell type.

2. Remove 50 μL of each sample and add 50 μL of lysis buffer to reach complete lysis. These samples will be used for normalization.

3. Further dilute the samples for normalization in lysis buffer. A fivefold dilution should give suitable concentrations for the Bradford assay.

4. Centrifuge the samples with lysis buffer at $11,000 \times g$ and use the supernatant to measure the protein concentration by using a Bio-Rad® Bradford assay. Based on the protein concentrations, further dilute the concentrated samples to obtain the same concentrations for experimental and control samples.

5. Use 1 mL of cell suspension for lipid extraction.

3.2.2   Lipid Extraction

1. Add 1 mL cell suspension in PBS into a 7 mL Dounce homogenizer.

2. Add 1 mL methanol and 2 mL chloroform (PBS–methanol–chloroform, 1:1:2).

3. Homogenize the sample (30 strokes/sample).

4. Transfer the homogenized solution to a vial (8 mL) and centrifuge for 10 min at $500 \times g$.

5. Remove the organic layer (bottom layer) with a glass Pasteur pipette and transfer to a clean vial (4 mL).

6. Pipette 1.5 mL of the organic solution to a new vial and remove the solvent using a rotary evaporator.

7. The dried content should be resuspended in 100 μL Solution C. *Note*: The use of Solution C instead of chloroform allows comparing the ionization efficiencies between different runs.

8. Transfer the chloroform solution to a 2 mL glass vial with a glass insert to decrease the dead volume of the vial.

9. These samples can be stored at −80 °C until LC-MS analysis.

3.3 LC-MS Run and Data Analysis

Optimal conditions should be determined for the HPLC and mass spectrometer that will be used. The following protocol has been optimized for an Agilent 6520 Accurate Mass ESI-QTOF system (Fig. 2).

1. Inject 30–40 µL of sample at a flow rate of 0.1 mL/min for 5 min with 100 % A. The flow through goes to waste (not mass spectrometer) during injection.

2. Once the injection is completed, increase the flow rate to 0.5 mL/min and start a gradient from 0 % B to 100 % B over 70 min.

3. Stay at 100 % B for 8 min before switching to 100 % A and equilibrate the system with 100 % A for 7 min prior to the next injection. The total analysis time is 90 min (*Note*: for best experimental reproducibility and to decrease false discovery rates, alternate between experimental and control samples during each run. For example, the order of running samples for an experiment where three experimental conditions and three control conditions are being analyzed should be as follows:

blank1/blank2 (or solution C)/control1/experimental1/control2/experimental2/control3/experimental 3

Note: (a) A blank run at the beginning is recommended to ensure the system is well-equilibrated and clean.

(b) Instead of the second blank sample (blank2), using Solution C to compare the ionization efficiency between runs is recommended.

4. Set the capillary voltage to 3,500 V.

5. Set the drying gas temperature and flow to 350 °C and 12 L/min.

6. Set the nebulizer pressure to 30 psi.

3.3.1 Conversion of Raw MS Data to XCMS Friendly Format and XCMS Profiling

Raw LC-MS files need to be converted into an XCMS friendly format. This can be achieved by converting .d directories to .mzXML using "trapper" software. Trapper can be downloaded form the following link:http://sourceforge.net/projects/sashimi/files/trapper%20(MassHunter%20converter)/

Once .mzXML files are obtained, they can be processed in XCMS. XCMS is a data analysis approach that relies on nonlinear retention time alignment, identification and matching of peaks in different samples. The instructions for downloading XCMS and running it under statistical suite R can be found at: http://metlin.scripps.edu/xcms. Once R is downloaded, XCMS package can be installed within R. To perform profiling by XCMS:

1. Create a folder that contains two separate folders with "experimental" and "control" .mzXML files. Experimental and control files need to be in separate folders.

2. Open R terminal and choose the parent folder that contains "experimental" and "control" folders and proceed with the comparison.

   Note: Alternatively, XCMS can also be run online via XCMS Online tool that can be found at https://xcmsonline.scripps. edu/. The instructions on how to upload files and obtain the XCMS output are explained in detail on this webpage [18, 19].

3. The output file can be opened and saved in Excel. The output file contains m/z's identified, their corresponding retention times, abundances in individual samples, fold changes, and p-values associated with them.

4. Sort the entries based on fold change and eliminate m/z's below a certain fold change value (we eliminate any species that change less than threefold).

5. OPTIONAL: You can also sort the entries based on p-values and eliminate m/z's with high p-values such as p-value > 0.05.

6. Group the entries based on isotopic peaks and eliminate all the entries for which there is no isotopic sister peak.

   Note: The isotopic distribution is observed due to naturally occurring different isotopes (^{12}C vs. ^{13}C or ^{1}H vs. ^{2}H). In a high-resolution mass spectrometry setting, depending on the atom composition of the molecule, m/z's that correspond to these different isotopes are observed rather than m/z for the monoisotopic mass. The presence of isotopic sister peaks is used as a diagnostic to rule out false positives.

3.3.2 Validation of XCMS Hits

1. Once the XCMS output file is sorted based on the above criteria, the remaining species should be manually checked for abundance, purity, and fold change.

2. Open the raw files (total ion chromatograms (TIC's)) in the MS analysis software.

3. Extract the m/z's and obtain extracted ion chromatograms (EIC's).

4. Locate the peaks that correspond to m/z with the given retention time and integrate the area under the curve to determine the abundance of the ion in different samples.

5. Recalculate fold changes based on the abundance in the previous step and eliminate all species with fold changes <3.

3.3.3 Assignment of Molecular Formulas and Candidate Lipid Assignment

Once the final list with m/z's has been obtained the next step is to find out what these species are. The species are most likely lipids due to the chloroform extraction performed. However, the presence of other hydrophobic metabolites cannot be ruled out.

Software that are used to analyze raw MS data usually are able to generate molecular formulas based on the high resolution m/z and ionization mode. As m/z's increase, the number of possibilities for generated molecular formulas also increases; thus, this approach is not very useful for high molecular mass lipids.

Another approach to find the candidate lipids for a given m/z is to conduct a search based on high resolution mass in the METLIN database. "Metabolite search" option that can be accessed at http://metlin.scripps.edu/metabo_search_alt2.php allows a search based on high resolution m/z once the charge of the species is indicated (Fig. 3).

- Enter m/z at "Mass" section.

- Define a tolerance for mass (±30 ppm is default).

- Choose the "Charge" (either positive or negative) and the corresponding adducts that are present for each mode.

    Note: Depending on the solvent system and modifiers used, different adducts can form in solution (e.g., $(M_1 + H)^+$ vs. $(M_2 + NH_4)^+$ or $(M_3 - H)^-$ vs. $(M_4 - OH)^-$). Since this affects the molecular composition, thus the high resolution mass, it is important to include adducts of different parent molecules that have the same m/z. For example, $(M_1 + H)^+$ and $(M_2 + NH_4)^+$ might have the same m/z due to the same molecular composition even though the parent lipids, M_1 and M_2, are different molecules.

- "Find metabolite" command will provide the metabolites that match with the high resolution m/z at the defined charge state.

This approach provides the identification of candidate lipids for m/z's identified. Following tandem MS experiments comparing the fragmentation patterns of the candidate lipid to that of m/z's of interest will confirm the nature of the molecule.

3.4 Tandem MS for Identification

The fragmentation patterns of m/z's should be compared to that of candidate lipids. In some cases, it is possible to obtain the exact candidate lipid for direct comparison. However, if the exact lipid is not commercially available, lipids from the same family can be used to compare the fragmentation of the backbone structures.

1. Obtain the lipid standard whose fragmentation pattern will be compared to m/z.

    Note: (a) Avanti Polar Lipids has a large selection of lipids belonging to different families. Small (1–10 mg) quantities can be obtained.

    (b) Isotope labeled standards are desired (if available) as their use allows additional experiments where the sample can be spiked with the isotope labeled lipids and direct comparison of retention times and fragmentation patterns on the same sample can be obtained. If an isotope labeled version is not available, unlabeled lipids can also be used.

2. Prepare a stock solution of the lipid standard. Chloroform or chloroform–methanol mixtures are usually suitable (1–10 mM depending on the lipid and amount purchased).

3. MS/MS experiments are carried out in a similar fashion as described in Subheading 3.3 but different collision energies need to be tested to find the optimal ionization. Fragmentation patterns can be observed at 15, 35, 55, 75, 95, and 115 V.

4. Compare the fragments obtained for m/z of interest at different collision energies to those of the lipid standard. If the lipid standard used is exactly the same, the same fragments are expected. If another lipid from the same family with a similar chemical structure is used, check for systematic differences that are consistent with the chemical structures of the fragments.

References

1. Fahy E, Sud M, Cotter D, Subramaniam S (2007) LIPID MAPS online tools for lipid research. Nucleic Acids Res 35(Web Server issue):W606–W612

2. van Meer G, de Kroon AI (2011) Lipid map of the mammalian cell. J Cell Sci 124(Pt 1): 5–8

3. Atilla-Gokcumen GE, Bedigian AV, Sasse S, Eggert US (2011) Inhibition of glycosphingolipid biosynthesis induces cytokinesis failure. J Am Chem Soc 133(26):10010–10013

4. Echard A (2012) Phosphoinositides and cytokinesis: the "PIP" of the iceberg. Cytoskeleton (Hoboken) 69(11):893–912

5. Canals D et al (2010) Differential effects of ceramide and sphingosine 1-phosphate on ERM phosphorylation: probing sphingolipid signaling at the outer plasma membrane. J Biol Chem 285(42):32476–32485

6. Kunkel GT, Maceyka M, Milstien S, Spiegel S (2013) Targeting the sphingosine-1-phosphate axis in cancer, inflammation and beyond. Nat Rev Drug Discov 12(9):688–702

7. Atilla-Gokcumen GE, Castoreno AB, Sasse S, Eggert US (2010) Making the cut: the chemical biology of cytokinesis. ACS Chem Biol 5(1): 79–90

8. Atilla-Gokcumen GE et al (2014) Dividing cells regulate their lipid composition and localization. Cell 156(3):428–439

9. Saghatelian A, McKinney MK, Bandell M, Patapoutian A, Cravatt BF (2006) A FAAH-regulated class of N-acyl taurines that activates TRP ion channels. Biochemistry 45(30): 9007–9015

10. Folch J, Lees M, Sloane Stanley GH (1957) A simple method for the isolation and purification of total lipides from animal tissues. J Biol Chem 226(1):497–509

11. Zhu ZJ et al (2013) Liquid chromatography quadrupole time-of-flight mass spectrometry characterization of metabolites guided by the METLIN database. Nat Protoc 8(3):451–460

12. Tagore R, Thomas HR, Homan EA, Munawar A, Saghatelian A (2008) A global metabolite profiling approach to identify protein-metabolite interactions. J Am Chem Soc 130(43): 14111–14113

13. Lv H (2013) Mass spectrometry-based metabolomics towards understanding of gene functions with a diversity of biological contexts. Mass Spectrom Rev 32(2):118–128

14. Gerl MJ et al (2012) Quantitative analysis of the lipidomes of the influenza virus envelope and MDCK cell apical membrane. J Cell Biol 196(2):213–221

15. Saghatelian A et al (2004) Assignment of endogenous substrates to enzymes by global metabolite profiling. Biochemistry 43(45): 14332–14339

16. Smith CA et al (2005) METLIN: a metabolite mass spectral database. Ther Drug Monit 27(6):747–751

17. Tautenhahn R et al (2012) An accelerated workflow for untargeted metabolomics using the METLIN database. Nat Biotechnol 30(9): 826–828

18. Patti GJ et al (2013) A view from above: cloud plots to visualize global metabolomic data. Anal Chem 85(2):798–804

19. Tautenhahn R, Patti GJ, Rinehart D, Siuzdak G (2012) XCMS online: a web-based platform to process untargeted metabolomic data. Anal Chem 84(11):5035–5039

Chapter 10

Imaging Membrane Order Using Environmentally Sensitive Fluorophores

G.W. Ashdown and Dylan M. Owen

Abstract

In the lipid raft hypothesis, ordered and disordered lipid membranes are responsible for regulating the distribution, dynamics, and interactions of membrane associated proteins. Ordered and disordered bilayers may be distinguished by the degree of order in their acyl tails (the order parameter) which in turn affects lipid mobility and lipid packing. Low density lipid packing in the disordered phase allows polar water molecules to penetrate into the usually non-polar bilayer interior. Transition to the ordered phase causes condensation of the membrane, tighter lipid packing, and more complete exclusion of polar water. This process can be measured and quantified using polarity sensitive fluorophores embedded within the bilayer which then have different emission properties depending on membrane phase. Two examples of these are Laurdan and di-4-ANEPPDHQ which can be used to image membrane order distributions in live cells via a variety of microscopy techniques.

Key words Laurdan, Di-4-ANEPPDHQ, Lipid rafts, Membrane microdomains, Fluorescence

1 Introduction

The lipid raft hypothesis is critical to our understanding of how membrane lipids may play a regulatory role during cell signalling and protein trafficking [1, 2]. One of the key tenants of this hypothesis is that in the presence of cholesterol, lipids with saturated acyl tails such as sphingolipids may form a liquid-ordered phase which can coexist with the bulk disordered phase mainly containing unsaturated phospholipids. In the ordered phase lipid lateral mobility is decreased, lipid packing density (the inverse area per lipid) is higher and the membrane bilayer is thicker. Membrane associated proteins may favor partitioning into the ordered or disordered phase depending on the type of lipid anchor they possess. In this way, protein interactions and distributions may be regulated by the distribution of ordered phase membrane lipid microdomains. In this context, high order membrane domains have been shown to play a role in a wide variety of membrane-related phenomena [3–7].

Dylan M. Owen (ed.), *Methods in Membrane Lipids*, Methods in Molecular Biology, vol. 1232,
DOI 10.1007/978-1-4939-1752-5_10, © Springer Science+Business Media New York 2015

One of the most notable examples of ordered lipid domain function is the regulation of cell signalling such as during T cell or B cell activation (formation of an immunological synapse). In T cells, loading the cells with the cholesterol analogue 7-ketocholesterol which disrupts ordered phase domains was shown to cause the mislocalization of key signalling proteins and consequently to decrease signalling efficiency [8]. Here, high membrane order is also observed in sub-synaptic T cell vesicles [4]. It has recently been shown by super-resolution microscopy that such vesicles may be important in the trafficking of key signalling molecules, such as Linker for activation of T cells (LAT) at the synapse [9]. Membrane order has also been studied in the context of cell migration where high order domains have been shown to accumulate at the leading edge of migrating cells as well as at focal adhesions [10]. Membrane order also plays an important role in membrane trafficking. Here, membrane components such as lipids and proteins are sorted between the apical and basolateral surfaces of many polarized cell types, such as MDCK cells [5]. Here, the apical surfaces, which face the harsher environment of the lumen, are enriched with ordered domains as well as specialized signalling proteins which are trafficked to the apical surface via high membrane order vesicles. Once at the apical surface, these components are prevented from diffusing into the basolateral areas by tight junctions. Finally, high membrane order has been shown to be important in a number of pathological processes. These include the observation that the HIV and the influenza viruses require high order domains for entry and budding during their life cycle [11, 12].

In recent years, numerous methods have been used to study lipid microdomains in cells but one of the most fruitful methods has been the use of fluorescence microscopy due to the combination of high spatial resolution, specificity and its minimally invasive nature [13, 14]. In particular, membrane order can be probed using a number of environmentally sensitive fluorescent membrane-partitioning dyes. These are small molecule dyes which reside within the bilayer itself and change their fluorescent properties in response to their local chemical or physical environment. Many such environmentally-sensitive probes exist for both membranes as well as bulk solutions allowing the probing of biophysical parameters such as solvent polarity, pH, viscosity, ion concentrations, temperature, and more. In the context of lipid microdomains and membrane order, those sensing solvent polarity have been the most widely used. In the ordered phase, the density of lipid packing is increased causing greater exclusion of polar water molecules from the otherwise non-polar bilayer interior. In the disordered phase, water penetration is increased and the internal environment of the bilayer becomes more polar.

The two most common membrane dyes to work in this way are Laurdan [15, 16] and di-4-ANEPPDHQ [17, 18]. Laurdan, which is excited in the UV, shifts its fluorescence emission spectrum from a peak around 450 nm when the dye is resident in the ordered phase to a peak of around 500 nm when the dye is in disordered phase bilayers. The order of the bilayer can then be measured by a multi-color ratiometric measurement between two spectral channels corresponding to fluorescence derived from dye in the ordered and disordered phase. Similarly, the probe di-4-ANEPPDHQ, excited at 488 nm, shifts its emission spectrum from 560 nm in the ordered phase to 620 nm in the disordered phase, and can thus be imaged and analyzed in the same manner as Laurdan. There is now a wide palate of such membrane probes available with various spectral properties [19]. Note also that these dyes also change their fluorescence lifetime between the two environments, displaying a longer fluorescence lifetime in the ordered phase. This allows the use of fluorescence lifetime imaging (FLIM) to quantify membrane order [18, 20, 21].

2 Materials

2.1 Cell Culture and Staining

1. Serum-free DMEM (for HeLa cells) or RPMI1640 (for Jurkat T cells) without Phenol Red.

2. Phosphate buffered saline (PBS).

3. Flask of HeLa or Jurkat T cells.

4. Trypsin–EDTA.

5. DMSO.

6. 37 °C incubator.

7. Laurdan or di-4-ANEPPDHQ made up to a 5 mM stock in laboratory grade ethanol.

8. #1.5 glass coverslips or coverslip bottomed dishes.

2.2 Imaging and Analysis

1. Confocal microscope capable of imaging in at least two spectral channels and exciting at 405 nm (or multiphoton at 800 nm) for Laurdan or at 488 nm for di-4-ANEPPDHQ. In this case a Nikon AR1 inverted confocal microscope will be used.

2. Computer with image analysis software such as ImageJ and the ImageJ GP analysis plugin installed [22].

3 Methods

3.1 Sample Preparation

1. Warm the DMEM media and PBS to 37 °C using a water bath.

2. Take a flask of HeLa cells, check the percentage confluence and remove and discard the media.

3. Wash the cells gently with PBS.

4. Add a few milliliters of Trypsin to the flask to cover the cells and incubate at 37 °C for 2–3 min or until the cells detach.

5. Re-suspend the cells in DMEM media.

6. Plate the cells into a glass coverslip-bottomed microscope dish at a confluence of around 50 % in around 2 ml of media.

7. Incubate overnight at 37 °C in a 5 % CO_2 atmosphere.

8. Remove the media and discard. Replace with pre-warmed serum free DMEM media without phenol red, or PBS. *See* **Note 1**.

9. Add the 5 mM Laurdan or di-4-ANEPPDHQ stock solution to the media to generate a final concentration of 5 μM (this will typically be a 1:1,000 or 1:500 dilution). For Jurkat T cells simply add cells to the glass-bottomed dish and stain under the same conditions as HeLa cells.

10. Incubate the cells for 30 min at 37 °C in a 5 % CO_2 atmosphere.

11. Make a second sample of Laurdan or di-4-ANEPPDHQ diluted in DMSO at a concentration of 1:10–1:100 for calibration purposes.

3.2 Image Acquisition

1. Turn on all the microscope equipment, including lasers. If required, also turn on the microscope 37 °C incubation chamber and allow the temperature to equilibrate.

2. Select a high-NA water-immersion objective lens, place a drop of pure water on the lens and add the live cell dish to the microscope stage.

3. Using the brightfield lamp and eye-pieces, find a region of interest and focus the cells.

4. In the microscope control software select two spectral channels to image; corresponding to the dye emission in the ordered and disordered phases. For Laurdan, ordered phase emission is in the range 400–460 nm and disordered phase emission is in the range 470–530 nm. For di-4-ANEPPDHQ, ordered phase emission is in the range 500–580 nm and disordered phase emission is in the range 620–750 nm (Fig. 1a). *See* **Note 2**.

5. Select an excitation wavelength. For Laurdan this will typically be at 405 nm and for di-4-ANEPPDHQ it will be 488 nm. *See* **Note 3**.

6. Ensure that the Channel Series is off such that a single excitation wavelength is present for both emission channels.

7. Start scanning. Increase the detector gains and laser power such that fluorescence is visible in both channels at approximately the same intensity. *See* **Note 4**.

8. Set the image size, scan speed and zoom and check again that the pixel intensities are appropriate.

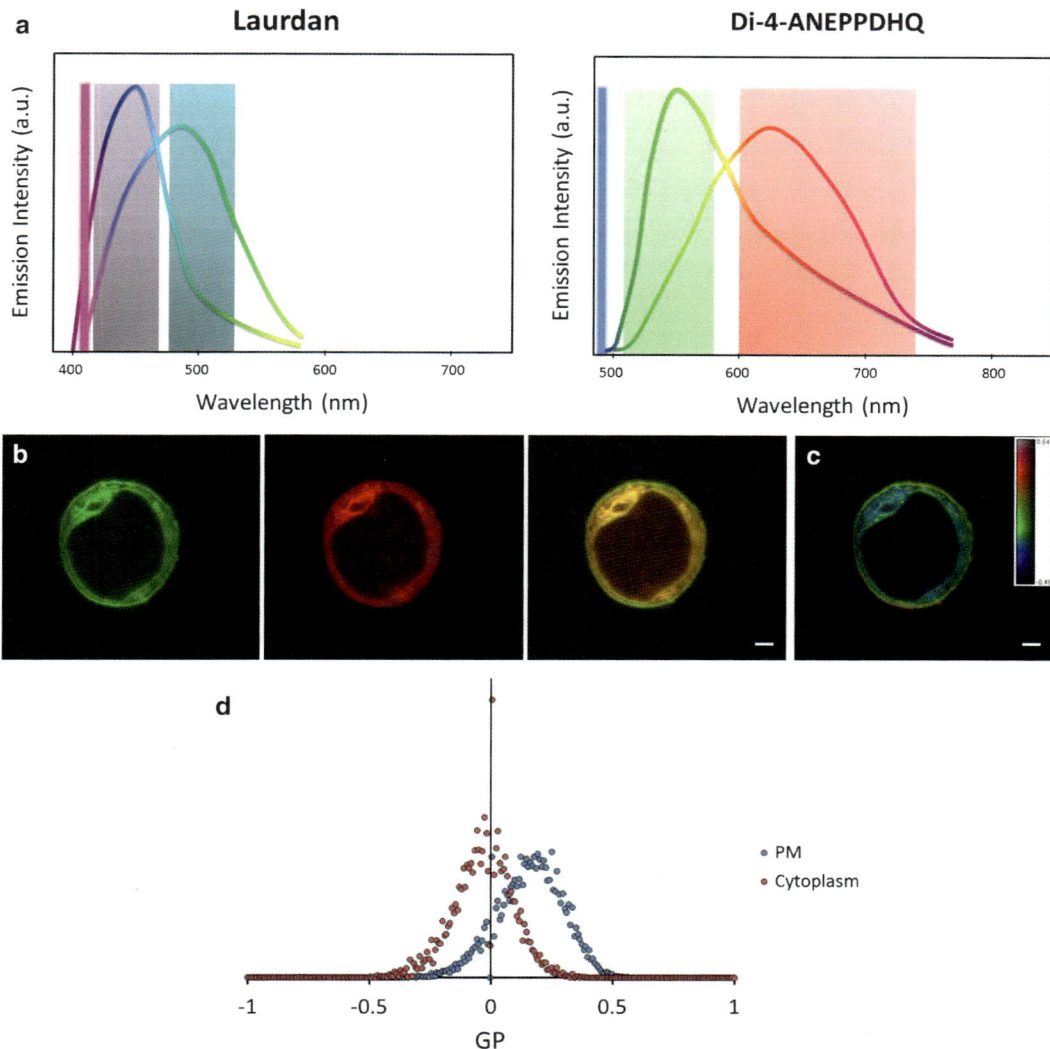

Fig. 1 (**a**) Schematic spectra of Laurdan (*left*) and di-4-ANEPPDHQ (*right*) showing recommended spectral imaging regions. (**b**) *Raw green* and *red* channel fluorescence images of di-4-ANEPPDHQ in live Jurkat T cells together with merged image. (**c**) Calculated GP value image. (**d**) Extracted GP histograms from pixels comprising the plasma membrane and intracellular membranes showing high membrane order (GP) at the plasma membrane

9. Select a higher level of line averaging if required. Higher line averaging will increase signal-to-noise but increase acquisition times.

10. Record the image. *See* **Note 5**.

11. Save the data as Nikon .nd2 files which also contain the metadata and export the two images as 8-bit .tif files such that they can be processed into membrane order maps.

12. Replace the sample with the DMSO calibration sample and image with exactly the same microscope settings. *See* **Note 6**.

3.3 Image Analysis and Quantification

1. Open the two images in ImageJ and ensure there are no saturated pixels, that the images are the same size and are saved as 8-bit .tif files (Fig. 1b).

2. Ensure that the files names match and the end of the file name reads _ch00 for the ordered channel and _ch01 for the disordered channel, for example Image1_ch00.tif and Image1_ch01.tif. *See* **Note 7**.

3. Using the ordered and disordered calibration images, measure the mean intensity in each one using ImageJ.

4. Using these values, calculate the GP value of Laurdan or di-4-ANEPPDHQ in DMSO using the equation:

$$GP = \frac{I_{ordered} - I_{disordered}}{I_{ordered} + I_{disordered}}$$

5. Calculate the G factor calibration value using the equation:

$$G = \frac{GP_{REF} + GP_{REF}GP_{DMSO} - GP_{DMSO} - 1}{GP_{DMSO} + GP_{REF}GP_{DMSO} - GP_{REF} - 1}$$

Using a value of GP_{REF} of 0.207 for Laurdan and -0.85 for di-4-ANEPPDHQ.

6. Run the plugin from ImageJ and navigate to the folder containing the cell images. For the GP color scale select "greys" and select an appropriate threshold. Enter the calculated value for the G factor and select process. The plugin will then calculate the GP values using the equation:

$$GP = \frac{I_{ordered} - GI_{disordered}}{I_{ordered} + GI_{disordered}}$$

7. Select an appropriate color scale for the pseudo-colored GP image. This step does not alter the quantification of the data (Fig. 1c).

8. Once the plugin has completed processing, open the created folder. Here, the "HSB" images contain the pseudo-colored membrane order maps and the "GP" folder contains the quantifiable grey-scale membrane order maps.

9. To quantify membrane order, open the grey-scale GP image in ImageJ and select an appropriate region-of-interest (ROI). Here, the membrane order (GP) is encoded on an 8-bit greyscale with low membrane order (GP = −1) colored black (Intensity = 0) and high membrane order (GP = +1) colored white (Intensity = 255).

10. Display the intensity histogram from the ROI and convert the 8-bit greyscale values to GP values using the equation: GP = Grey/127.5 − 1. Higher GP values indicate higher membrane order (Fig. 1d).

4 Notes

1. If long-term imaging of cells is required, imaging can be performed in full media with serum. This however results in a slight increase in background fluorescence.

2. The exact spectral bands used for each dye are flexible and can be changed to suit available emission filters. If the bands selected are narrower and further apart, contrast between the ordered phase and disordered phase will be greater but signal-to-noise will be worse as less fluorescence signal will be collected. Conversely, if the selected channels are broader of closer together, contrast will be lowered but signal-to-noise increased. A calibration image can be acquired if it is required to quantitatively compare data taken with different spectral ranges.

3. The choice of excitation wavelength is also somewhat flexible. For example di-4-ANEPPDHQ can be efficiently excited anywhere between 450 and 500 nm. Laurdan can also be used with a range of excitation wavelengths and can also be conveniently excited using multi-photon illumination around 800 nm from common, femtosecond pulsed Ti:Sapphire laser sources. Multiphoton excitation has the advantage of greater penetration depth through thick samples and reduced out-of-plane photobleaching. If multi-photon excitation is used, remember to open the confocal pinhole.

4. For optimal signal-to-noise and quantifiable data in the processed images, the intensities in each channel should be set to be approximately equal. There should also be no underflow or saturated pixels which can be checked using a saturated pixel look-up-table.

5. Laurdan and di-4-ANEPPDHQ imaging can also be used with multi-point imaging, time-series and z-stacks if required. If it is required to quantitatively compare images that have significantly different intensities, the laser power can be changed to compensate; however, the detector gains must always be kept constant as these will change the ratio between the two channels and hence the membrane order value.

6. If the calibration sample is too bright, the laser power can be reduced. Di-4-ANEPPDHQ is very dim in DMSO and it may be necessary to use higher concentrations or to perform frame accumulation to obtain the calibration image.

7. The ImageJ plugin can batch process all files in a folder so long as they are labelled in pairs as in this example.

References

1. Simons K, Ikonen E (1997) Functional rafts in cell membranes. Nature 387:569–572
2. Lingwood D, Simons K (2010) Lipid rafts as a membrane-organizing principle. Science 327(5961):46–50
3. Gaus K et al (2005) Condensation of the plasma membrane at the site of T lymphocyte activation. J Cell Biol 171(1): 121–131
4. Owen DM et al (2010) High plasma membrane lipid order imaged at the immunological synapse periphery in live T cells. Mol Membr Biol 27(4–6):178–189
5. van Meer G et al (1987) Sorting of sphingolipids in epithelial (Madin-Darby canine kidney) cells. J Cell Biol 105(4):1623–1635
6. Carter GC et al (2009) HIV entry in macrophages is dependent on intact lipid rafts. Virology 386(1):192–202
7. Takeda M et al (2003) Influenza virus hemagglutinin concentrates in lipid raft microdomains for efficient viral fusion. Proc Natl Acad Sci 100(25):14610–14617
8. Rentero C et al (2008) Functional implications of plasma membrane condensation for T cell activation. PLoS One 3(5):e2262
9. Williamson DJ et al (2011) Pre-existing clusters of the adaptor Lat do not participate in early T cell signaling events. Nat Immunol 12(7):655–662
10. Gaus K et al (2006) Integrin-mediated adhesion regulates membrane order. J Cell Biol 174(5):725–734
11. Khurana S et al (2007) Human immunodeficiency virus type 1 and influenza virus exit via different membrane microdomains. J Virol 81(22):12630–12640
12. Carrasco M, Amorim MJ, Digard P (2004) Lipid raft-dependent targeting of the influenza A virus nucleoprotein to the apical plasma membrane. Traffic 5(12):979–992
13. Lagerholm BC et al (2005) Detecting microdomains in intact cell membranes. Annu Rev Phys Chem 56:309–336
14. Owen DM et al (2007) Optical techniques for imaging membrane lipid microdomains in living cells. Semin Cell Dev Biol 18(5):591–598
15. Parasassi T et al (1998) Laurdan and Prodan as polarity-sensitive fluorescent membrane probes. J Fluoresc 8(4):365–373
16. Parasassi T et al (1991) Quantitation of lipid phases in phospholipid vesicles by the generalized polarization of Laurdan fluorescence. Biophys J 60(1):179–189
17. Jin L et al (2005) Cholesterol-enriched lipid domains can be visualized by di-4-ANEPPDHQ with linear and nonlinear optics. Biophys J 89(1):L04–L06
18. Owen DM et al (2006) Fluorescence lifetime imaging provides enhanced contrast when imaging the phase-sensitive dye di-4-ANEPPDHQ in model membranes and live cells. Biophys J 90(11):L80–L82
19. Kwiatek JM et al (2013) Characterization of a new series of fluorescent probes for imaging membrane order. PLoS One 8(2):e52960
20. Owen DM et al (2012) Sub-resolution lipid domains exist in the plasma membrane and regulate protein diffusion and distribution. Nat Commun 3:1256
21. Golfetto O, Hinde E, Gratton E (2013) Laurdan fluorescence lifetime discriminates cholesterol content from changes in fluidity in living cell membranes. Biophys J 104(6):1238–1247
22. Owen DM et al (2011) Quantitative imaging of membrane lipid order in cells and organisms. Nat Protoc 7(1): 24–35

Chapter 11

3D Super-Resolution Imaging by Localization Microscopy

Astrid Magenau and Katharina Gaus

Abstract

Fluorescence microscopy is an important tool in all fields of biology to visualize structures and monitor dynamic processes and distributions. Contrary to conventional microscopy techniques such as confocal microscopy, which are limited by their spatial resolution, super-resolution techniques such as photoactivated localization microscopy (PALM) and stochastic optical reconstruction microscopy (STORM) have made it possible to observe and quantify structure and processes on the single molecule level. Here, we describe a method to image and quantify the molecular distribution of membrane-associated proteins in two and three dimensions with nanometer resolution.

Key words 2D and 3D super-resolution microscopy, PALM, STORM, Cluster analysis

1 Introduction

In the past decades fluorescence microscopy has been a popular tool in all life sciences. Its popularity stems from the versatility of its applications. It can be utilized to image a wide range of targets—from whole organisms to single cells, organelles, and more recently, single molecules. The development of new hardware and software has reduced the resolution limit of conventional fluorescence light microscopy techniques to the boundaries of the refraction limit; typically ~250 nm in lateral direction and ~600 nm in axial direction. With the dimension of proteins in order of 1–20 nm, the ultimate goal has always been to circumvent this limitation and allow the visualization of structures and dynamic processes below these limits. A major step towards this goal was the development of super-resolution techniques such as PALM, STORM and other related techniques [1–4]. The success of localization-based super-resolution techniques relies on the ability to distinguish between fluorophores whose point spread functions overlap and cannot be separated by conventional fluorescence microscopy techniques. Localization microscopy techniques temporally segregate fluorescence events by reducing the number of excited fluorescent

Dylan M. Owen (ed.), *Methods in Membrane Lipids*, Methods in Molecular Biology, vol. 1232,
DOI 10.1007/978-1-4939-1752-5_11, © Springer Science+Business Media New York 2015

molecules to one or a few molecules per diffraction limited volume. To achieve this, PALM takes advantage of the photo-physical properties of photo-activatable or photo-switchable fluorescent proteins [5], while STORM utilizes the stable dark state of small molecule dyes for photo-switching [6].

Commonly used fluorescent proteins for PALM include PS-CFP2 [7] and mEOS2 [8]. PS-CFP2 is a cyan-to-green photo-convertible protein, while mEOS2 photo-converts from green to red. Both proteins are photo-converted by illumination with 405 nm light. In the case of PS-CFP2, imaging and bleaching of the photo-converted state is achieved by exposure to 488 nm light, whereas in the case of mEOS2 561 nm light is used.

In the related technique, direct (d)STORM, conventional organic dyes are used in combination with a reducing buffer [3, 6]. The oxygen-reduced environment allows small organic dyes to be switched on and off by pushing them into a long-lived dark state. Popular dyes for (d)STORM include Alexa 647 and Cy5 [2, 3, 6], but many other dyes have been reported to be suitable for (d) STORM imaging [9]. To improve spatial segregation and achieve an optimal signal-to-noise ratio, many commercial systems use total internal reflection fluorescence (TIRF) illumination and hence restrict the number of molecules than can be activated in z-direction to about 100 nm [10]. This technique achieves the highest signal-to-noise levels for single molecule imaging, and hence the highest possible resolution by minimizing background interference from out-of-focus light. Fluorescent events are subsequently recorded by a sensitive camera, most commonly an EMCCD camera. Initially, super-resolution imaging has been done on fixed samples, and hence lacking dynamic information. Improvement of hardware and software solutions has enabled researchers to produce super-resolution images of live cells.

In recent years, 3D approaches have completed the toolbox of single molecular localization techniques. A number of different 3D super-resolution have been developed achieving varying resolutions, penetration depths, and ranges. The first approaches to 3D super-resolution imaging was Biplane PALM developed by Juette et al. [11] and astigmatic spot analysis developed by Huang et al. [12]. A further approach was invented by Pavani et al. [13] and Baddeley et al. [14], who engineered a double-helix PSF approach. All three approaches exploit the z-dependency of the PSF by defocusing the emission of a single fluorescent molecule above or below the in-focus image plane. For Biplane imaging this is achieved by simultaneously imaging two different focal planes, one above the ideal focal plane and one below [11]. Consequently, the shape of the PSF in each plane can be used to deduct its axial position. This approach results in a z-resolution of about 75 nm across an imaging depth of ~1 μm. Astigmatic spot analysis is based on the introduction of a cylindrical lens between the objective and the camera [12].

The resulting optical astigmatism causes the PSFs to be elliptical. The ratio between the x- and y-dimensions of the PSFs reveals the axial position of single molecules, achieving a z-resolution of 50 nm with an imaging range of ~600 nm. The third approach based on PSF shape modification employs either a phase mask in the detection path (double-helix PSF imaging, DH-PSF) [13] or a glass wedge in front of the objective resulting in a phase ramp (phase ramp imaging localization microscopy, PRILM) [14]. Both approaches yield a twisted double-helical PSF with two prominent lobes. The relative orientation and the distance of the two PSF lobes to each other are exploited to deduct the axial position of the single fluorescent molecules, resulting in a z-resolution of about 40 nm with a z-range of up to 2 μm.

In general, with all techniques described above, the fluorescent molecules detected in each frame are imaged until they are bleached, followed by a new cycle of activation and bleaching. This process is repeated many times requiring several thousand frames to compile a complete fluorescence image. Once the raw image time series has been recorded, single molecule point-spread functions (PSFs) are identified over the background noise. The center of identified PSFs has to be calculated and recorded. A common approach was developed by Thompson et al. [15]. It is based on the assumption that the shape of a PSF follows a 2D Gaussian distribution. Hence, the center of mass of each PSF can be calculated using a Gaussian mask after smoothing with an idealized Gaussian function. This returns the position of the molecule, i.e., the center the coordinate, and the localization precision. In the case of N recorded photons, background variance b, PSF width s, and camera pixel size relative to the PSF width a, the localization precision is given by:

$$\text{Precision} = \sqrt{\left(\frac{s^2}{N}\right) + \left(\frac{\dfrac{a^2}{12}}{N}\right) + \left(\frac{8\pi s^4 b^2}{a^2 N^2}\right)}$$

The centers of the PSFs can also be determined by fitting them to 2D Gaussian distributions using a least-squares approach. A number of solutions now exist for this computationally more involved approach.

Special consideration has to be given to re-excitation events of single fluorophores. (d)STORM experiments are only possible due to the repeated excitation and bleaching cycles of organic dyes. But also photo-activatable fluorescent proteins have now been shown to undergo multiple blinking cycles. For example, with mEOS2 approximately 50 % of molecules undergo re-excitation [16]. To eliminate possible clustering artifacts [17], an additional processing step has to be introduced. The use of an off-gap allows for

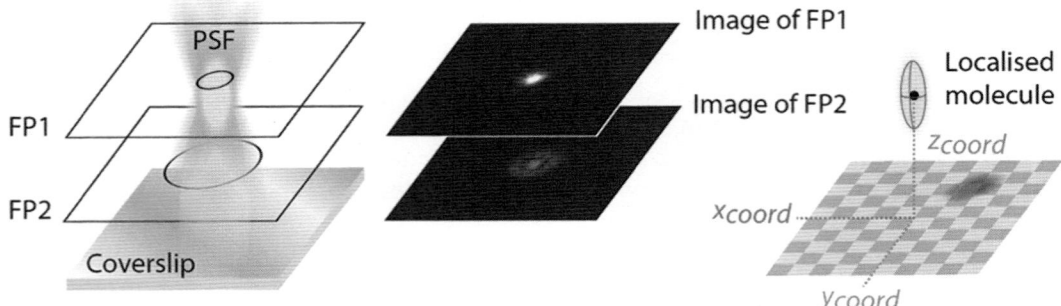

Fig. 1 3D superresolution—Biplane imaging. A single molecule PSF is imaged on two focal planes (FP) simultaneously. This results in differently shaped PSFs in each image plane. The z-coordinate of each PSF can consequently be deducted from the shape of the PSF in each plane in comparatively to a theoretical 3D PSF

temporal grouping of blinking events of a single molecule and hence corrects for clustering artifacts caused by repeated excitation and bleaching cycles [18].

Once the centers of each PSF have been determined and re-excitation events have been accounted for, an image is generated showing the exact coordinate for each molecule. Several display options are available for this. Typically, each molecular coordinate is convolved with a standardized Gaussian PSF with a width equalling the localization precision. However, it should be taken into account that the resolution of such a super-resolution image not only depends on the localization precision but also on the density of the localized molecules.

The molecular distribution map generated by the super-resolution image can now be used for quantitative analysis. Statistical methods based on pair-correlation analysis [19], Ripley's K-function analysis [20] and related techniques investigate molecular clustering and cluster distributions. Ripley's K-function analysis is the most commonly used point-pattern analysis method [21] and has been applied to analyze cluster behavior and distribution of proteins in the T-cell synapse [22–24]. Ripley's K-function analysis outputs one clustering value for an entire region of interest. The related Getis and Franklin's L-function [25] assigns a clustering value to each individual molecule, rather than to a region in its entirety. With this method, color-coded cluster maps can be generated, which then can be further used to extract various cluster parameters [22, 23]. These routines have so far been used for analysis of 2D super-resolution datasets, but have also now been applied to 3D PALM datasets [26].

In this chapter, we will describe a protocol for Biplane 3D super-resolution imaging (Fig. 1) and quantitative data analysis of the

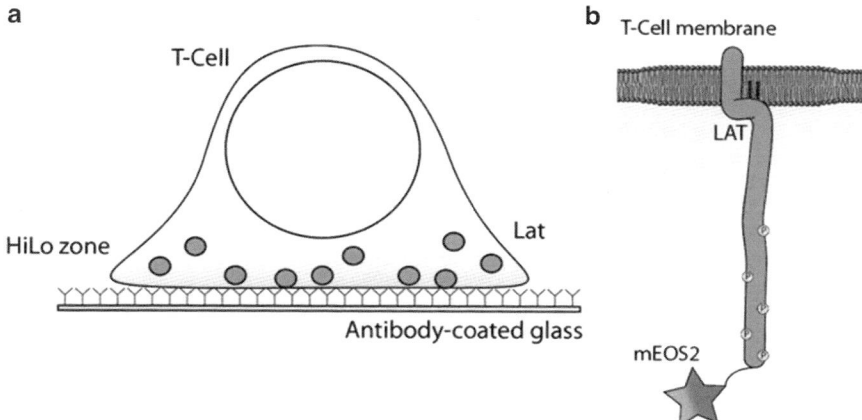

Fig. 2 T-Cell activation and Lat-membrane association. (**a**) To image Lat vesicles (*blue circles*) adjacent to the plasma membrane of T-cells and membrane-bound Lat, T-cells are activated by adhesion to an antibody-coated glass coverslip. Only Lat molecules that are inside the HiLo zone, which typically penetrates ~1 μm beyond the glass/cell interface, are detected. (**b**) Membrane-Lat is dually palmitoylated and is tagged with the green-to-red photo-convertible fluorescent protein mEOS2 at its C-terminus to allow visualization by super-resolution imaging (Color figure online)

resulting 3D dataset. As an example we use the integral membrane protein Linker for Activation of T-cells (Lat) in the synapse of an activated T-cell. Lat was tagged with the fluorescent green-to-red photo-switchable protein mEOS2 (Fig. 2a). Lat is dually palmitoylated targeting the protein to lipid-ordered domains in the plasma membrane of T-cells [27, 28] (Fig. 2b). The size of lipid-ordered domains range from 10 to 200 nm [29]. Since the size of lipid domains is well below the resolution limit of conventional fluorescence microscopy techniques, they are an ideal target for super-resolution microscopy analysis. The methods applied here to investigate the distribution of LAT are also applicable to other integral membrane proteins, membrane-associated proteins in the inner and outer leaflet as well as fluorescent membrane lipid dyes such as diI.

2 Materials

2.1 Reagents

1. Jurkat T-cell line (American Type Culture Collection, ATCC).
2. RPMI media supplemented with 10 % FCS, 2 % L-Glutamine.
3. For T-cell activation: mouse anti-CD3 (OKT3) and mouse anti-CD28 antibody.
4. 0.22 μm filtered phosphate buffered saline (PBS).
5. Sterile water.
6. DNA plasmid Lat-mEOS2.
7. 100 nm diameter gold beads as fiducial markers.

8. 18 mm diameter Chamlide microscopy chamber (LCI Live Cell Instrument) and compatible #1.5 (0.17 mm) TIRF-suitable glass coverslips.

9. 16 % paraformaldehyde solution.

2.2 Microscopy Equipment

This chapter assumes the user is familiar with basic instrumentation for wide-field and TIRF microscopy.

1. An inverted TIRF microscope with a 405 and 561 nm laser capable of simultaneous excitation. The imaging laser (561 nm) should have an output power of at least 50 mW. The microscope should have a sensitive, cooled EMCCD camera with at least 256×256 pixels and should allow imaging frame rates of 2–5 ms. The total magnification of the system should be set so that each pixel records 100 nm × 100 nm of the sample. Detection of mEOS2 fluorescence requires dichroics and filters in the range of approximately 580–650 nm while exciting at 405 and 561 nm simultaneously. To minimize background noise and sample drift, it is of benefit to enclose the microscope inside a light-proof and temperature-controlled incubation chamber.

2. An oil-immersion objective suitable for TIRF illumination with 60–100× magnification was used here. Typically, objectives with a numerical aperture (NA) of at least 1.45 are recommended for TIRF illumination.

3. A computer with the capacity to handle large datasets and rapid data transfer rates to allow sensible acquisition and date processing size of 3D PALM datasets that are typically around 5 GB in size is required. A typical configuration is at least a 1 Tb hard—drive, 3 GHz quad—core processor and at least 16 GB of fast RAM. Installation of Matlab (Mathworks Inc.) is also advisable.

2.3 Cell Transfection Equipment

1. Neon electroporation transfection system (Invitrogen) including buffer E, buffer R, and Neon pipettes (10 µl).

3 Methods

3.1 Antibody-Coating of Coverslips

1. Clean coverslips for 30 min in an UV/ozone cleaner. (*See* **Note 1**).

2. Rinse coverslips with sterile water. (*See* **Note 2**).

3. Apply antibodies to the coverslips: use mouse anti-CD3 and mouse anti-CD28 at 20 µg/ml. (*See* **Note 3**).

4. Incubate at 37 °C for at least 1 h or overnight in a humidified chamber to prevent antibodies from drying out.

5. Wash coverslips three times with sterile 1× PBS.

6. (Optional) Block with 1 % BSA/PBS for 1 h at room temperature.

7. Wash three times with sterile 1× PBS.

3.2 Electroporation of Jurkat T-Cells with Neon Transfection System (Invitrogen)

1. Use cells at a confluency of $1–2 \times 10^6$ cells/ml and below passage no 12.

2. Warm up buffers E and R to RT.

3. Set up Neon machine, prepare Neon pipette and tips, and add 3 ml buffer E to buffer tube.

4. Resuspend cells to disrupt clumps and transfer to a 15 ml tube.

5. Count cells to determine the total number of cells in the flask.

6. Spin 15 ml tube at $200 \times g$ for 5 min at room temperature.

7. Transfer supernatant to a new 15 ml tube (this can be used as recovery media).

8. If preparing recovery media, dilute the old media 1:1 with fresh media and top up to 20 % FBS. (Fresh media can also be used instead.)

9. Plate the media into the recovery plate (e.g., 1.0 ml/well of a 12-well plate) and place at 37 °C to equilibrate.

10. Resuspend cell pellet in 15 ml PBS.

11. Spin again and discard supernatant.

12. Resuspend cell pellet in enough PBS to make a 20×10^6 cells/ml suspension.

13. Transfer 50 μl of PBS cell suspension to a 1.5 ml Eppendorf tube.

14. Spin Eppendorf tube at $277 \times g$ for 5 min at RT.

15. During the centrifugation, dispense the Lat-mEOS2 DNA. Use 1 μg per 10 μl transfection.

16. Retrieve recovery plate from the 37 °C incubator.

17. Remove PBS supernatant from Eppendorf tube and resuspend pellet in 50 μl Buffer R. Mix very gently. (*See* **Note 4**).

18. Transfer 10 μl of cells/buffer R mixture into the DNA and mix very gently.

19. Fit a neon-tip and aspirate 10 μl of cells. Avoid formation of bubbles.

20. Electroporate cells with 2×30 ms pulses at 1,150 V.

21. Transfer the cells into the recovery wells and rock the plate gently to wash them.

22. When reusing the tip do not flush the tip with media, but go straight to aspirating the next 10 μl of cells. (*See* **Notes 5** and **6**).

23. Place cells at 37 °C and incubate overnight.

3.3 T-Cell Activation on Coverslips and Fixation for PALM Imaging

1. Dilute T-cell suspension to 1–2×10^6 cells/ml in complete RPMI (10 % FCS, 2 % L-Glutamine).

2. Transfer 1 ml of T-cell suspension to coated coverslips and incubate for 10 min at 37 °C.

3. Add 1/3 of 16 % PFA solution to total volume in each well and mix very gently.

4. Wash three times in 1× PBS.

5. Transfer to microscope and image. (*See* **Note 7**).

3.4 Microscope Setup and Data Acquisition

1. Turn on microscope. Activate 405 and 561 nm lasers. Set the EMCCD camera temperature to –60 °C and the EM gain to ~150 V. (*See* **Note 8**).

2. Assemble sample holder and coverslip. Add ~500 µl 1× PBS (or enough volume to sufficiently cover coverslip surface). Add ~ 30 µl of fiducial markers to the sample. Wait for ~10 min to allow fiducials to settle on the glass surface.

3. Place a drop of immersion oil on the objective and place the sample on the microscope stage. Make sure that the stage and sample are level. Use transmitted light and oculars to locate and focus on the cells.

4. Select epifluorescence illumination with an appropriate filter covering the green-to-red range to identify a transfected cell.

5. Switch the detection pathway to the camera. Set the camera integration time to 30 ms and the excitation wavelength (561 nm) to ~15 mW at the sample. Select a filter with an appropriate detection band in the orange range (using a filter which allows simultaneous excitation at 405 nm). Identify the transfected cell and focus on the basal membrane. Make sure that several fiducial markers are inside the field of view.

6. Activate 3D imaging mode and move Biplane prisms and field stop into detection path. Set the focal plane offset for Biplane imaging to an appropriate offset (typically between 372–516 nm). Switch the microscope to TIRF illumination mode. Change the TIRF illumination angle to highly inclined and laminated optical sheet (HiLo) illumination. Set the focal plane between the upper and lower focal plane of the Biplane prisms. (*See* **Note 9**).

7. Set the microscope to record a time-series of 15,000–20,000 frames and begin recording. Very few PSFs should be observed per frame, blinking with a lifetime of a small number of frames. (*See* **Note 10**).

8. Turn on the 405 nm laser at very low power, typically at <10 µW at the sample. The number of blinking PSFs should increase. It is crucial to keep the density of blinking molecules to low to avoid overlapping PSFs.

9. The number of viable photo-convertible molecules decrease over the course of the acquisition. It may be necessary to gently increase the 405 nm laser power to increase the number of recorded PSFs. Re-focusing may also be necessary to compensate for axial drift caused by thermal fluctuations or hardware instabilities.

10. Save the raw data in a format that is compatible with the available data processing software.

3.5 Data Analysis

1. Calculate the center of single molecule PSFs and the localization precision. The following manual gives an example using the commercially available software Zen 2010 by Carl Zeiss Microscopy. However, there are a few open-source software packages available for 3D-superresolution processing [30, 31]. Zeiss Zen applies Gaussian and Laplace filtering to the raw time series. Single molecule events are identified when $I - M > 6S$, with I being the intensity, M the mean image intensity, and S the standard deviation of the image intensity. The center of each PSF is then calculated by fitting to a three-dimensional theoretical or experimental PSF. To generate an experimental PSF, typically a z-stack with z-steps of about ~10 nm is acquired by imaging sparsely distributed small fluorescent beads (for example 0.1 μm Tetraspecks). Zeiss Zen also calculates the localization precision and has an option for drift correction based on fiducial marker recognition. The output text file contains a list of x–y–z coordinates and the localization precision of all detected molecules.

2. Select a representative region of interest (ROI) excluding cell edges to avoid edge artifacts in further analysis steps. The ROI is typically a region of 2–5 μm² with a depth of roughly 1 μm. The optimal size depends on the number of molecules detected per region. Robust cluster analysis requires at least 300–500 localized molecules per 1 μm³.

   The ROI can be chosen either in Zeiss Zen on the reconstructed 3D PALM image or by plotting a scatter plot of the molecules' coordinates for example in Matlab. The complete dataset is then cropped to the selected ROI and points outside the ROI are discarded. (*See* **Note 11**).

3. Calculate the Ripley's K-function for the ROI. The Ripley's K-function evaluates the degree of clustering of a spatial point pattern relative to a random distribution. It counts the number of encircled points within concentric circles of increasing radii centered on each point. The K-function $(k(r))$ at a radius r, for a volume V, containing n points is calculated as follows:

$$K(r) = V \sum_{i=1}^{n} \sum_{j=1}^{n} \left(\frac{\delta_{ij}}{n^2} \right) \text{ where } \delta_{ij} = 1 \text{ if } d(\text{point } i, \text{ point } j) < r, \text{ otherwise } 0$$

where δ_{ij} is the distance between points i and j. For a random distribution $K(r)$ scales linearly with the volume. Linearise the K-function to generate the L-function:

$$K(r) = V\sum_{i=1}^{n}\sum_{j=1}^{n}\left(\frac{\delta_{ij}}{n^2}\right) \text{where } \delta_{ij} = 1 \text{ if } d(\text{point } i, \text{point } j) < r, \text{otherwise } 0$$

such that it scales linearly with the radius.

4. Plot the value of $L(r) - r$ versus r. $L(r) - r = 0$ for a spatially random distribution at all radii. Positive $L(r) - r$ values indicated clustering above what would be expected for a random distribution. The greater the value of $L(r) - r$ at any given r, the greater the degree of clustering at that spatial scale. Confidence intervals (e.g., 99 %) can be generated by simulating randomly distributed points with the same number and region size of the data. Depending on the number of points, 100–500 random distributions should be generated and analyzed to generate reliable statistics.

5. Calculate the Getis and Franklin's (G&F's) L-function for local point pattern analysis. While the Ripley's K-function produces a mean value for the ROI as a whole, the G&F's L-function generates a value for the clustering level of each point individually based on the density in 3D. This counts the number of points encircled by a sphere of a specific radius centered on that point (relative to the value expected of a random distribution). Choose a radius (cluster scale) with which to interrogate the pattern for clustering. Appropriate values might be the size of expected clusters, or the spatial scale at which the Ripley's K-function curve peaks. Calculate G&F's L-function for each molecule, i, at that radius (for example 50 nm) using:

$$L_j(50) = \sqrt[3]{\frac{3}{4\pi}V\sum_{i=1}^{n}\left(\frac{\delta_{ij}}{n}\right)} \text{where } \delta_{ij} = 1 \text{ if } d(\text{point } i, \text{point } j) < r, \text{otherwise } 0.$$

6. Use the x and y coordinates and the G&F's $L_i(r)$ value for each point to generate a cluster heat map: This can be achieved in several packages and programming languages but is simple in Matlab. First, generate three 1D arrays covering the range of the x, y, and z values of the ROI. Typically arrays with 10 nm resolution are sufficient. Use the meshgrid command to convert the arrays into a 3D grid on which to interpolate the surface. Use the TriScatteredInterp function to interpolate the surface from the coordinates, $L_i(r)$ values and the 3D grid. Plot the surface using the contour function (typically with the Jet color scale) and apply the desired cosmetic changes.

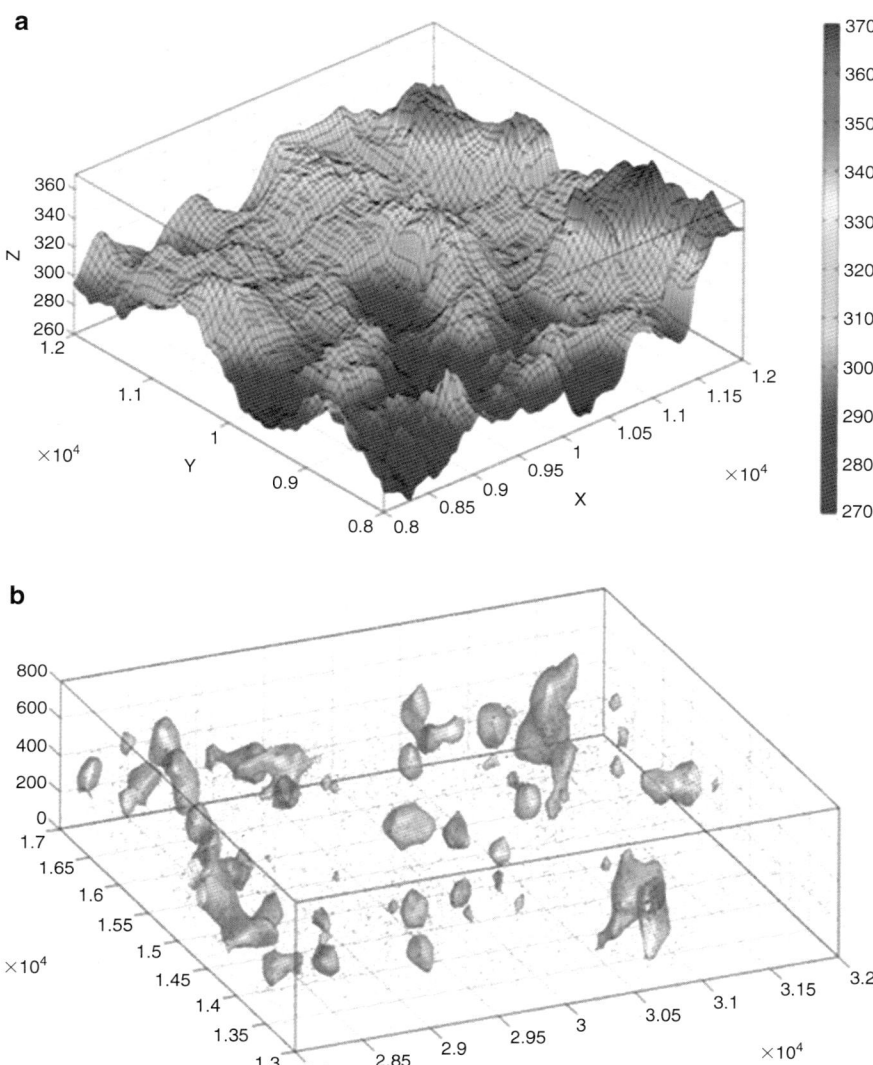

Fig. 3 3D topographical cluster map and an isosurface plot of Lat distribution in 3D. DiI and Lat-mEOS2 was imaged in HiLo mode and its 3D protein distribution analyzed. (**a**) *Color-coded* height map showing the distribution of DiI molecules in *x, y*, and *z*-directions in the plasma membrane. (**b**) Isosurface plot of Lat demonstrating the distribution of Lat clusters in 3D (Color figure online)

This volume can be sliced into various views in Matlab. Be careful to apply the maximum and minimum color values appropriately so as not to saturate the color scale. Save the plot as both a space.mat file and as a TIFF file (Fig. 3).

7. Generate a binary map from the heat map by plotting the data using the isosurface command and an appropriate $L(r)$ threshold. Save the plot as both a space.mat file and as a TIFF file (Fig. 3).

8. Extract clustering parameters using Matlab.

4   Notes

1. Cleaning coverslips in an UV/Ozone sterilizer gives the glass a hydrophilic surface charge. Therefore, less antibody solution needs to be used to coat coverslips, since the solution is able to cover the glass surface evenly without being affected by surface tension.

2. Rinse uncoated coverslips with water in a TC plate (i.e., 18 mm coverslips fit into wells of 12-well plates). Leave the plate uncovered in the TC hood air stream for a few minutes until the excess water has evaporated from the top surface. Evaporation from under the coverslip will seal it to the plate surface, making it easier to work with. During later washing steps the coverslip will be released for removal, although it may need to be moved gently with some fine tweezers.

3. Save antibody by spotting out small amounts of antibody out instead of covering the entire coverslip (~50 µl will create a ~5 mm diameter spot) or use a Dako Pen to mark boundaries on coverslips.

4. Cells must be used within 15 min. Otherwise cell viability will be affected.

5. Each Neon tip can be used twice. Therefore it is sensible to prepare duplicate samples.

6. See Invitrogen.com for additional protocols for other cells types using the Neon electroporation system. Other transfection procedures, such as Lipofectamine may also be used depending on the cell type.

7. Image as soon after fixation as possible to avoid decrease in sample integrity. If imaging on the same day is not feasible, store at room temperature in the dark overnight and image on the next day.

8. Turning on the incubation chamber, set to 25 °C, several hours before imaging, equilibrates microscope hardware temperature and reduces thermal drift during the acquisition. Sample drift will reduce the resolution and can lead to detection artifacts.

9. Both focal planes, i.e., the upper and the lower focal plane, should appear de-focused to the same extent.

10. Imaging can be continued until no more blinking molecules are observed. Getting an accurate, quantitative measurement of the number of molecules in the field of view is difficult however. Repeated activation of fluorescent molecules can be compensated for [16–18], but the contribution of noise, the fraction of molecules detected above noise, the fraction of correctly matured fluorescent proteins, etc. are usually unknown.

11. The described cluster analysis method requires edge-compensation for the ROIs because molecules located very close to the edge will appear less clustered as they have less nearby neighbors. While the ROI does not have to be square, square regions suffer less from edge effects than rectangles, as the area–perimeter ratio is highest. Edge compensation can be achieved by analyzing a slightly larger region than desired and then cropping the generated cluster map, or by weighting edge points appropriately to compensate for this effect. Some point-pattern analysis packages such as SpPack offer a variety of edge correction options [32].

References

1. Betzig E et al (2006) Imaging intracellular fluorescent proteins at nanometer resolution. Science 313:1642–1645

2. Rust MJ, Bates M, Zhuang X (2006) Sub-diffraction-limit imaging by stochastic optical reconstruction microscopy (STORM). Nat Methods 3:793–795

3. Heilemann M et al (2008) Subdiffraction-resolution fluorescence imaging with conventional fluorescent probes. Angew Chem Int Ed 47:6172–6176

4. Eggeling C et al (2009) Direct observation of the nanoscale dynamics of membrane lipids in a living cell. Nature 457:1159–1162

5. Lippincott-Schwartz J, Patterson GH (2009) Photoactivatable fluorescent proteins for diffraction-limited and super-resolution imaging. Trends Cell Biol 19:555–565

6. Heilemann M, Margeat E, Kasper R, Sauer M, Tinnefeld P (2005) Carbocyanine dyes as efficient reversible single-molecule optical switch. J Am Chem Soc 127:3801–3806

7. Chudakov DM et al (2004) Photoswitchable cyan fluorescent protein for protein tracking. Nat Biotechnol 22:1435–1439

8. Mckinney SA, Murphy CS, Hazelwood KL, Davidson MW, Looger LL (2009) A bright and photostable photoconvertible fluorescent protein. Nat Methods 6:131–133

9. van de Linde S et al (2009) Multicolor photo-switching microscopy for subdiffraction-resolution fluorescence imaging. Photochem Photobiol Sci 8:465–469

10. Axelrod D (2001) Total internal reflection fluorescence microscopy in cell biology. Traffic 2:764–774

11. Juette MF et al (2008) Three-dimensional sub-100 nm resolution fluorescence microscopy of thick samples. Nat Methods 5:527–529

12. Huang B, Wang W, Bates M, Zhuang X (2008) Three-dimensional super-resolution imaging by stochastic optical reconstruction microscopy. Science 319:810–813

13. Pavani SRP et al (2009) Three-dimensional, single-molecule fluorescence imaging beyond the diffraction limit by using a double-helix point spread function. Proc Natl Acad Sci 106:2995–2999

14. Baddeley D, Cannell MB, Soeller C (2011) Three-dimensional sub-100 nm super-resolution imaging of biological samples using a phase ramp in the objective pupil. Nano Res 4:589–598

15. Thompson RE, Larson DR, Webb WW (2002) Precise nanometer localization analysis for individual fluorescent probes. Biophys J 82:2775

16. Annibale P, Scarselli M, Kodiyan A, Radenovic A (2010) Photoactivatable fluorescent protein mEos2 displays repeated photoactivation after a long-lived dark state in the red photoconverted form. J Phys Chem Lett 1:1506–1510

17. Annibale P, Vanni S, Scarselli M, Rothlisberger U, Radenovic A (2011) Identification of clustering artifacts in photoactivated localization microscopy. Nat Methods 8:527

18. Annibale P, Vanni S, Scarselli M, Rothlisberger U, Radenovic A (2011) Quantitative photo activated localization microscopy: unraveling the effects of photoblinking. PLoS One 6:e22678

19. Sengupta P et al (2011) Probing protein heterogeneity in the plasma membrane using PALM and pair correlation analysis. Nat Methods 8:969–975

20. Owen DM et al (2010) PALM imaging and cluster analysis of protein heterogeneity at the cell surface. J Biophotonics 3:446–454

21. Ripley BD (1977) Modelling spatial patterns. J R Stat Soc B (Methodol) 39(2):172–212

22. Williamson DJ et al (2011) Pre-existing clusters of the adaptor Lat do not participate in early T cell signaling events. Nat Immunol 12:655–662

23. Rossy J, Owen DM, Williamson DJ, Yang Z, Gaus K (2012) Conformational states of the kinase Lck regulate clustering in early T cell signaling. Nat Immunol 14:82–89

24. Lillemeier BF et al (2010) TCR and Lat are expressed on separate protein islands on T cell membranes and concatenate during activation. Nat Immunol 11:90–96

25. Getis A, Franklin J (1987) Second-order neighborhood analysis of mapped point patterns. Ecology 68:473–477

26. Owen DM et al (2013) Quantitative analysis of three-dimensional fluorescence localization microscopy data. Biophys J 105:L05–L07

27. Tanimura N et al (2003) Dynamic changes in the mobility of LAT in aggregated lipid rafts upon T cell activation. J Cell Biol 160: 125–135

28. Zech T et al (2009) Accumulation of raft lipids in T-cell plasma membrane domains engaged in TCR signalling. EMBO J 466–476. doi:10.1038/emboj.2009.6

29. Pike LJ (2006) Rafts defined: a report on the keystone symposium on lipid rafts and cell function. J Lipid Res 47:1597–1598

30. Henriques R et al (2010) QuickPALM: 3D real-time photoactivation nanoscopy image processing in ImageJ. Nat Methods 7:339–340

31. Brede N, Lakadamyali M (2012) GraspJ: an open source, real-time analysis package for super-resolution imaging. Opt Nanosc 1:11–17

32. Perry GL (2004) SpPack: spatial point pattern analysis in excel using visual basic for applications (VBA). Environ Model Software 19: 559–569

Chapter 12

Electron Microscopy Methods for Studying Plasma Membranes

Alison J. Beckett and Ian A. Prior

Abstract

Electron microscopy allows direct visualization of the underlying organization of cell surface components on a nano-scale. Immuno-gold labelling of isolated plasma membranes generates point patterns that enable mapping of protein and lipid distributions. 2D spatial statistics reveals the extent to which these distributions are clustered or dispersed and allows the extent of co-localization between different cell surface components to be precisely determined. This approach has been successfully applied to the study of signalling network organization and the consequences of physiological changes in modulating cell surface function.

Key words Ras, Electron microscopy, Nanodomains, Signalling, Plasma membrane

1 Introduction

The plasma membrane is organized into subdomains over different length scales. Some such as caveolae, clathrin-coated vesicles, cytoskeletal networks, and junctional complexes between cells or the cell surface and subcellular organelles can be clearly visualized using high-resolution fluorescence imaging or electron microscopy techniques [1–4]. The majority of the cell surface is characterized by morphologically featureless areas where nano-compartmentalization of proteins and lipids is also proposed to occur [5, 6]. This organization is thought to promote protein–protein and protein–lipid interactions that facilitate effective cell signalling, adhesion, and membrane trafficking. Due to their small size and ephemeral nature they preclude direct observation resulting in debate about their existence and characteristics [7].

Conventional resin processing of cells following by ultramicrotomy generates 1-D views of the cell surface that are useful for observing the membrane and the underlying cellular reference space. When cells are processed in situ in the cell culture dishes,

Dylan M. Owen (ed.), *Methods in Membrane Lipids*, Methods in Molecular Biology, vol. 1232,
DOI 10.1007/978-1-4939-1752-5_12, © Springer Science+Business Media New York 2015

this method makes it easy to determine positioning of features of interest on the apical or basolateral membrane. A variant of this method involves sectioning of cells *en face* from the basal layer that was in contact with the cell culture dish. The first section taken from the block gives large 2D views of the plasma membrane together with the underlying 30–70 nm cytosolic space—similar to an area imaged in a TIRF microscope.

2D views are necessary for mapping the spatial arrangements of cell surface components. We have developed a combined EM-statistics approach for 2D spatial mapping of proteins and lipids within the apical cell surface [8]. The method produces areas of pure plasma membrane that are not contaminated by organelles or cytosol. The technique described in detail below and related variants has been successfully used to map signalling pathway and lipid organization within the plasma membranes of a variety of adherent and non-adherent cell types including HeLa, BHK, adipocytes, endothelial cells, and immune cells [9–13].

2   Materials

Prepare all solutions using ultrapure water (dH$_2$O; 18 MΩ cm at 25 °C). Typically we use dedicated glassware or disposable plasticware to prepare and store all solutions. Take care when handling hazardous materials to comply with local safety and waste disposal regulations.

2.1   Gold Preparation Components

1. 1 % gold chloride: add 50 ml of dH$_2$O to 0.5 g of AuCl$_3$, rotate the tube to mix until dissolved and then syringe-filter. Store at 4 °C.

2. 1 % trisodium citrate: weigh 0.5 g of trisodium citrate, add to 50 ml of dH$_2$O, and stir until dissolved. Store at room temperature.

3. 25 mM potassium carbonate: weigh 0.172 g of K$_2$CO$_3$, add to 50 ml of dH$_2$O, and stir until dissolved. Store at room temperature.

4. 1 % tannic acid: weigh 0.5 g of tannic acid, add to 50 ml of dH$_2$O, and stir until dissolved and then syringe-filter. Store at room temperature.

2.2   Plasma Membrane Rip-Offs/ Labelling

1. 10× stock KOAc solution: 0.25 M Hepes pH 7.4 (KOH), 1.15 M potassium acetate, 25 mM magnesium chloride. Weigh 5.958 g Hepes, 11.286 g potassium acetate, 0.238 g MgCl$_2$. Add to 80 ml of dH$_2$O, stir until dissolved, adjust

volume to 100 ml with dH_2O and pH 7.4 using potassium hydroxide solution. Prepare 20 ml aliquots and store working stock at 4 °C and spare aliquots at -20 °C. When making 1× stock solution for use in experiments add 1 ml of 10× stock to 9 ml of dH_2O.

2. 10× stock phosphate buffered saline (PBS): 1.37 M NaCl, 27 mM KCl, 100 mM Na_2HPO_4, 18 mM KH_2PO_4, pH 7.4. Weigh 8.006 g NaCl, 0.201 g KCl, 1.420 g Na_2HPO_4, 0.245 g KH_2PO_4. Add to 80 ml of dH_2O, stir until dissolved, adjust volume to 100 ml with dH_2O and pH 7.4 using HCl solution. Filter-sterilize and store at 4 °C (*see* **Note 1**). When making 1× stock solution for use in experiments add 1 ml of 10× stock to 9 ml of dH_2O.

3. 10 % Bovine serum albumin (BSA) solution: Add 1 g of BSA to 9 ml of 1× PBS, rotate tube to mix until dissolved, and adjust volume to 10 ml with 1× PBS. Store at 4 °C.

4. 10 % Fish skin gelatin (FSG) solution: Add 1 g of FSG to 9 ml of 1× PBS, rotate tube to mix until dissolved, and adjust volume to 10 ml with 1× PBS. Store at 4 °C.

5. 0.25 M glycine stock solution: Weigh 0.938 g of glycine, add to 50 ml of 1× PBS, and stir until dissolved. Store at room temperature.

6. Fixatives: Electron microscopy grade, 10×10 ml aliquots of 25 % glutaraldehyde and 16 % paraformaldehyde stored in glass ampoules are purchased from EM consumables suppliers. Store at room temperature (*see* **Note 2**).

7. 2 % methyl cellulose: Dissolve 2 g of 25 cP methyl cellulose in 98 ml of dH_2O heated to 90 °C. Cool on ice whilst continuing stirring. Store in a sealed container for 72 h at 4 °C. Centrifuge at $100,000 \times g_{av}$ for 90 min. Decant supernatant and store at 4 °C.

8. Uranyl acetate: Weigh 1.5 g of uranyl acetate and add to 50 ml of dH_2O. Store at 4 °C (*see* **Note 3**).

9. Poly-L-lysine: 0.1 % (w/v) poly-L-lysine in dH_2O is purchased from Sigma.

10. Coated EM grids. We prepare Pioloform film coated 200 mesh hexagonal copper grids. These are coated with carbon prior to use. Alternatively formvar film coated grids can be purchased from EM consumable suppliers.

11. Antibodies. We have successfully used primary antibodies that have been affinity purified or polyclonal antisera. All antibodies must be characterized prior to use to ensure specificity and sensitivity.

3   Methods

3.1 Preparing Colloidal Gold

This method is taken from Slot and Geuze [14].

3.1.1 Size-Dependent Colloid Formation

50 ml of colloidal gold is typically prepared for conjugation to a single antibody. For univariate labelling prepare 5.5 nm; for bivariate labelling prepare 3.5 and 5.5 nm.

1. Mix 0.5 ml 1 % gold chloride solution with 39.5 ml dH_2O in a 250 ml conical flask.

2. In a separate 250 ml conical flask, mix together 2 ml of 1 % trisodium citrate plus variable amounts of 1 % tannic acid and 25 mM K_2CO_3 dependent on the gold size required (*see* Table 1). Make up to 10 ml final volume using dH_2O.

3. Heat both solutions to 60 °C using a stirring hotplate. Add the reducing solution to the gold solution and stir vigorously. Boil for 5 min. The solution should turn deep red once colloids have formed. Cool on ice and pH to 8.5 (NaOH) (*see* **Notes 4–7**).

3.1.2 Titrating Antibody-Gold Binding

The gold colloid will aggregate in the presence of salt unless stabilized by protein on the surface. This can be used as an assay for determining the optimal antibody concentration for coating the gold.

4. Prepare eleven 1.5 ml microfuge tubes containing 0–5 µg of antibody (0.5 µg increments between tubes) in total volumes of 20 µl (in dH_2O) (*see* **Note 8**). Add 250 µl of gold colloid. Incubate for 5 min at room temperature.

Table 1
Gold sizing components

Gold diameter (nm)	1 % tannic acid (ml)
3.5	2.50[a]
4.0	1.25[a]
4.5	0.75[a]
5.5	0.50[a]
6.0	0.25
7.5	0.13
10.0	0.04
11.5	0.025
14.0	0.013

[a]Add an equal volume of 25 mM K_2CO_3

5. Add 100 µl of 10 % NaCl. Look for lowest amount of antibody that stabilizes gold, i.e., where no color change to blue occurs (typically 6–10 µg of antibody per ml of gold, i.e. 1.5–2.5 µg per 250 µl used in the titration) (*see* **Note 9**).

3.1.3 Preparing the Final Antibody-Gold Solution

6. Incubate gold with the optimal antibody concentration for 10 min at room temperature.

7. Whilst incubating the antibody-gold solution, microfuge 0.6 ml of 10 % BSA for 5 min at maximum speed. Add the pre-spun BSA to the antibody-gold solution to a final concentration of 0.1 % (*see* **Note 10**).

8. Centrifuge to concentrate the gold and remove unbound antibody. $120,000 \times g_{av}$ for 3.5 nm or $100,000 \times g_{av}$ for 5.5 nm for 1 h at 4 °C.

9. Immediately at the end of the spin collect the tubes. A hard pellet will be visible with a loose streak of gold sliding down into the bottom of the centrifuge tube. Collect the loose part of pellet with a 200 µl pipette. Approximately 300–500 µl will be collected in total from a 50 ml gold preparation. Store at 4 °C (*see* **Note 11**).

3.1.4 Gradient Purification to Make Gold for Double Labelling Experiments

10. For further size resolution, load the gold onto a 10–30 % glycerol gradient prepared with PBS, 0.1 % BSA. Gradient fractions are prepared by combining glycerol, 10 % BSA, 10× PBS and water to the required final concentrations (*see* **Note 12**). For a 14 ml tube for SW40 Ti rotor we use a step gradient comprising 2.65 ml fractions of 30, 25, 20, 15, and 10 % glycerol in PBS, 0.1 % BSA. All of antibody-gold solution from a 50 ml preparation (approx 300–500 µl) is loaded onto the top and then spun at $200,000 \times g_{av}$ for 30 min at 4 °C. Collect twenty 200 µl fractions from the top half of the gradient. When preparing 2–3 and 5 nm gold for double labelling we run both samples at the same time. The two sizes of gold will be differentially distributed within their respective gradients (*see* **Note 13**).

11. Gold near the top of the gradient will be dispersed. For double labelling choose 2–3 nm fractions nearest the top that are darkest red. For 5.5 nm gold, choose the nearest fractions to the top that did not show significant overlap with the 2–3 nm gold gradient run in parallel (Fig. 1).

12. Pipette 5 µl of each fraction to be tested onto a clean piece of Parafilm (*see* **Note 14**). Place a pioloform-coated grid face down onto the gold droplet and incubate for 1 min at room temperature. Transfer the grid to a 50 µl drop of 2 % methyl cellulose. Pick up the grids in 5 mm copper wire loops that are commonly used for immuno-EM labelling (*see* **Note 15**).

1. Prepare 2 nm and 5 nm colloidal gold
2. Conjugate to antibody (7-10μg antibody per ml gold)
3. Wash by ultracentrifugation for single labelling studies
4. Gradient purify for double labelling studies

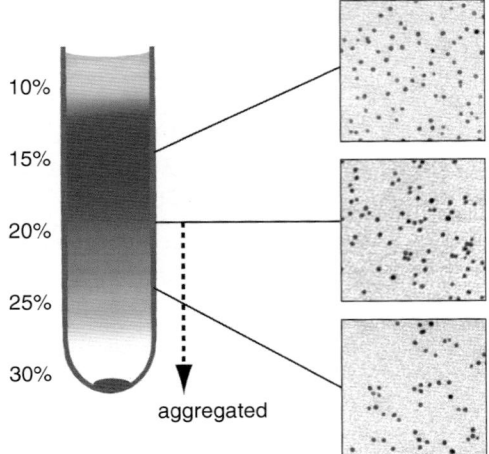

5. Calibrate size distribution of the gold particles

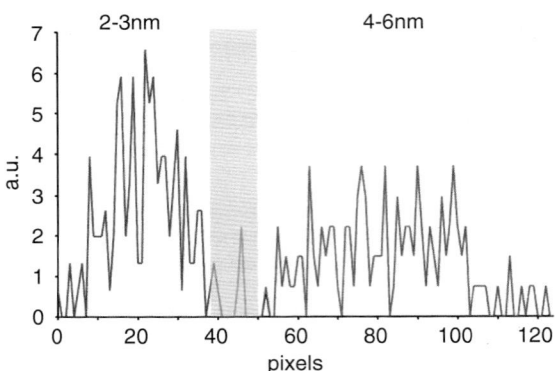

Fig. 1 Preparation and sizing of gold fractions. Glycerol gradients allow enrichment for non-overlapping size variants from 3.5 and 5.5 nm gold preparations. Aggregated gold found in the denser fractions should not be used. Calibration of gold sizes allows a sizing window to be determined to exclude sizes from the overlapping region

Drain excess mixture on filter paper and leave to dry for at least 10 min. Carefully remove the grids from the dried loop and store in a grid box.

13. Take two images of each fraction in a transmission electron microscope (TEM) to allow determination of gold size distributions. Use the same magnification for all images; typically we use 87,000× on a 120 kV FEI Tecnai G2.

*3.1.5 Gold Size
Calibration*

3.5 and 5.5 nm gold preparations contain a significant population of particles of intermediate size that cannot be used in any co-localization analysis because of the ambiguity of assignment. To identify particle size ranges where there is little cross contamination of sizes the fractions need to be calibrated. The following procedure involves manually highlighting individual gold particles by drawing around them and then removing the background and then determining the gold size distributions using Image J. Each person is unique in how closely they crop the particles therefore to ensure consistency, it is important that the gold fractions are individually calibrated by each user. These parameters are then applied to the relevant datasets obtained by the user with the fractions.

1. Using Adobe Photoshop or an equivalent program, select and crop a square area comprising 200–400 particles.

2. Using the Lasso tool, draw around each nanoparticle. Depress the Alt key whilst drawing around each gold particle to allow multiple gold particles to be added. Avoid those that are cut off at the edge of the area. Select the inverse of the area so that now the background is selected and Fill with white. You will now see only the nanoparticles on a white background. Save as a .tif file.

3. Open the file using Image J. Threshold the image using the range 0–170.

4. Modify Set Measurements to record center of mass to eight decimal places.

5. Modify Analyze Particles to count the number of gold particles at different sizes from a minimum of 2 pixels to a maximum of 150 pixels.

6. The curves generated from the size distributions allow a counting window of 5–6 pixels to be determined where most overlap occurs between the different sizes of gold (Fig. 1). This counting window will be used in co-localization studies. Typically for 2–3 and 5 nm gold double labelling, approximately 80–90 % of the total gold will be in the non-overlapping size ranges, with ≤ 5 % of the counts representing contamination of the other gold.

**3.2 Preparing
and Processing
Plasma Membrane
Rip-Offs**

1. Incubate pioloform/carbon-coated grids with 1 mg/ml poly-L-lysine for 10 min. Rinse briefly in dH_2O, 2×1 min, then air dry and store in a dust-free environment.

2. Culture cells on 10 mm glass coverslips; transfect if necessary. Try to ensure that on the day of the experiment the cells are 70–80 % confluent for optimal rip-offs (*see* **Notes 16** and **17**).

3. Place grids onto a clean piece of filter paper in a group (typically two per coverslip) with the polylysine-coated pioloform/carbon side facing up.

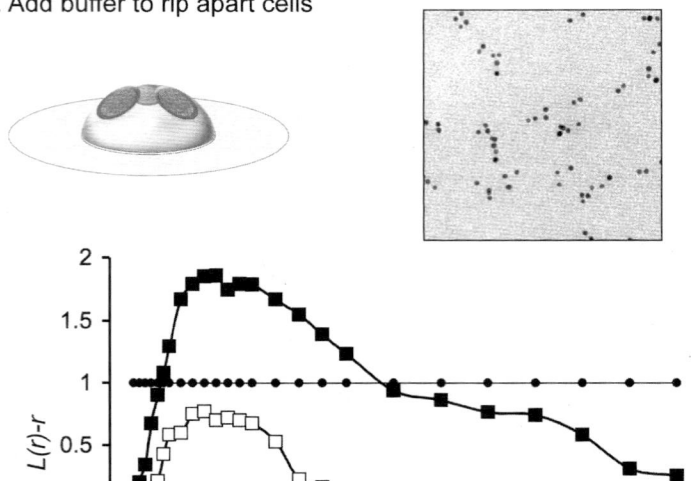

1. Wash coverslip in PBS   4. Fix, wash, label

2. Press onto EM grids   5. Image & determine co-ordinates

3. Add buffer to rip apart cells

Fig. 2 Plasma membrane preparation and analysis. Cells on coverslips are pressed onto EM grids; addition of buffer rips the cells apart to produce plasma membrane sheets. After fixation and labelling, areas of plasma membranes are visualized in a TEM. Images are processed to enable gold co-ordinates to be identified using ImageJ. Ripley's k-function analysis determines the extent to which patterns deviate from a random distribution. Values above $L(r) - r$ of 1 are significantly clustered over the indicated distance (nm)

4. Immediately before use wash cells by dipping the coverslip into 1× PBS to remove any debris. Touch the edge of the coverslip with filter paper to remove excess solution. Place the coverslip cell-side down onto the grids. Place filter paper on top of coverslip and exert pressure using a silicon bung for 1–2 s (*see* **Note 18**).

5. Remove the filter paper and turn the coverslip over so that grids are uppermost. Carefully pipette 100 μl of 1× KOAc buffer in between and around the grids so the surface tension causes them to float up onto the top of the drop. A slight touch on the edge with fine forceps may be needed to encourage them to lift up. Avoid wetting the backs of the grids (Fig. 2).

6. Wash grids by transferring them cell side down on to a 100 μl drop of 1× KOAc buffer on Parafilm for 10 s. All subsequent

steps are carried out at room temperature on 100 µl drops arrayed in lines on Parafilm unless otherwise stated. Grids are transferred from drop to drop with fine forceps.

7. Fix for 10 min with 4 % paraformaldehyde, 0.1 % glutaraldehyde in 1× KOAc buffer. Fixative is made up immediately prior to starting the processing.

8. Quench free aldehyde groups with 25 mM glycine in KOAc for 3 × 3 min.

9. Block non-specific antibody binding sites with blocking solution (0.2 % FSG, 0.2 % BSA in KOAc) for 10 min.

10. Dilute the antibody gold probes into blocking solution and microfuge for 1 min at full speed. Incubate each grid with 5 µl of the antibody-gold solution for 30–60 min.

11. Wash the labelled grids with blocking solution (3 × 1 min) and then de-ionized water (3 × 1 min) (*see* **Note 19**).

12. Whilst the grids are washing, prepare an ice-bucket with a piece of Perspex on top covered by a piece of Parafilm. Prepare 1 ml of final concentration 1.8 % methyl cellulose, 0.3 % uranyl acetate (MC/UA), avoiding making bubbles in the solution when mixing together by pipetting. Pipette 2 × 200 µl drops of MC/UA mix and then 50 µl drops for each pair of grids from each condition.

13. Transfer each of the grids via the two wash MC/UA drops before incubating on the final drop for 10 min.

14. Pick up the grids in 5 mm copper wire loops, drain excess mixture on filter paper and leave to dry for at least 10 min in a dust free environment. Remove the grids taking care not to break the MC/UA film; the grids are now ready for viewing (*see* **Note 20**).

3.3 Imaging Rip-Offs and Processing Negatives for Image Analysis

1. Using a transmission electron microscope, scan for rip-offs at low magnification (4,000–6,000×); ensure that the field is bright since rip-offs are not well contrasted against the background (*see* **Note 21**). Take photographs at higher magnification (typically 80,000–100,000×). Ensure that the areas in the photos are completely filled with a membrane sheet and do not contain 3D features such as large vesicular structures or cellular debris. Use the same magnification for all images to make later computer analysis easier. Typically 15–20 negatives are taken of each condition. For solutions to difficulties in finding rip-offs *see* **Note 22**. For solutions to observation of low labelling *see* **Note 23**. *See* **Note 24** for information on how to identify nonspecific labelling and how to reduce it.

2. Using Adobe Photoshop select and crop a square area (we use an area of 725 × 725 nm) and save as a new .tif file.

3. Remove the background from the image using the same method described for the gold sizing calibration above. For single labelled samples you can manually spot the gold in a new layer to produce a map and then delete the image to just leave behind the spotted points against a white background. Save as a .tif file.

3.4 Ripley's k-Function Analysis

The 2D spatial statistics underpinning this analysis are extensively discussed elsewhere [8, 9, 15–18].

1. Using predetermined values to identify gold of the correct size, locate the center of mass of every gold particle to 8-decimal places using Image J as described in the gold calibration section. Save as a .txt file. For bivariate analysis of co-localization, two data sets will be obtained corresponding to small gold and large gold.

2. Run the Ripley's univariate (single labelling) or bivariate (double labelling) *k*-function analysis program. The Excel VisualBASIC macro is available from the authors upon request and has been described elsewhere. Care has to be taken to ensure that that pixel to nanometer conversion factor is correct before running the macro. Instructions for checking this are provided with the macro.

3. Normalize each dataset by dividing by the average confidence interval (99 % C.I.) determined by the program (univariate); or 95 % C.I. from Monte Carlo simulations (bivariate). This allows analyses from different images to be combined into a single scale allowing cross-comparison.

4. Calculate a mean of the normalized values. Values above 1.0 indicate significant clustering at that value, values of 0 indicate a random distribution and values below −1.0 indicate dispersal (Fig. 2). Confounding non-specific clustering of gold probes can be caused by ultracentrifugation, *see* **Note 25** for information on how to characterize it and how to remove the influence of this from the data analysis.

4 Notes

1. We typically buy pre-made 10× stock solutions of PBS.

2. Fixative in ampoules are opened and added to buffers in a fume cupboard. Only freshly opened paraformaldehyde is used to fix EM samples. Spare paraformaldehyde is frozen in aliquots for use in immuno-fluorescence experiments. 25 % glutaraldehyde can be aliquoted into microfuge tubes and stored at −20 °C indefinitely before being used to fix samples for EM. We have a dedicated fridge and freezer to store fixatives,

fixed samples and other EM stains to avoid the potential of fixing biologically sensitive samples such as antibodies and other proteins that we store in the fridge-freezer.

3. Uranyl acetate is a mildly radioactive and very toxic if ingested, inhaled or in contact with the skin; it has the potential to cause cumulative damage from long-term exposure. Frequently, dried power precipitate is present around the neck of the storage container. Extreme care should be taken when handling and to clean the area afterwards of microscopic particles or spills.

4. We make 3.5 and 5.5 nm gold since larger sizes are too big to efficiently label multiple proteins within a single nanocluster; this is possibly a consequence of steric hindrance.

5. Dedicate glassware, centrifuge tubes, thermometers and other equipment for the use of gold colloid production to prevent contamination.

6. Tannic acid is polyphenol with an indeterminate molecular weight. Different sources will vary in their composition. Historically, gallnut Alleppo tannin from Mallinckrodt was used to make gold for EM. This source is no longer available; therefore, it may be necessary to calibrate the amount of tannic acid needed to make the required gold size.

7. When adding the solutions together, an immediate color change will be observed. For larger gold sizes, ie. ≥ 5 nm, the boiling is needed to reach the required size and the color will develop more slowly with larger particles. For 3.5 nm particles we only briefly boil (≤ 1 min) to keep all of the particles as small as possible.

8. Commercial antibodies usually contain stabilizing BSA that will compete with the antibody for gold binding. Sodium azide may also be present; this is incompatible with gold-antibody conjugation. In either case, purify the antibody before conjugation. For polyclonal antisera either determine the protein concentration or dilute 1:10 in dH_2O and titrate with 0–5 µl in the binding assay.

9. For very small 2–3 nm gold this color change will be very subtle. If in doubt use 10 µg of antibody per ml of gold.

10. BSA is present in all of the antibody-gold solutions and washes after the initial conjugation to ensure that the gold remains stable. Over time, the antibody will desorb from the gold and be replaced by BSA. Effective shelf-life depends upon the specific protein that is bound. We have had anti-GFP-gold conjugates remain effective for >3 years after synthesis however routinely we make fresh gold when starting a new set of nanoclustering experiments and especially when performing double labelling co-localization analysis.

11. Harvest the pellet as soon as the spin stops to reduce re-suspension of the more compact pellet that contains gold aggregates.

12. Glycerol is measured by weighing into a tube taking into account the density of 1.26 g/ml. i.e. to make 10 ml of a 30 % glycerol solution we weigh 3.78 g of glycerol. 1 ml of 10× PBS, 0.1 ml of 10 % BSA, and 5.9 ml of dH_2O are then added.

13. Avoid fractions from the bottom half of the gradient; these contain significant proportions of aggregated gold (Fig. 1). Typically, useable 2–3 nm gold will be in the first 2–5 fractions where gold is observed. Useable 5 nm gold will be a further 5–10 fractions from the top of the gradient than the 2–3 nm gold.

14. Parafilm provides a flat, clean hydrophobic surface for incubating EM grids with solutions. Spray the bench with some water, add the Parafilm on top and gently flatten to remove air bubbles. Remove the protective paper covering. Do not touch the Parafilm directly with fingers otherwise significant contamination on the grids will be seen in the TEM. A transparent plastic cover is typically used to prevent accidental disruption of incubations; this cover should not touch any part of the Parafilm.

15. Loops are made by winding 8–10 cm of 0.1 mm copper wire around 5 mm drill bits and twisting the free ends to make a single stem. This is then glued onto wooden sticks or preferably plastic sticks that can fit into the holes of a 200 µl pipette tip box or other suitable stand.

16. Higher confluency increases the chances of observing a ripped off plasma membrane. However, if the cells are aggregated in clumps (sometimes caused by over-diluting cells during splitting), or fully confluent they will detach en masse and produce few plasma membrane rip-offs. Similarly, if confluency is low there will be fewer opportunities for a membrane sheet to be produced. Typically we find that 50–80 % confluency with evenly dispersed cells works well.

17. Typically we seed cells at two densities into dishes containing coverslips. The next day a coverslip from each dish is then taken and put into a single 3 cm dish or well of a 6-well plate prior to transfection. On the day of the EM processing, the coverslips are inspected and the one with 70–80 % confluency is selected for rip-off processing, the other is used for fluorescent imaging to check transfection efficiency.

18. Corks, rubber bungs or any similar device can be used. The pressure exerted needs to be calibrated for each cell type and the confluence of the cells. Typically we stand up and use our body weight to push down with both arms straight for 1 s.

19. Uranyl acetate precipitates in the presence of phosphate ions, ensure that the tweezers are also washed in de-ionized water in between each water wash.

20. Uranyl acetate precipitates in the presence of sunlight, ensure that the incubations and subsequent drying of the looped grids are not performed in bright light near a window. MC/UA dries to produce a supporting film that prevents air-drying artifacts and produces a homogenous negative stain across the sample.

21. Sheets can often be several microns in diameter with a well-defined edge contrasted against the grid background (Fig. 3). Where lots of cell debris is present on the grid the membrane sheets often stand out as looking 'clean' against this cluttered background. A common place to find these areas is adjacent to semi-intact cells where the procedure has successfully disrupted a small section of the cell margin.

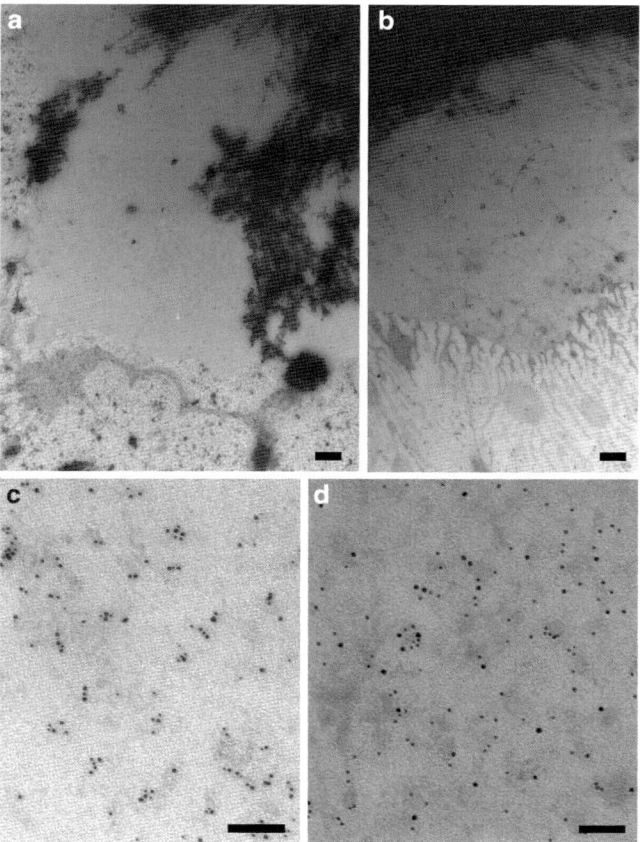

Fig. 3 Identifying plasma membrane sheets. Plasma membrane sheets often present a homogenous appearance against the grainy grid background (**a**). In many cases they are found at the margins of cells (**b**). Representative examples of singly (**c**) and double (**d**) labelled plasma membrane sheets. Bars = 500 nm (**a**, **b**) and 50 nm (**c**, **d**)

22. Under optimal conditions, a typical grid will contain only 10-20 areas and 20 % of grids will contain no obvious rip-offs. For most cell types, cell confluence is an important determinant of the number of rip-offs produced. It is important to be systematic in scanning the whole grid and not missing any areas. Typically a grid will take 1–2 h to examine properly. Rip-offs have very little contrast against the background, try searching at magnifications of 6,000–8,000× and keep the brightness high to see them more easily. At least 50 % of rip-offs can be found still attached to whole cells, look carefully around the periphery of each cell (Fig. 3).

23. Before performing any immuno-EM, use immuno-fluorescence to check that the antibody works with glutaraldehyde fixed cells. Most antibodies need to be used at 10× higher concentration for immuno-EM compared to immuno-fluorescence. If the labelling is for a transfected protein ensure that the transfection efficiency is ≥30 %. Increase the concentration of the gold-antibody conjugate by diluting less in blocking solution. Gold harvested from a gradient is more dilute and will give lower labelling than non-gradient purified gold so may not need dilution. Touch the side of the grid on a piece of filter paper directly before incubating on the gold to remove excess block which will dilute the gold further. The antibody-gold conjugate is relatively unstable; the desorbed free antibody competes for antigen binding. Either make fresh antibody-gold conjugate or wash away free antibody by resuspending the gold in a 10× volume of PBS, 0.1 % BSA and re-centrifuging. It is normal if the majority of rip-offs that you find do not have label on them. Typically, ≤5 separate plasma membrane sheets are found with clear labelling on a grid even when transfection efficiency is good.

24. Non-specific labelling of the plasma membrane, and gold aggregates producing non-specific clustering both affect the analysis to varying degrees. For transfected cells, the extent of non-specific labelling can be determined by labelling non-transfected rip-offs. Changing the blocking conditions and length of antibody labelling time can reduce this background. For antibodies that we routinely use, there is negligible background equivalent to ≥1 % of specific labelling.

25. Centrifugation of the gold-antibody conjugate induces aggregation such that even the loose part of the pellet contains aggregates. Gradient purification reduces the amount present but also reduces the labelling efficiency. The contribution of non-specific clustering to an analysis can be removed by performing Ripley's k-function analysis on a grid containing only the gold-antibody conjugate. The resultant curve normalized to the 99 % C.I. can be subtracted from any data curves.

References

1. Kusumi A, Fujiwara TK, Chadda R, Xie M, Tsunoyama TA et al (2012) Dynamic organizing principles of the plasma membrane that regulate signal transduction: commemorating the fortieth anniversary of Singer and Nicolson's fluid-mosaic model. Annu Rev Cell Dev Biol 28:215–250

2. Parton RG, del Pozo MA (2013) Caveolae as plasma membrane sensors, protectors and organizers. Nat Rev Mol Cell Biol 14:98–112

3. Soboloff J, Rothberg BS, Madesh M, Gill DL (2012) STIM proteins: dynamic calcium signal transducers. Nat Rev Mol Cell Biol 13:549–565

4. Insall RH, Machesky LM (2009) Actin dynamics at the leading edge: from simple machinery to complex networks. Dev Cell 17:310–322

5. Laude AJ, Prior IA (2004) Plasma membrane microdomains: organization, function and trafficking. Mol Membr Biol 21:193–205

6. Simons K, Sampaio JL (2011) Membrane organization and lipid rafts. Cold Spring Harb Perspect Biol 3:a004697

7. Munro S (2003) Lipid rafts: elusive or illusive? Cell 115:377–388

8. Prior IA, Muncke C, Parton RG, Hancock JF (2003) Direct visualization of Ras proteins in spatially distinct cell surface microdomains. J Cell Biol 160:165–170

9. Prior IA, Parton RG, Hancock JF (2003) Observing cell surface signaling domains using electron microscopy. Sci STKE 2003:PL9

10. Zhang J, Leiderman K, Pfeiffer JR, Wilson BS, Oliver JM et al (2006) Characterizing the topography of membrane receptors and signaling molecules from spatial patterns obtained using nanometer-scale electron-dense probes and electron microscopy. Micron 37:14–34

11. Wilson BS, Steinberg SL, Liederman K, Pfeiffer JR, Surviladze Z et al (2004) Markers for detergent-resistant lipid rafts occupy distinct and dynamic domains in native membranes. Mol Biol Cell 15:2580–2592

12. Cho KJ, Kasai RS, Park JH, Chigurupati S, Heidorn SJ et al (2012) Raf inhibitors target ras spatiotemporal dynamics. Curr Biol 22:945–955

13. Chapkin RS, Wang N, Fan YY, Lupton JR, Prior IA (2008) Docosahexaenoic acid alters the size and distribution of cell surface microdomains. Biochim Biophys Acta 1778:466–471

14. Slot JW, Geuze HJ (1985) A new method of preparing gold probes for multiple-labeling cytochemistry. Eur J Cell Biol 38:87–93

15. Besag JE (1977) Contribution to the discussion of Dr Ripley's paper. J R Stat Soc B39:193–195

16. Diggle PJ (1986) Displaced amacrine cells in the retina of a rabbit: analysis of a bivariate spatial point pattern. J Neurosci Methods 18:115–125

17. Ripley BD (1979) Tests of randomness for spatial point patterns J. J R Stat Soc B41:368–374

18. Ripley RD (1977) Modelling spatial patterns. J R Stat Soc B39:172–192

Chapter 13

Measuring Cytoskeleton and Cellular Membrane Mechanical Properties by Atomic Force Microscopy

Charles Roduit, Giovanni Longo, Giovanni Dietler, and Sandor Kasas

Abstract

Atomic force microscope is an invaluable device to explore living specimens at a nanometric scale. It permits to image the topography of the sample in 3D, to measure its mechanical properties and to detect the presence of specific molecules bound on its surface. Here we describe the procedure to gather such a data set on living macrophages.

   Key words AFM, Macrophages, CD40 ligand, Stiffness tomography, Chemical mapping

1 Introduction

Almost all cellular physiological functions depend on the mechanical properties of diverse cellular components [1, 2]. The filamentous 3D networks of the cytoskeleton play an important role in the mechanical properties of cells. It permits cellular motion, intracellular displacement of organelles and plays a cornerstone role in cell division. The function of this highly organized network mostly depends on its mechanical properties that are unfortunately still very poorly known today. Similarly, the mechanical properties of lipid membranes that delimit cells and organelles play an important role in numerous cellular processes. Beside physiology, recent studies have pointed out that cellular mechanical properties are strongly modified in numerous pathological conditions such as cancer [3, 4] or muscular dystrophy [5] opening novel avenues to innovative diagnostic tools or complementary investigation techniques. The exploration of the cytoskeleton and lipid membrane structure is possible nowadays by means of various optical or electronic microscopy techniques. While these instruments can reveal very precisely the 3D structure and the organization of the cytoskeletal filaments and cellular membranes, they have no access to their mechanical properties and often are usable only in non-physiological conditions. This type of investigation domain is therefore addressed by

Dylan M. Owen (ed.), *Methods in Membrane Lipids*, Methods in Molecular Biology, vol. 1232,
DOI 10.1007/978-1-4939-1752-5_13, © Springer Science+Business Media New York 2015

sophisticated instrumentation [6] such as optical or magnetic tweezers [7], acoustic microscopy [8], PDMS patterned substrates, or
micropipette punching. Among the instruments that can explore at
a micro or a nano scale the mechanical properties of living samples,
the atomic force microscope (AFM) is probably the most readily
accessible one [9]. The simplicity of its setup and its capacity to
simultaneously record topographic or chemical information in parallel to the mechanical probing of the sample, in almost physiological environment, made this instrument very popular among scientists
involved in cellular biophysics.

In this chapter we will describe the procedure that permits to
simultaneously record chemical and stiffness maps of the cell surface, the 3D distribution of the mechanical properties as well the
3D topography of a single living cell by using the so-called force
volume imaging mode. In this mode, the AFM tip is periodically
indented (i.e., pushed) into the sample. The study of the deformation of the cantilever as a function of the indented depth permits
the 3D characterization of its mechanical properties [10]. An additional data processing step, referred to as stiffness tomography [11,
12], goes further and isolates the mechanical contribution of the
cellular membrane [13, 14]. It also detects the presence of subcellular structures located underneath the surface according to their
mechanical properties. The attachment of a ligand onto the AFM
tip that specifically interacts with molecules present in the cell
membrane reconstructs a biochemical mapping of the sample
whereas the localization of the spot where the AFM tip comes in
contact with the cell's surface allows the determination of the 3D
topography of the cell. All the data processing to extract valuable
information from the force volume files is accomplished by the
open source software called OpenFovea [15].

2 Materials

2.1 Cantilever Preparation

Silicon nitride cantilevers (*see* **Note 1**) are processed according to
the following procedure:

1. Clean the cantilever holding chip in a plasma cleaner (*see* **Note 2**).

2. Activate the tip by immersing the chip into a solution of glutaraldehyde 10 % v/v (*see* **Note 3**).

3. Rinse in bi-distillated water at least 3 times.

4. Incubate the chip with the ligand solution for 1 h (*see* **Note 4**).
   A droplet (2 μl) deposited on the cantilever is usually enough
   for a good coating. The concentration of the ligand should be
   in the order of 1 μm/ml in the case of antibodies.

5. Wash the chip thoroughly with phosphate buffer saline (PBS).

6. Chips can be stored few days at 4 °C for a delayed use.

3 Methods

3.1 Cell Culture Preparation

Since protocols usually depend on the used cell line please refer to the appropriate procedure (*see* **Notes 5** and **6**). For THP-1 macrophages the protocol is as follows:

1. Maintain THP-1 macrophages (National Center for Cell Science, www.nccs.res.in) in minimal essential medium (Invitrogen) supplemented with 10 % fetal bovine serum, and 100 U/ml penicillin/streptomycin (Invitrogen) to inhibit microbial growth. The cell culture must be incubated in a humidified atmosphere at 37 °C containing 5 % CO_2.

2. Treat plastic Petri dishes (35 mm in diameter, Falcon) (*see* **Note 7**) with poly-D-Lysin (Sigma (0.1 mg/ml)) for 5 min and air-dry them for minimally 2 h.

3. Transfer the cell culture in Falcon tubes for centrifugation at $3,000 \times g$ for 10 min.

4. Remove the supernatant and replace it with 2 ml of fresh medium.

5. Deposit 10 µl in the polylysine pretreated Petri dish and add 2 ml of fresh medium supplemented with Phorbol 12-myristat 13-acetate (PMA, Sigma) at a final concentration of 50 nm/ml to promote cell adhesion for 24 h.

6. Wash and incubate the cell culture in standard DMEM 48 h before AFM experimentation.

3.2 AFM Imaging

1. Mount the previously ligand coated chip in the cantilever holder of the AFM (*see* **Note 8**).

2. Deposit the cells containing Petri dish on the inverted microscope of a biologically oriented AFM.

3. Choose a strongly adhering cell (*see* **Note 9**) with the inverted optical microscope.

4. Engage the AFM tip on a cell-free spot close to the target cell.

5. Proceed to the sensitivity calibration of the AFM and record at least one force volume file for further processing with imaging parameters as close as possible (except the scan size) to those used to image the target cell (scan size 0 µm, scanning speed 5 Hz, pulling speed 20 µm/s, trigger threshold set to relative).

6. Withdraw the AFM tip and move it above the target cell.

7. Engage the AFM tip onto the target cell by maintaining a scan size of 0 µm in order to prevent cell damage and tip contamination.

8. Check the retraction curve to asses that the tip detaches off the cell surface during force volume data acquisition, *see* Fig. 1.

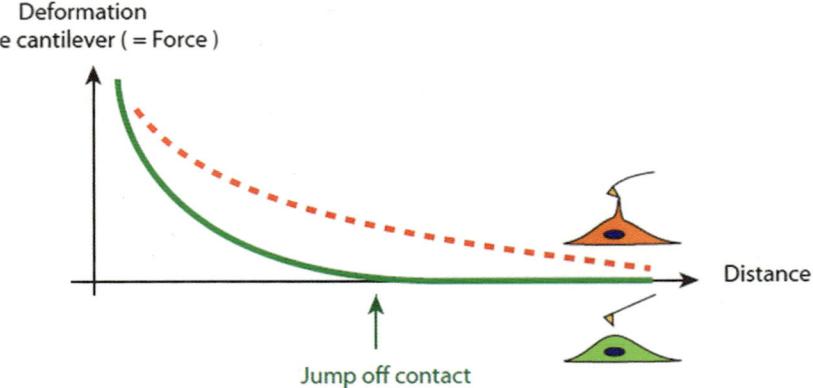

Fig. 1 In case the AFM tip remains attached to the cell membrane at the end of the retraction process, the force distance curve adopts a shape that resembles to the *red dashed line*

9. Fine tune the force distance parameters (z scan range and trigger threshold) in order to obtain force distance curves that have at least 1/3 of their length off contact.

10. Set the scan size to 2 μm or any other value and record your data.

3.3 Data Processing

1. Open the data processing software (*see* **Note 10**).

2. Load the force volume data sets in OpenFovea by selecting **New** option in the **File** menu.

3.4 Calculating Stiffness Maps

1. Adjust the point of contact detection sensitivity in the **Parameters** menu.

2. Insert the model (sphere or cone) as well the requested tip and material parameters.

3. In the **Compute** menu select the resolution of stiffness tomography images (number and size of force distance curve segments (*see* **Note 11**)). Press **Compute Stiffness** to start the automatic data processing.

4. In the **Display** menu choose **Topography**, **Stiffness**, or **Stiffness tomography** options. An example of stiffness tomography data is depicted in Fig. 2.

3.5 Detecting Unbinding Events

1. Adjust the sensitivity of the detector by pressing **Adjust sensitivity** in the **Parameters** menu. Choose the **Event fit model**, set the **Event distance threshold** and **Event length threshold** parameters.

2. Press the **Detect Events** button. To display the events position map press **Events Array** in the **Display menu**. A typical event position array is depicted in Fig. 3.

3. To display the properties of unbinding events choose **Force**, **Loading rate**, or **Distance** histograms.

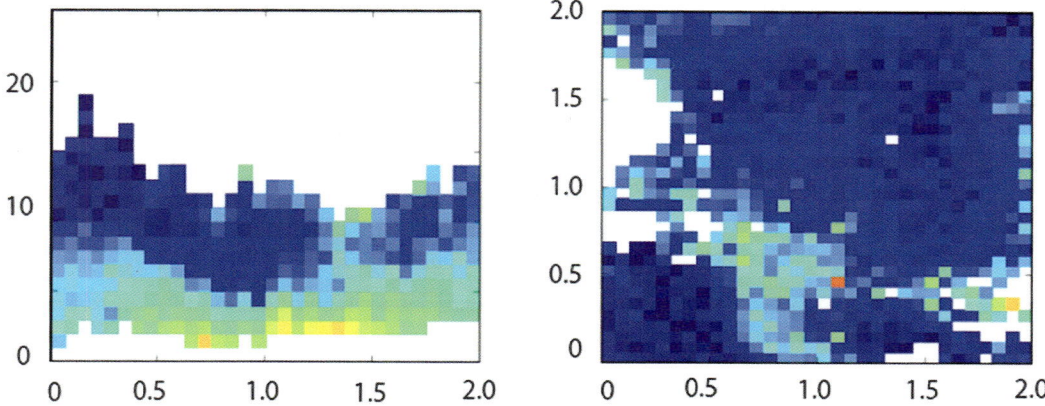

Fig. 2 Stiffness tomography slice (*left*) and stiffness map (*right*) of a living macrophage. False colors code for the stiffness (*blue*=soft, *red*=hard)

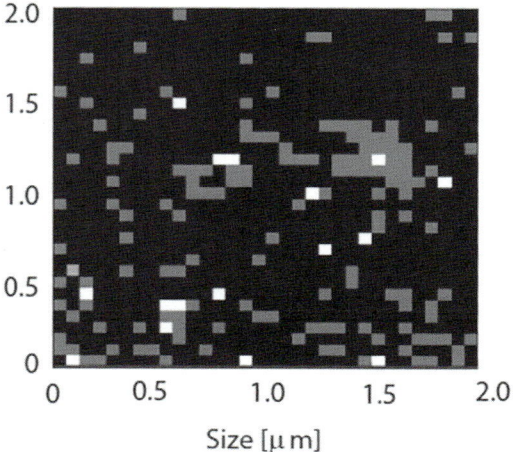

Size [μm]

Fig. 3 Same area as in Fig. 2. *Gray pixels* indicate the presence of one unbinding event whereas the *white pixels* indicate the presence of two unbinding events on the same location

3.6 Correlating Cellular Membrane Stiffness and Unbinding Events

In the **Ev. Histo**. menu choose the parameters to correlate (for example, to correlate the unbinding force between the molecule present on the tip and the receptor in the cell membrane vs. the stiffness of the cell membrane where the unbinding event occurred choose **Force vs Stiffness** option).

4   Notes

1. Fresh silicon nitride cantilever should be used. In order to prevent cell damage, choose soft, rectangular or triangular cantilevers ($k \leq 0.06$ N/m), avoid sharpened tips, and prefer those with a large radius of curvature (above 20 nm).

2. Typical parameters should be 1,011 electrons/cm^3 and a temperature of 104 K during 5 min.

3. Other, more sophisticated chemical linkers can be used such as polyethylene glycol (PEG); however, in our experience glutaraldehyde is sufficient in most cases.

4. Efficient binding to the AFM tip can already be obtained for a vast majority of proteins after 10 min of incubation.

5. Cells should be non-confluent to prevent overlapping and permit AFM sensitivity calibration on a rigid substrate.

6. Adherent cells such as epithelial cells do not require polylysine pre-coating. In the case of non-adherent cells use polylysine-coated surfaces.

7. The cell culture should be completed in a recipient compatible with the microscope used.

8. Avoid drying by depositing few microliters of imaging buffer on its surface.

9. The adherence of the cell can be monitored by gently hitting by hand the body of the inverted microscope.

10. OpenFovea can be downloaded for free at http://lpmv.epfl.ch/openfovea, and is compatible with Mac, Windows and Linux operating systems. The user manual of the software can be downloaded from the http://www.freesbi.ch/fr/openfovea.

11. In order to distinguish the contribution of the cell membrane using the stiffness tomography mode, increase the number of segments (>10) and decrease their size (5–10 nm).

Acknowledgments

This work was supported by Swiss National Science Foundation grants no 200021-144321 and no 3213-130676.

References

1. Zemel A, De R, Safran SA (2011) Mechanical consequences of cellular force generation. Curr Opin Solid State Mater Sci 15:169–176

2. Jaqaman K, Grinstein S (2012) Regulation from within: the cytoskeleton in transmembrane signaling. Trends Cell Biol 22: 515–526

3. Plodinec M, Loparic M, Monnier CA, Obermann EC, Zanetti-Dallenbach R, Oertle P, Hyotyla JT, Aebi U, Bentires-Alj M, Lim RYH, Schoenenberger C-A (2012) The nanomechanical signature of breast cancer. Nat Nanotechnol 7:757–765

4. Lekka M, Pogoda K, Gostek J, Klymenko O, Prauzner-Bechcicki S, Wiltowska-Zuber J, Jaczewska J, Lekki J, Stachura Z (2012) Cancer cell recognition—mechanical phenotype. Micron 43:1259–1266

5. Puttini S, Lekka M, Dorchies OM, Saugy D, Incitti T, Ruegg UT, Bozzoni I, Kulik AJ, Mermod N (2009) Gene-mediated restoration of normal myofiber elasticity in dystrophic muscles. Mol Ther 17:19–25

6. Oddershede LB (2012) Force probing of individual molecules inside the living cell is now a reality. Nat Chem Biol 8:879–886

7. Ou-Yang HD, Wei M-T (2010) Complex fluids: probing mechanical properties of biological systems with optical tweezers. Annu Rev Phys Chem 61:421–440

8. Hagiwara Y, Saijo Y, Ando A, Chimoto E, Suda H, Onoda Y, Itoi E (2009) Ultrasonic intensity microscopy for imaging of living cells. Ultrasonics 49:386–388

9. Webb HK, Vi Khanh T, Hasan J, Crawford RJ, Ivanoya EP (2011) Physico-mechanical characterisation of cells using atomic force microscopy—current research and methodologies. J Microbiol Methods 86:131–139

10. Weisenhorn AL, Kasas S, Solletti JM, Khorsandi M, Gotzos V, Romer DU, Lorenzi GP (1993) Deformation observed on soft surfaces with an AFM. Proc SPIE 1855:26–34

11. Kasas S, Wang X, Hirling H, Marsault R, Huni B, Yersin A, Regazzi R, Grenningloh G, Riederer B, Forro L, Dietler G, Catsicas S (2005) Superficial and deep changes of cellular mechanical properties following cytoskeleton disassembly. Cell Motil Cytoskeleton 62:124–132

12. Roduit C, Sekatski S, Dietler G, Catsicas S, Lafont F, Kasas S (2009) Stiffness tomography by atomic force microscopy. Biophys J 97:674–677

13. Yersin A, Hirling H, Kasas S, Roduit C, Kulangara K, Dietler G, Lafont F, Catsicas S, Steiner P (2007) Elastic properties of the cell surface and trafficking of single AMPA receptors in living hippocampal neurons. Biophys J 92:4482–4489

14. Roduit C, van der Goot FG, De Los Rios P, Yersin A, Steiner P, Dietler G, Catsicas S, Lafont F, Kasas S (2008) Elastic membrane heterogeneity of living cells revealed by stiff nanoscale membrane domains. Biophys J 94: 1521–1532

15. Roduit C, Saha B, Alonso-Sarduy L, Volterra A, Dietler G, Kasas S (2012) OpenFovea: open-source AFM data processing software. Nat Methods 9:774–775

Chapter 14

Fluorescence Linear Dichroism Imaging for Quantifying Membrane Order

Richard K.P. Benninger

Abstract

The plasma membrane of a cell is an ordered environment, giving rise to anisotropic orientations and restricted motion of constituent lipids and proteins. The membrane environment is also dynamic and heterogeneous, which is important for the regulation of membrane-localized signaling. A number of fluorescent microscopy approaches enable the membrane order to be quantified with high spatial and temporal resolution. A polarization-resolved fluorescence method, termed fluorescent linear dichroism (fLD) imaging, can quantify the orientation of membrane bound fluorophores which allows spatially resolved measurement of membrane order and sub-resolution membrane topology (ruffling). Here we describe the detailed methods for performing fLD imaging in biological membrane environments such as the plasma membrane of living cells. This includes the preparation of the sample with appropriate fluorescent dyes, the requirements of the microscope system, the data collection protocol, and post-acquisition image processing, analysis, and interpretation.

Key words Fluorescence, Microscopy, Polarization anisotropy, Plasma membrane, Orientation, Order, Topology

1 Introduction

The plasma membrane of a cell is an ordered environment of lipid molecules, proteins and other signaling molecules. The membrane environment is also heterogeneous; consisting of different classes of lipid molecules which determine properties of the membrane environment such as order, fluidity, polarity and mobility, and other factors. In addition to providing a barrier between the cytosol and extracellular space of a cell, the membrane can also act as a signaling platform for a diverse range of cellular signaling processes. These include exo/endocytosis, ion channel function and excitability, the function of GPCRs and other receptors or transporters, as well as intercellular communication. The regulation of many signaling molecules underlying these processes is recognized to be highly dependent on the membrane environment, such as

Dylan M. Owen (ed.), *Methods in Membrane Lipids*, Methods in Molecular Biology, vol. 1232,
DOI 10.1007/978-1-4939-1752-5_14, © Springer Science+Business Media New York 2015

the presence of ordered lipid phases or domains. Therefore the membrane environment is an important parameter that affects cellular function.

Several optical measurements have been developed to allow non-invasive measurement and quantification of the membrane environment. These include measurements of membrane polarity with dyes such as Laurdan and di-4-ANEPPDHQ [1] which can reflect changes in the environment of the membrane interior. Alternatively, measuring the fluidity of the membrane can be achieved by measuring the mobility of membrane bound fluorophores such as with FRAP [2] or fluorescence correlation spectroscopy [3].

We have previously developed a complimentary method for measuring membrane order which involves measuring the orientation of lipophilic fluorophores, where the chromophore resides on the fatty acid tail within in the plasma membrane interior. Due to the ordering of the lipids within the membrane, the chromophore is also ordered with anisotropic orientation (Fig. 1), and this orientation can be measured using polarized fluorescence microscopy [4]. A reduction in the order of the membrane interior, such as through reduced cholesterol content, leads to a greater range of orientations (tilt angles) of the fatty acid tails; therefore changing the tilt angle the chromophore makes relative to the membrane surface. Thus by measuring the average orientation of a fluorophore within the membrane, we can measure the membrane order.

The approach to measure the fluorophore orientation requires polarized excitation and polarization resolved fluorescence detection. Fluorophores have an absorption dipole aligned along some axis (Fig. 1a) and the more aligned the absorption dipole is with respect to the excitation polarization, the stronger the absorption and therefore the stronger the fluorescence emission. Furthermore, the more aligned the fluorophore emission dipole is with respect to the excitation polarization, the more polarized the fluorescence emission. A quantity called the fluorescence polarization anisotropy (r) represents the dependence of the detected fluorescence on emission polarization, which reflects both the fluorophore orientation and motion of the fluorophore between the absorption and emission process [5]. However fluorescence polarization anisotropy is generally used in samples with isotropic orientations to examine fluorophore mobility. A quantity called the reduced linear dichroism (LD^r) represents the dependence of fluorophore absorption on excitation polarization, which reflects just the fluorophore orientation [4]. These quantities can be compared to theoretical calculations based on geometrical considerations of fluorophore orientation distributions and dipole positions, which includes the relative orientation of the plasma membrane with respect to the excitation and detection polarization (Fig. 1b). For example, if a fluorophore dipole is perfectly orientated to be collinear with an

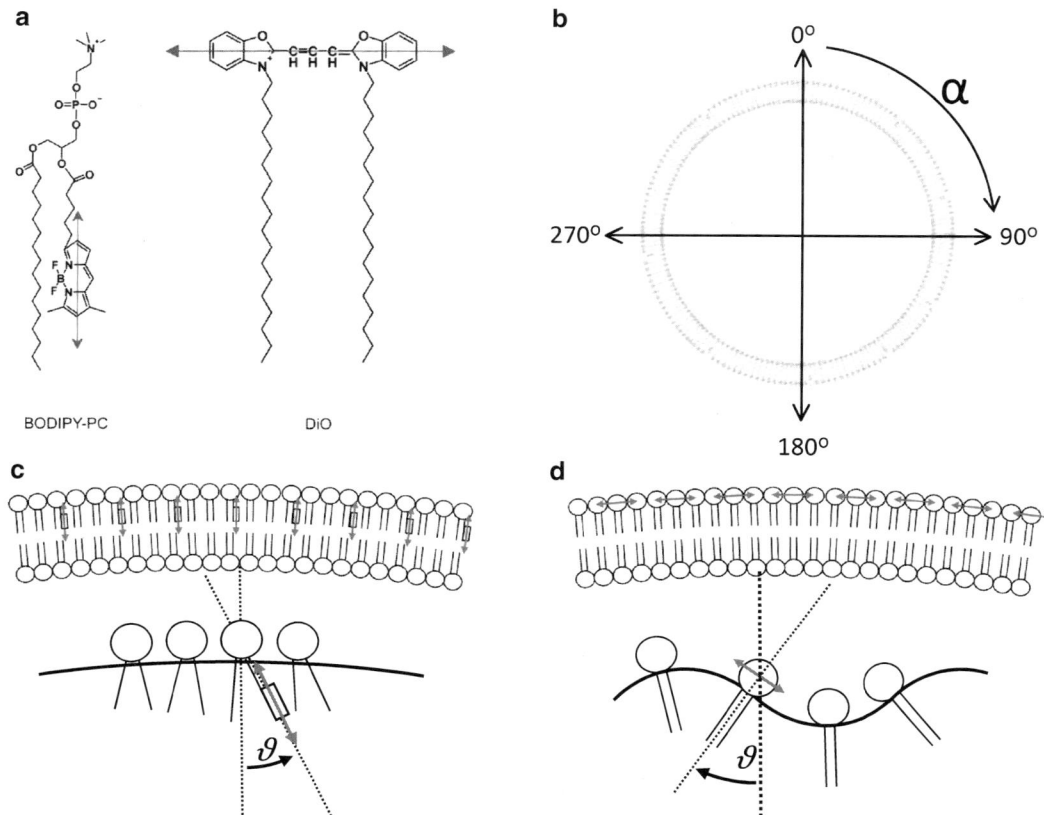

Fig. 1 (**a**) Chemical structures of the two lipophilic fluorophores utilized in fLD imaging. Highlighted in *red* are the absorption dipoles of the chromophores, localized to the fatty acid tail for BODIPY-PC and to the head group for DiO. Note the opposite orientation of the dipole. (**b**) Orientation of the plasma membrane, where α represents the angle of the normal to the plasma membrane surface with respect to the vertical axis. (**c**) Schematic of BODIPY-PC in the plasma membrane of a cell with anisotropic orientation due to the ordered lipid environment. The orientation of the chromophore located in the tail with respect to the membrane normal reflects order of the membrane interior, which can tilt by an average angle $\langle \vartheta \rangle$. (**d**) Schematic of DiO in the plasma membrane of a cell. The orientation of the chromophore located in the head group with respect to the membrane normal reflects local topology of the membrane, where the local normal can tilt by an average angle $\langle \vartheta \rangle$ (Color figure online)

experimental axis, it will have a LD^r of 1, whereas if it shows isotropic orientation it will have an LD^r of 0.

One advantage of this approach is that the position of the chromophore within the fluorescent lipid can be used to measure different orientation properties of the membrane environment. Within the membrane environment, the orientation of the probe will not be solely determined by the position of the plasma membrane (Fig. 1b). As highlighted above, a chromophore positioned on the fatty acid tail will tilt according to the order of the membrane interior (Fig. 1c); which can become disordered upon cholesterol disruption [4], leading to greater title of the fatty acid tail. Changes in head group can localize the fluorophore to different lipid phases in the membrane. However, the chromophore can also be positioned in the head group and the orientation at this position is largely independent of the order in membrane interior. Instead, the chromophore positioned in the head group will tilt according to the local membrane topology (Fig. 1d). For example if the membrane shows a convoluted, "ruffled" topology, there will be a greater range of orientations (tilt angles) that the fluorophore can take, thus leading to a change in the polarization-dependence of absorption and fluorescence emission [6]. A powerful approach therefore is to measure the relative orientation of multiple fluorophores with different chromophore positions, to reflect changes in different properties of the membrane environment.

It should be noted that the flexibility of polarization resolved fluorescence imaging for measuring fluorophore orientation has applications beyond measuring properties of membrane environment. Other applications include membrane-bound biosensor measurements of calcium activity and GPCR activation [7]; the rigidity of membrane-bound proteins [8] and their interaction [5], and include protein complexes bound to other cellular membrane environments such as the nuclear pore complex [9].

In this protocol we will describe the measurement of membrane order and membrane topology in a cell line using lipophilic fluorophores with a chromophore that differentially localizes to the head group or tail group. We will describe the required microscope system, the preparation of the sample, the acquisition of time-lapse images, and the post-acquisition image processing, analysis, and interpretation.

2   Materials

All reagents for use with live cells should be sterile at the time of use. Appropriate institutional safety approval should be sought if primary human or animals cells are used.

*2.1   Equipment
for Imaging*

1. A confocal microscope or other optical sectioning microscope (*see* **Note 1**) with appropriate excitation source (*see* **Note 2**) and emission filters and detector (*see* **Note 3**), (Fig. 2).

2. Linear excitation polarizer and control (*see* **Note 4**).

3. Linear emission polarizer and control (*see* **Note 5**).

4. Heated chamber for maintaining cell viability and function (*see* **Note 6**).

5. Power meter.

6. High-quality linear polarizer (e.g., from Thorlabs).

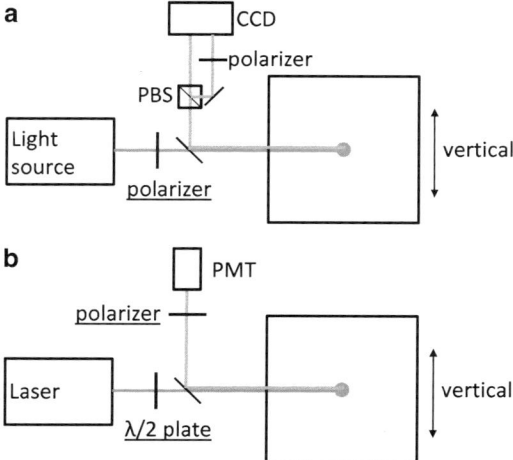

Fig. 2 Schematic of two possible microscope systems used for polarized fluorescence imaging. (**a**) In a representative wide field configuration, the excitation polarization is controlled by a rotating linear polarizer. The emission polarization is controlled by a polarizing beamsplitter imaging system, simultaneously imaging each emission polarization on a CCD camera. (**b**) In a laser scanning microscope, the excitation polarization is controlled by a half wave plate (λ/2). The emission polarization is controlled by a rotating linear polarizer. In each case *underlined* components are required to be sequentially rotated/adjusted

*2.2 Reagents
for Staining*

1. DiO fluorophore (*see* **Note 7**), from Molecular Probes division of LifeTechnology.

2. β-BODIPY FL C5-HPC (BODIPY-PC) fluorophore (*see* **Note 8**), from Molecular Probes division of LifeTechnology.

3. Phosphate-buffered saline (PBS).

4. Scattering solution of milk diluted in buffer 1:100.

5. 10 µM solution of fluorescein in buffer.

6. Chambered cover glass, for example Labtex wells (Nunc).

2.3 Other

1. Software for image analysis such as Matlab (as will be detailed here), ImageJ or Metamorph, etc. which can handle image arithmetic such as image addition, subtraction, multiplication, and division.

3 Methods

3.1 Calibration

1. First the excitation polarization is orientated. Place a linear polarizer just above the microscope objective parallel with the microscope stage, and orientate it such that its polarization is vertical within the image plane. The alignment of this polarizer will dictate the axes of the fLD analysis. Place a power meter

just above the polarizer such that it detects the excitation illumination. Rotate the excitation polarization (or waveplate) until the detected signal is maximized—this is the "vertical" excitation polarization setting (Fig. 2). Now rotate the excitation polarization (or waveplate) until the detected signal is minimized—this is the "horizontal" excitation polarization setting (Fig. 2). These two polarization settings should be 90° with respect to each other.

2. Place 500 µl of scattering solution in one well of the chambered coverglass and place on microscope stage. Bring the wall of one chamber into the field of view, where focusing on the interface between the wall and coverslip will indicate the z position (axial position) of the bottom surface of the well. Refocus to a z position above this, 25 µm into the well solution.

3. Set the microscope acquisition in free running mode, and select the vertical excitation polarization. Adjust the detection polarization settings (see **Note 9**) such that the signal in one channel is maximized, and the signal in the other channel is minimized. These channels will be "vertical" and "horizontal" detection channels respectively (Fig. 2), (see **Note 10**).

4. Next, to calculate the G and E factor, place 500 µl of fluorescein solution in a separate well of the chambered cover glass. Focus to a position as in **step 2**.

5. Collect images of the uniform fluorescein sample under all combinations (vertical, horizontal) of excitation and detection polarization. These are correction images C_{VV}, C_{VH}, C_{HV}, C_{HH} (Fig. 3), where the first subscript refers to excitation polarization, the second subscript the detection polarization, "H" represents horizontal and "V" represents vertical.

6. Under same acquisition settings, collect background images upon no excitation C_{OV}, C_{OH} (Fig. 3) (see **Note 11**).

3.2   Acquisition

1. Prepare a staining solution of 0.5 µM of DiO in 1 ml of 1× PBS. Prepare a staining solution of 0.5 µM of BODIPY-PC in 1 ml imaging buffer. Take two dishes or wells of plated cells (see **Note 12**) and wash twice with PBS at room temperature. Incubate cells with the staining solution at 4 °C for 30 min. After incubation, again wash cells twice with 1× PBS and store on ice until ready for imaging (see **Note 13**).

2. Place the dish of cells stained with DiO on the microscope stage and focus on a healthy cell or set of cells. Switch to free-running fluorescence imaging and focus at the mid-way axial position of a cell.

3. Retaining the z position, set an acquisition to collect images of the sample under all combinations (vertical, horizontal) of

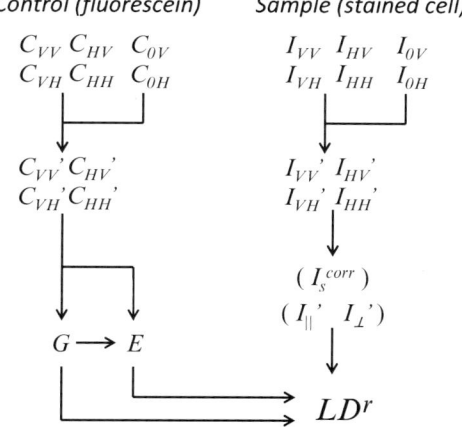

Fig. 3 Work flow for the images required for calculating the reduced linear dichroism. "*C*" refers to control (uniform sample) images. "*I*" refer to sample images. *First subscript* refers to excitation polarization direction, *second subscript* to emission polarization direction. "V" is vertical and "H" is horizontal polarization with respect to an axis in the sample. "0" refers to no excitation. *G* is the *G*-factor and "*E*" is the excitation factor

excitation and detection polarization. These are sample images I_{VV}, I_{VH}, I_{HV}, and I_{HH} (Fig. 3), where the first subscript refers to excitation polarization, the second subscript the detection polarization, "H" represents horizontal and "V" represents vertical.

4. Under same acquisition settings, collect sample background images upon no excitation I_{0V}, and I_{0H} (Fig. 3).

5. To examine changes over time, set up a time-lapse acquisition to acquire each set of polarized images every 10 s for the desired time duration (*see* **Note 14**).

6. Acquire images, noting whether the object of interest is maintained within the field of view.

7. Repeat **steps 2–6** for cells stained with BODIPY-PC.

8. Check that the appropriate images were collected with appropriate polarizations. For BODIPY-PC staining, under vertical polarized excitation and vertical polarized detection the cell membrane should be brighter where they are aligned perpendicular to the vertical axis, at the top and bottom of the cell (the pole) where the chromophore dipole is parallel with the vertical axis (*see* Fig. 4). For DiO staining the areas with increased intensity should be rotated 90° with respect to the pattern observed under BODIPY-PC staining, for a given excitation polarization.

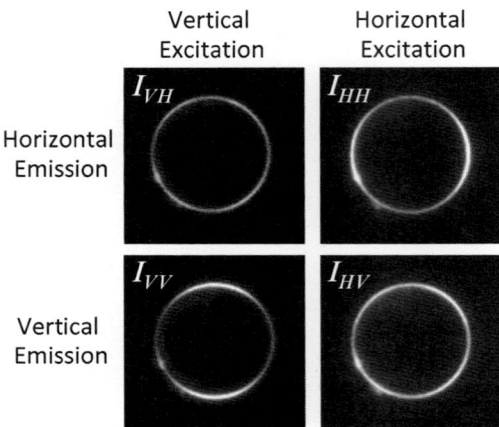

Fig. 4 Example images of BODIPY-PC stained cell acquired under different excitation and emission polarizations. For vertical excitation and detection there is increased intensity at the *top* and *bottom* of the cell image where the chromophore is aligned vertically. For horizontal excitation and detection there is increased intensity at the sides of the cell image where the chromophore is aligned horizontally

3.3 Analysis

1. Open Matlab, or other image analysis software, and load the set of acquired correction images and sample images for DiO-stained cells.

2. Background subtract all correction images: $C_{VV}' = C_{VV} - C_{OV}$; $C_{VH}' = C_{VH} - C_{OH}$; $C_{HV}' = C_{HV} - C_{OV}$; $C_{HH}' = C_{HH} - C_{OH}$ (Fig. 3) (*see* **Note 15**).

3. The *G*-factor corrects for differences in detection efficiency for the different detection polarization channels, to avoid fluorescence polarization-independent artifacts. The *G*-factor image represents this correction for the relative detection efficiency across the field of view for vertical and horizontal polarization (which are, respectively, parallel and perpendicular to vertical excitation). This is calculated by combining the correction images in **step 2** according to the following formula (Fig. 3) (*see* **Note 16**):

$$G = \sqrt{\frac{C_{VV}'C_{HV}'}{C_{VH}'C_{HH}'}}$$

4. The *E*-factor corrects for differences in the excitation power/efficiency at the sample for the different excitation polarizations, to avoid absorption-polarization independent artifacts. The *E*-factor image represents this correction for the relative excitation efficiency across the field of view for horizontal and

vertical detection polarizations. This is calculated by combining the correction images according to the following formula (Fig. 3) (*see* **Note 16**):

$$E = \frac{C_{VV'} + 2GC_{VH'}}{C_{HH'} + 2GC_{HV'}}$$

5. If the *G* factor and *E*-factor do not vary substantially across the field of view, the average *G* factor in the center of the image can be taken.

6. Background subtract all sample images: $I_{VV'} = I_{VV} - I_{OV}$; $I_{VH'} = I_{VH} - I_{OH}$; $I_{HV'} = I_{HV} - I_{OV}$; $I_{HH'} = I_{HH} - I_{OH}$ (*see* **Note 15**).

7. If the microscope objective has a high numerical aperture (NA), calculate the depolarizing effects on the images (*see* **Note 17**).

8. The linear dichroism (LD) is the difference in absorption following orthogonally polarized excitation, where absorption is calculated from the total fluorescence (fLD). The reduced linear dichroism (LD^r) scales the LD to the isotropic absorption. The LD^r image is formed by combining the sample images as follows (Fig. 5) (*see* **Note 18**):

$$LD^r = \frac{\left(I_{VV'} + 2GI_{VH'}\right) - E\left(I_{HH'} + 2G^{-1}I_{HV'}\right)}{\left(I_{VV'} + 2GI_{VH'}\right) + 2E\left(I_{HH'} + 2G^{-1}I_{HV'}\right)}$$

9. For multiple points in the cell, calculate the LD^r value over a small region of interest approximately equivalent to the $1\ \mu m \times 1\ \mu m$ area, i.e., a few pixels (Fig. 6).

10. For each point calculate the angle the local membrane makes with the vertical orientation, corresponding to the vertical excitation polarization. This can be achieved by drawing a line that makes contact and is parallel with the membrane at which the region of interest is taken (Fig. 6). Measure the vertical and horizontal distance of this line. The angle with respect to the vertical is calculated as

$$\vartheta = \tan^{-1}\frac{\text{horizontal}}{\text{vertical}}$$

11. Repeat sampling the LD^r and the corresponding membrane orientation angle for multiple points in the cell (*see* **Note 19**).

12. Plot the LD^r value as a functions of its corresponding angle α (Figs. 6 and 7) (*see* **Note 20**).

13. Repeat this process for data acquired from BODIPY-PC-stained cells (Fig. 5).

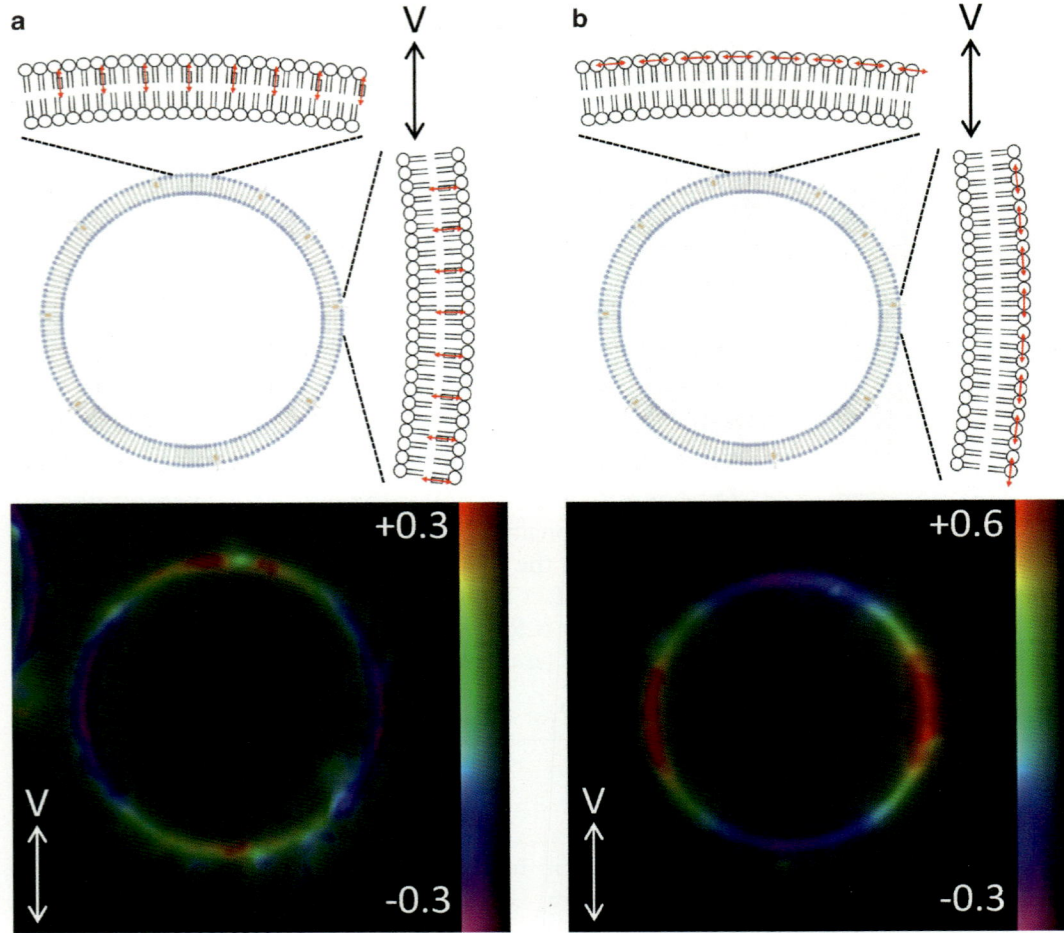

Fig. 5 (a) Schematic of BODIPY-PC chromophore aligned in different parts of the plasma membrane of a cell, together with a false color scale image of LD^r in a BODIPY-PC stained cell, where increased LD^r at the *top* of the cell is associated with the chromophore dipole being more parallel with the vertical excitation axis (indicated, V). **(b)** Schematic of DiO chromophore aligned in different parts of the plasma membrane of a cell, together with a false color scale image of LD^r in a DiO stained cell, where increased LD^r at the *side* of the cell is associated with the chromophore dipole being more parallel with the vertical excitation axis (indicated, V)

3.4 Interpretation

1. The LD^r vs. rotation angle is first fit to a model of the fluorophore orientation to calculate the mean tilt angle ϑ (*see* **Note 21**). First import the data set of BODIPY-PC LD^r values and DiO LD^r values as a function of the corresponding membrane orientation angle α.

2. For BODIPY-PC, the LD^r vs. alpha data is fit to the following function (*see* **Note 22**):

$$\text{LD}^r(\alpha) = \frac{\left(2\cos^2\alpha - 1\right)X}{1 + \left(1 - \cos^2\alpha\right)X}$$

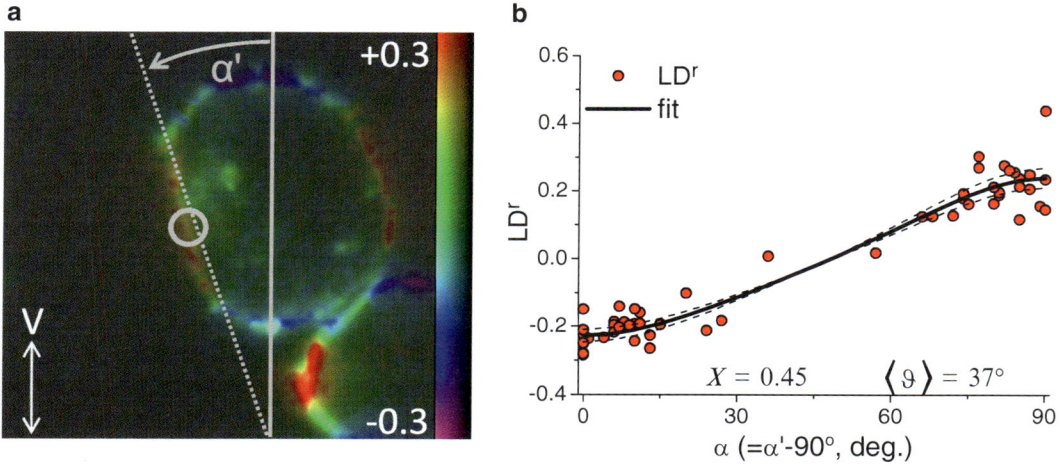

Fig. 6 Schematic showing how LD^r (from a DiO stained cell) can be calculated as a function of membrane orientation angle. (**a**) For a localized region-of-interest (*grey circle*), the average LD^r is taken. A *straight line* intersecting the membrane at the region-of-interest is taken, and the angle of this line with respect to the vertical is calculated (α'), and converted to the angle of the membrane normal ($\alpha = 90° - \alpha'$). (**b**) A plot of LD^r vs. angle (α) can then be fitted to the model to obtain the order parameter (X) and the average chromophore tilt $\langle \vartheta \rangle$

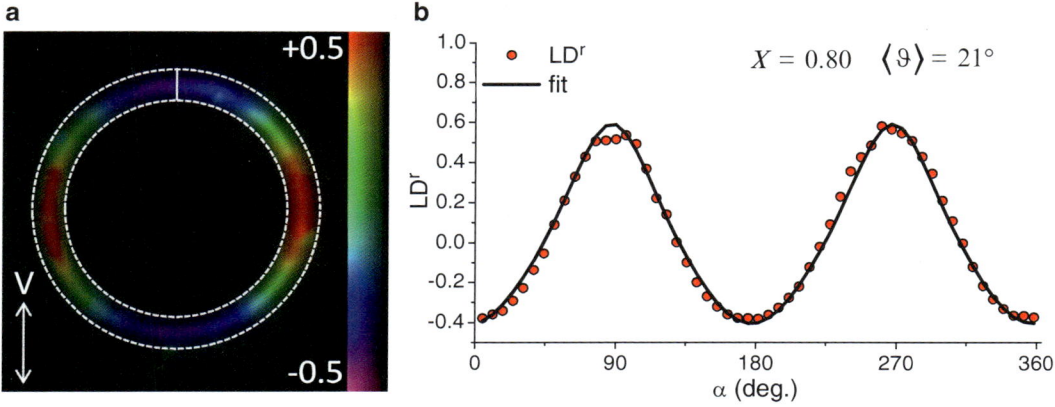

Fig. 7 Schematic showing an alternative method how LD^r (from a DiO stained cell) can be calculated as a function of membrane orientation angle. (**a**) If the cell shows a regular circular structure, the LD^r verses α can be extracted directly, by unwrapping an annular mask (*dotted line*). (**b**) This can also be fit to obtain the order parameter (X) and the average chromophore tilt $\langle \vartheta \rangle$

where the order parameter is represented by $X = \dfrac{3(\cos^2 \vartheta) - 1}{2}$

The maximum and minimum LD^r values are at $\alpha = 0°$ and $\alpha = 90°$, respectively, which are given as

$$\mathrm{LD}^r(0°) = X$$

$$\mathrm{LD}^r(90°) = \dfrac{-X}{1 + X}$$

3. For DiO, the LD^r vs. alpha data is fit to the following function (*see* **Note 22**):

$$LD^r(\alpha) = \frac{\frac{1}{2}(1 - 2\cos^2\alpha)X}{1 + \frac{1}{2}(\cos^2\alpha - 1)X}$$

where again the order parameter is represented by $X = \dfrac{3(\cos^2\vartheta) - 1}{2}$

Here, the maximum and minimum LD^r values are at $\alpha = 90°$ and $\alpha = 0°$, respectively, which are given as

$$LD^r(0°) = -X\!\big/\!2$$

$$LD^r(90°) = \frac{X\big/2}{1 - X\big/2}$$

4. In each case determine the order parameter X which best fits the experimental data, from which the value of $\langle\cos^2\vartheta\rangle$, and thus $\langle\vartheta\rangle$ can be obtained (Figs. 6 and 7) (*see* **Note 23**).

5. The greater the tilt angle (corresponding to a smaller modulation in LD^r vs. α), the more disordered the membrane in which the DiO or BODIPY-PC is orientated in. However, it should be noted that the measured tilt angle $\langle\vartheta\rangle$ represents the average tilt over a distribution. A fully disordered sample will show an isotropic distribution of orientations, from which $\langle\vartheta\rangle = 54.7°$, which is the magic angle. An angle lower than this indicates anisotropic orientation, with a minimum orientation of $\langle\vartheta\rangle = 0°$ indicating perfect orientation of the fluorophore within the membrane. However if $\langle\vartheta\rangle > 54.7°$, this corresponds to an ordered fluorophore orientated perpendicular to its preferred orientation, such as parallel with the surface of the membrane. Thus we should note, if the fluorophore happens to be perfectly orientated at the magic angle, this will not be distinguishable from an isotropic orientation.

6. The order of the BODIPY-PC fluorophore reflects the order of the internal portion of the plasma membrane, such as ordered and disordered phases. For example if cholesterol is extracted from the plasma membrane, the interior of the membrane becomes more disordered (corresponding to a smaller modulation in LD^r vs. α) and $\langle\vartheta\rangle$ approaches 54.7° (Fig. 8).

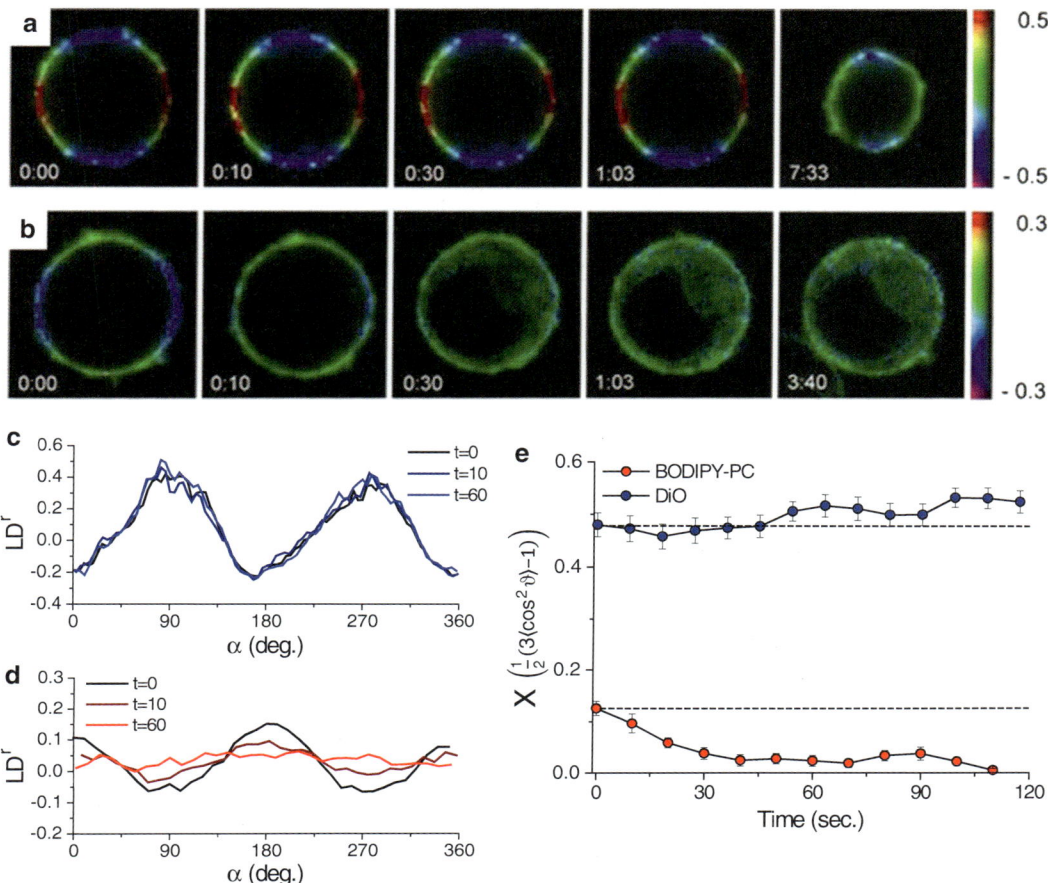

Fig. 8 Example of how the LD^r for BODIPY-PC (fatty acid tail chromophore), can be used to measure changes in order of the membrane interior. In this case, measuring the effect of altered cholesterol levels over time. (**a**) False color scale LD^r images of DiO stained cell at various time points (indicated) following cholesterol extraction at time 0 showing no change in LD^r. (**b**) False color scale LD^r image of BODIPY-PC stained cell as in **a** showing loss of LD^r modulation around the cell. (**c**) Profile of LD^r around DiO stained cell before, during and after cholesterol extraction showing no change. (**d**) Profile of LD^r around BODIPY-PC stained cell before, during and after cholesterol extraction showing loss of modulation. (**e**) Change in order parameter over time after cholesterol extraction. Despite an already reduced order, the order further reduces close to 0 (isotropic orientation) for BODIPY-PC stained cell; but remains unchanged for DiO stained cells indicating no change in membrane topology

7. The order of the DiO fluorophore reflects the order of the membrane surface topology. For example if membrane is highly ruffled at the edge of the cell-cell contact region in the immune synapse, the membrane surface is highly disordered (corresponding to a smaller modulation in LD^r vs. α) and $\langle \vartheta \rangle$ again approaches 54.7° (Fig. 9).

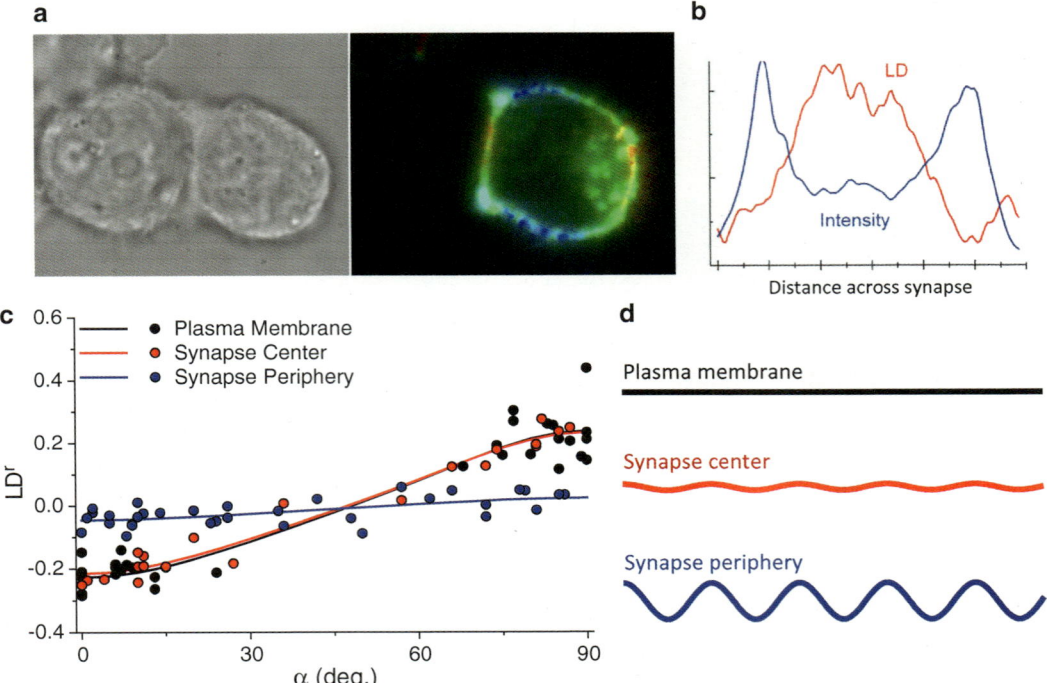

Fig. 9 Example of how the LD^r for DiO (head group chromophore) can be used to measure changes in membrane topology, in this case, measuring a spatially dependent increase in membrane ruffling across the cell-cell contact region during immune synapse formation. (**a**) Bright field and LD^r image of DiO-stained NK cell line interacting with a target cell. (**b**) Change in intensity and LD^r across the cell contact region indicating an increase in DiO fluorescence but reduction in LD^r at the edges of the interaction. (**c**) LD^r values vs. membrane normal orientation together with fits for plasma membrane outside the interaction zone ("plasma membrane," *black*), plus the center ("synapse center," *red*) and edge ("synapse periphery," *blue*) of the interaction zone. (**d**) Additional micro-scale membrane topology of center and periphery of interaction zone required to change the LD^r and chromophore tilt $\langle \vartheta \rangle$, compared to the plasma membrane outside the interaction zone; assuming no ruffling outside the interaction zone (*see* **Note 21**)

4   Notes

1. The microscope does not necessarily have to be a confocal microscope, but does need to have optical sectioning capability to reject out of focus light such that fluorescence from a specific location in the membrane is detected. This could be in the form of structured illumination, two-photon excitation, Yokogowa spinning disk, or other fluorescent optical-sectioning method. If two-photon microscopy or other non-linear microscopy is used, the theory becomes more complex as the absorption is the square of the excitation intensity [4].

2. This will depend on the microscope used, but must be suitable to excite the chosen fluorophore. BODIPY-PC and DiO have

excitation maxima at 480–500 nm, such that a 488 nm Ar + laser line or diode laser will be suitable.

3. This will also be dependent on the microscope, but must be suitable to detect fluorescence from the chosen fluorophore. BODIPY-PC and DiO have emission maxima at ~510 nm, such that a 500–550 nm band-pass filter will efficiently collect most fluorescence.

4. The control must be able to rotate the linear polarizer between horizontal and vertical alignments. In cases where the excitation source is already polarized (such as the laser source in confocal microscope), a half-wave plate orientated at 45° to the excitation polarization should be used instead to flip the polarization between two orthogonal positions.

5. The controller must be able to rotate the linear polarizer(s) to positions parallel and perpendicular to the excitation polarization direction (Fig. 2). In some microscope systems, such as a confocal or two-photon microscope, two detectors allow simultaneous detection. In this case a polarizing beam splitter and linear polarizers can be used. In a system utilizing a CCD camera, a "dual-view" system (Photometrics, Tucson AZ) containing a polarizing beam splitter and linear polarizers can be used to split an image into two adjacent images made up of orthogonal polarizations. In these of these two cases the polarizing beam splitter again must be aligned relative to the excitation polarization direction (Fig. 2).

6. Membrane environment (order, polarity, etc.) is highly dependent on temperature. Therefore maintaining the sample under a physiological temperature is necessary during the imaging process. Depending on the medium or buffer used, appropriate CO_2 levels may also be required.

7. DiO has the chromophore localized in the head group of the lipid. In principle other head group lipids such as DiI or DiD could also be used, albeit requiring different excitation and detection wavelengths.

8. BODIPY-PC is a PC lipid molecule with BODIPY tagged to one of the tail groups. This can preferentially localize to more disordered membrane phases. If the orientation in more ordered phases is to be measured, a chromophore-tagged lipid with an alternative head group, such as GM can be used. Other fluorophore-tagged lipids can also be used where the chromophore is lipid-tail localized.

9. This of course will depend exactly what form the detection channel will be in (Fig. 2). In the case of a polarization wheel, it must be rotated to maximize and minimize signal. In the case of a polarizing beam splitter, such as an optical insight system the beam splitter must be rotated.

10. Depending on the form of detection polarization used it may be important to account for the presence of polarization bleedthrough. Linear polarizers are generally highly efficient, however polarizing beamsplitters are not. In a polarizing beamsplitter, the p polarization is transmitted, and the s polarization reflected. Generally the transmitted light has very high purity (>99 %), but the reflected light less purity (~90 %). Therefore additional linear polarizers are important, particularly for the reflected light (s polarization) channel. In the absence of additional linear polarizers, the reflected s-polarization channel can be corrected:

$$I_s^{corr} = \frac{1}{9}\left(10I_s - I_p\right)$$

In this case whether s is horizontal or vertically polarized depends on the orientation of the detection polarizing beam splitter or polarizers. But this can be checked using the scattering solution and vertical excitation: vertical detection will be higher for vertical excitation and horizontal detection will be higher for horizontal excitation.

11. For best results in collecting correction images, acquire using a long integration time and average multiple image frames (for a laser scanning microscope, use a long pixel dwell time, and average multiple lines).

12. If cells do not require plating (for example, immune cell lines), these can be prepared through centrifugation and adding to chambered coverglass as a final step before imaging.

13. The protocol for staining cells with BODIPY-PC and DiO may be different for different types of cells—this should be optimized so as not to disrupt the function of the cell but achieve good fluorescence signal.

14. For additional 3D resolution, a z stack of LD^r images can be collected at each time point, by acquiring the set of polarization resolved images at multiple z positions. Set the upper and lower bounds of the z position by refocusing under free running mode to image the top of the cell and the bottom of the cell. Acquire a single stack before setting up the time lapse to check the appropriate range of z values is sampled for the object. If the system is too slow to achieve this in the time required to resolve changes in LD^r or motion of the cell, discount collection of z stacks.

15. Rather than subtracting the entire background image from the entire sample image on a pixel by pixel basis, an alternative is to calculate the average intensity for each background image and subtract that single value from all pixels of the equivalent polarized sample images.

16. An example Matlab code for forming the *G* and *E* factor can be found below, which includes an optional correction for the s-polarization bleedthrough, where in the example s-polarization corresponds to horizontal:

```
%Control images: CVV, CVH, CHH, CHV.
%Control background images: C0V, C0H.
%First subscript excitation pol. second subscript emission pol.
CVVcorr = CVV-C0V; %background correction
CVHcorr = CVH-C0H; %background correction
CHHcorr = CHH-C0H; %background correction
CHVcorr = CHV-C0V; %background correction


G = sqrt((CVVcorr.*CHVcorr)./(CVHcorr.*CHHcorr)); %G factor
G = mean(mean(G)); %Optional spatial average
E = (CVVcorr + 2*G*CVHcorr) ./(CHHcorr + 2*G*CHVcorr); %E factor
E = mean(mean(E)); %Optional spatial average


%Sample images: IVV, IVH, IHH, IHV.
%Control background images: I0V, I0H.
%First subscript excitation pol. second subscript emission pol.
IVVcorr = IVV-I0V; %background correction
IVHcorr = IVH-I0H; %background correction
IHHcorr = IHH-I0H; %background correction
IHVcorr = IHV-I0V; %background correction


%Optional poalrization correction for polarizing beam splitter (PBS).
%If reflected light from PBS is 'Verticle' polarization:
IVVcorrcorr = (1/9)*(10*IVVcorr-IVHcorr);
IHVcorrcorr = (1/9)*(10*IHVcorr-IHHcorr);
```

17. A high numerical aperture can induce mixing of the polarizations. For emission polarizations, the effect of this mixing is calculated as

$$I_{\parallel} = K_a I_x + K_b I_y + K_c I_z$$

$$I_{\perp} = K_a I_x + K_c I_y + K_b I_z$$

where I_x is aligned parallel to the optical axis and perpendicular to the excitation polarization, I_y is aligned perpendicular to the optical axis and perpendicular to the excitation polarization, and I_z is aligned perpendicular to the optical axis and parallel to excitation polarization. In the limit of low NA $I_\| = I_z$ and $I_\perp = I_y$. Under high NA K_a dominates as >95 % of the depolarization, therefore the relevant polarizations are corrected as

$$I_\|' = I_\| - \frac{K_a}{K_a + K_c} I_\perp$$

$$I_\perp' = \frac{K_c}{K_a + K_c} I_\perp$$

In our notation, for vertical excitation (V) $I_\| = I_{VV}$ and $I_\perp = I_{VH}$. For horizontal excitation (H) $I_\| = I_{HH}$ and $I_\perp = I_{HV}$. In each case I_x is approximated as $I_\perp$. K_a, K_c are calculated as follows, where $\vartheta = \sin^{-1} \frac{NA}{n}$:

$$K_a = \frac{1}{3}\left(2 - 3\cos\vartheta + \cos^3\vartheta\right).$$

$$K_c = \frac{1}{4}\left(5 - 3\cos\vartheta - \cos^2\vartheta - \cos^3\vartheta\right)$$

18. An example Matlab code for forming the reduced linear dichroism (LD^r) image:

```
%Corrected polarized images: IVV, IVH, IHH, IHV
%First subscript excitation pol. second subscript emission pol.
%Correction factors G, E

LDr_top = (IVV + 2*G.*IVH) - E.*(IHH + 2*(1./G). *IHV);
LDr_bottom = (IVV + 2*G.*IVH) + 2*E.*(IHH + 2*(1./G). *IHV);
LDr = LDr_top./LDr_bottom;
```

19. If the cell image is close to a circular shape, an annular region of interest can be taken (*see* Fig. 7). Some image analysis software can unwrap the annulus to a linear array of pixels, which represents the LD as a function of angle around the cell.

20. Qualitatively, the greater the modulation of the LD^r with angle α, then the greater the level of order. For BODIPY-PC this corresponds to a more ordered membrane interior. For DiO this corresponds to a more ordered (less ruffled) membrane topology.

21. Here, we present the data fitting for the mean tilt angle $\langle\vartheta\rangle$ assuming that the membrane is orientated at an angle α. However alternate LD^r models for fitting parameters describing the presence of vesicles, membrane nanotubes, and for quantifying the membrane ruffling can be generated.

This is beyond the scope of this chapter but examples are described in [4, 6].

22. If the order parameters and mean title angles are very close between samples, then one option is to use a more accurate model which accounts for assumptions made in the main model where the formation of the isotropic fluorescence represents polarized light absorption. As a result of this assumption, high NA-induced depolarization and hindered motion of the fluorophore in the membrane can have a small (<5 %) effect on the calculated LD^r. The numerically solved model is beyond the scope of this protocol but is described in detail in [4].

23. The mean tilt angle $\langle \vartheta \rangle$ will be averaged over the distribution of tilt angles. There may be a single fixed angle or a distribution of angles. $\langle \cos^2 \vartheta \rangle$ is calculated from an arbitrary tilt distribution as

$$\cos^2 \vartheta = \frac{\int\limits_0^\pi \cos^2 \vartheta N(\vartheta) \sin \vartheta \, d\vartheta}{\int\limits_0^\pi N(\vartheta) \sin \vartheta \, d\vartheta}$$

As a result, if $\langle \vartheta \rangle = 57°$ as a single tilt, this will be equivalent to, and indistinguishable from, isotropic orientation.

Acknowledgements

We would like to thank Bjorn Őnfelt (Karolinska Institutet) and Tim Lei (University of Colorado) for critical reading of the manuscript. This work was supported by NIH grant DK085145 (RKPB).

References

1. Owen DM et al (2012) Quantitative imaging of membrane lipid order in cells and organisms. Nat Protoc 7:24–35

2. Kenworthy AK (2007) Fluorescence recovery after photobleaching studies of lipid rafts. Methods Mol Biol 398:179–192

3. Ries J et al (2009) Accurate determination of membrane dynamics with line-scan FCS. Biophys J 96:1999–2008

4. Benninger RK et al (2005) Fluorescence imaging of two-photon linear dichroism: cholesterol depletion disrupts molecular orientation in cell membranes. Biophys J 88:609–622

5. Blackman SM et al (1996) The orientation of eosin-5-maleimide on human erythrocyte band 3 measured by fluorescence polarization microscopy. Biophys J 71:194–208

6. Benninger RK et al (2009) Live cell linear dichroism imaging reveals extensive membrane ruffling within the docking structure of natural killer cell immune synapses. Biophys J 96:L13–L15

7. Lazar J et al (2011) Two-photon polarization microscopy reveals protein structure and function. Nat Methods 8:684–690

8. Rocheleau JV et al (2003) Intrasequence GFP in class I MHC molecules, a rigid probe for fluorescence anisotropy measurements of the membrane environment. Biophys J 84:4078–4086

9. Kampmann M et al (2011) Mapping the orientation of nuclear pore proteins in living cells with polarized fluorescence microscopy. Nat Struct Mol Biol 18:643–649

Chapter 15

Scanning Fluorescence Correlation Spectroscopy on Biomembranes

Eduard Hermann, Jonas Ries, and Ana J. García-Sáez

Abstract

Fluorescence correlation spectroscopy (FCS) is a powerful quantitative method to study dynamical properties of biophysical systems. It exploits the temporal autocorrelation of fluorescence intensity fluctuations originating from a tiny volume ($\sim fL$). A theoretical model function can be then fitted to the measured auto-correlation curve to obtain physical parameters such as local concentration and diffusion time. However, the application of FCS on membranes is coupled to several difficulties like accurate positioning and stability of the set-up.

In this book chapter, we explain the theoretical framework of point FCS and Scanning FCS (SFCS), which is a variation especially suitable for membrane studies. We present a list of materials necessary for SFCS studies on Giant Unilamellar Vesicles (GUVs). Finally, we provide simple protocols for the preparation of GUVs, calibration of the microscope setup, and acquisition and analysis of SFCS data to determine diffusion coefficients and concentrations of fluorescent particles embedded in lipid membranes.

Key words Fluorescence correlation spectroscopy, Giant unilamellar vesicles, Scanning FCS, Diffusion coefficient, Membrane

1 Introduction

Fluorescence correlation spectroscopy (FCS) has become a powerful technique to determine local concentrations, rotational and translational diffusion coefficients, molecular weights, association and dissociation constants, chemical rate constants, and photodynamics by using an extended confocal laser scanning microscope [1, 2]. The principle of FCS is based on a statistical analysis of intensity fluctuations that arise from particle number fluctuations in a sufficiently small ensemble, as it is the case for molecules diffusing through a sub-micrometer detection volume.

The first FCS implementation goes back to the work by Magde, Elson, and Webb in 1972 [3]. After first successful applications in life sciences and further development [3–9], its use was then extended to the investigation of two-dimensional diffusion in artificial

Dylan M. Owen (ed.), *Methods in Membrane Lipids*, Methods in Molecular Biology, vol. 1232,
DOI 10.1007/978-1-4939-1752-5_15, © Springer Science+Business Media New York 2015

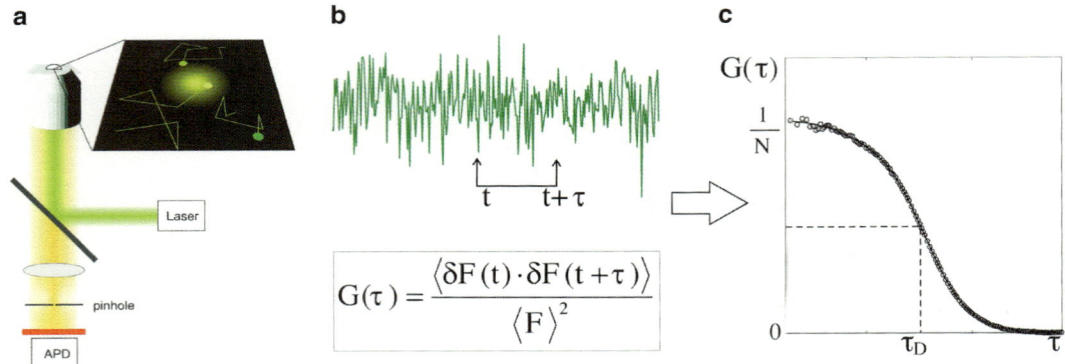

Fig. 1 Principle of FCS. Focused laser light excites diffusing fluorescent particles in a detection volume (**a**). The diffusion causes particle number fluctuations and therefore also fluorescent fluctuations (**b**) which are correlated resulting in an autocorrelation curve (**c**). Its intersection with the horizontal axis corresponds to the inverse mean particle number in the detection volume and its characteristic decay time corresponds to the diffusion time of the particles. Ries, J.; Schwille, P. New concepts for fluorescence correlation spectroscopy on membranes. Phys Chem. Chem. Phys. 2008, 10, 3487–3497. Reproduced by permission of the PCCP Owner Societies

supported lipid bilayers [10]. Measurements on native cell membranes and free-standing artificial lipid bilayers [11, 12] demonstrated the immense potential of FCS for membrane studies [13–20]. A widespread commercialization of FCS as an add-on to confocal microscopes has converted it into a well-established technique since then.

1.1 Principle of FCS

The most common variant of an FCS system is confocal FCS (Fig. 1a). A confocal detection volume of typically $\sim fL$ is exposed to laser light that induces fluorescence in the present molecules. The fluorescence light is detected by avalanche photo diodes (APD). Every excited fluorophore contributes to the overall fluorescence intensity. Particles permanently diffuse between the confocal volume and the surrounding reservoir leading to the fact that the particle number and thus the overall fluorescence intensity fluctuates around an average value [21]. The fluctuations are therefore directly connected to the particle dynamics. Mathematically, fluctuations can be quantified by autocorrelating the signal (Fig. 1b). The autocorrelation curve measures the self-similarity of the signal and is defined as

$$G(\tau) = \frac{\langle \delta F(t) \cdot \delta F(t+\tau) \rangle}{\langle F(t) \rangle^2}. \qquad (1)$$

$F(t)$ is the overall fluorescence signal, also called fluorescence trace, from the confocal volume as a function of time, τ is the lag or correlation time, $G(\tau)$ is the autocorrelation function and the angular

Table 1
Model functions for FCS autocorrelation curves

Type of diffusion	Model function
3D diffusion	$G_{3D}(\tau) = \dfrac{1}{N}\left(1 + \dfrac{\tau}{\tau_D}\right)^{-1}\left(1 + \dfrac{\tau}{S^2\tau_D}\right)^{-\frac{1}{2}}$
2D diffusion	$G_{2D}(\tau) = \dfrac{1}{N}\left(1 + \dfrac{\tau}{\tau_D}\right)^{-1}$
2D diffusion for two components	$G_{2D+2C}(\tau) = \dfrac{1}{N_{total}}\dfrac{q_f^2 \Upsilon_f G_{Df}(\tau) + q_s^2 \Upsilon_s G_{Ds}(\tau)}{(q_f \Upsilon_f + q_s \Upsilon_s)^2}$
2D diffusion with triplet	$G_{2D+T}(\tau) = \left[1 + \dfrac{T}{1-T}\exp\left(-\dfrac{\tau}{\tau_T}\right)\right]G_{2D}(\tau)$
2D diffusion with blinking	$G_{2D+bl}(\tau) = \left[1 + \dfrac{c_{dark}}{c_{bright}}\exp\left(-k_{bl}\tau\right)\right]G_{2D}(\tau)$
2D with elliptical-Gaussian profile	$G_{2DG}(\tau) = \dfrac{1}{N}\left(1 + \dfrac{\tau}{\tau_D}\right)^{-\frac{1}{2}}\left(1 + \dfrac{\tau}{S^2\tau_D}\right)^{-\frac{1}{2}}$

G is the autocorrelation curve, N is the average number of particles in the detection volume, τ is the lag time, τ_D is the diffusion time, S is the structure parameter of the detection volume, q is the molecular brightness of the fast (f) and slow (s) diffusing components, Υ refers to their molecular fraction, T is the fraction of fluorophores in the triplet state, τ_T corresponds to the triplet time, c denotes the concentration of fluorophores in the dark and bright state, k_{bl} is the blinking rate. The correction factors for two components, triplet and blinking are also valid for 3D diffusion

brackets denote averaging over time. $\delta F(t)$ is the fluorescence fluctuation at time t, defined as the deviation of the signal from its average over time $\delta F(t) = F(t) - \langle F(t)\rangle$

In a typical FCS experiment, the fluorescence trace is recorded and autocorrelated by a software or hardware correlator. This way, we obtain the experimental autocorrelation curve. On the other hand, a mathematical model function can be derived and includes the crucial dynamical parameters of the system under study. These parameters are then obtained by fitting the theoretical to the measured curve (Fig. 1c). Apart from the dynamic properties, an appropriate model function takes into account the size and shape of the confocal volume, the excitation profile and the molecular brightness of the fluorophore and its concentration as a function of position and time. Some widely used model functions are listed in Table 1.

An example for a very simple model is 2D Brownian diffusion within membranes. If the particles move within the x–y plane perpendicular to the optical axis, the theoretical autocorrelation function is given by

$$G_{2D}(\tau) = \frac{1}{N}\left(1 + \frac{\tau}{\tau_D}\right)^{-1} \qquad (2)$$

where N is the number of particles in the confocal volume and τ_D is the diffusion time related to the confocal volume. Usually, the confocal volume has Gaussian form with w_0 being the waist ($1/e2$-*radius*) in the *x–y* plane and w_z the waist in *z* direction. The effective 3D detection volume is then given by $V_{eff} = \pi w_0^2 w_z$ and the effective detection area on the equator plane is $A_{eff} = \pi w_0^2$. The area concentration of the particles C can be obtained using the relation

$$N = C\pi w_0^2. \tag{3}$$

Note that for $\tau = 0$, the autocorrelation function is just the reciprocal particle number:

$$G_{2D}(0) = \frac{1}{N}. \tag{4}$$

The other parameter in the formula, τ_D, is the diffusion time which corresponds to the time the particles need in average to cross the distance w_0. The diffusion coefficient D, which mainly characterizes the dynamic properties of a particle, is related to the diffusion time through the expression

$$\tau_D = \frac{w_0^2}{4D}. \tag{5}$$

Independently of how the particular model looks like, the basic idea always consists of fitting a theoretical correlation function to the measured data by varying the parameters of interest. Fitting is therefore a crucial step in FCS data analysis. Detailed derivation of the fitting functions can be found in [3, 17]. In Table 1 we present model functions for some important cases, taking into account the processes that take place in the sample under study. A comprehensive discussion of FCS theory and application can be found in reference [22].

A variety of photophysical effects can occur in the sample under study. Examples of photophysical effects which may affect the autocorrelation curve include blinking and triplet state excitation. In case of blinking, the molecule has a fluorescent and non-fluorescent state depending on its chemical environment. For example, proteins like GFP change their fluorescent properties as a function of pH. Excited molecules can also transition from singlet to metastable triplet state where they stay for some microseconds and remain dark during that time. If these effects play a role in the experimental setup, they have to be considered in the model, too.

One of the most interesting and challenging applications of FCS is the study of membrane dynamics. Biological membranes confine cellular compartments and also control the exchange of energy, matter and information between the inside of the cell and its external medium. In an aqueous environment they organize into continuous lipid bilayer sheets which exhibit low permeability to polar molecules. A variety of proteins that act as receptors, channels or transporters are embedded in biological membranes to enable their role as specific permeability barriers. The mobility and dynamic processes of the proteins as well as their structural conformation are essential to understand their function in the membrane. However, studying membranes is often connected with enormous challenges. It is very difficult to purify and reconstitute membrane proteins into artificial lipid bilayers. Several reviews and practical courses have addressed these problems and proposed various approaches to study membranes and their components. One interesting case is the existence of lipid domains or rafts, in which fluid/fluid phase separation in membranes leads to clustering of lipids and proteins which then perform specific functions. It is essential to study in detail both structural and functional dynamics to understand lipid rafts. This information can be obtained by FCS as evidenced by recent studies, but several difficulties have to be overcome when generally measuring FCS on membranes. First, the focal volume has to be positioned on the membrane and maintained there until the measurement process is completed. This is especially challenging if one considers the fact that the slow dynamics within membranes require long measurement times, typically 10,000 times longer than the expected diffusion time [23]. Due to thermal undulations and drifts, the membrane may leave the focal volume and it is technically extremely difficult to let the laser focus follow the membrane on the time scale of its undulations. In addition, the long measurement time immensely increases the probability of photobleaching. This problem can be partially reduced by lowering the laser power or considering the effect in the theoretical models [18]. However, a special technical variant, scanning fluorescence correlation spectroscopy (SFCS), has been proved to be more appropriate in membrane studies than point-FCS. Instead of a stationary confocal volume, SFCS continuously scans the detection volume across the membrane plane and thus measures the fluorescent trace along a defined scan path [18, 24]. It can be achieved using the scanning unit of a laser-scanning microscope (LSM). In the next section, we focus in detail on this technique.

1.2 Scanning FCS

In SCFS, the focal volume is scanned with a constant velocity along a defined path while recording the fluorescence intensity. When applied to membranes, the scan path is oriented perpendicular to the membrane plane (Fig. 2a). With this orientation, the focal volume is repeatedly scanned along a linear path and passes through the membrane only at specific time points, which reduces the residence time of fluorophores in the detection volume and thus minimizes the probability of photobleaching. Long measurement times, as required for membrane studies, are realized by a high

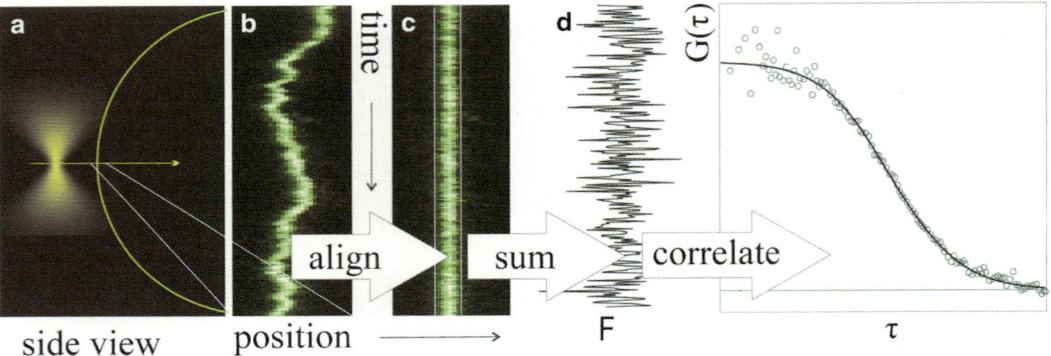

Fig. 2 Scanning FCS perpendicular to the membrane. (**a**) The detection volume is scanned repeatedly on a line perpendicular to the membrane. (**b**) The pseudo-image shows the membrane clearly but its position does not remain constant in time due to undulations and/or translational movements, for which reason each intensity trace is shifted in the way that the membrane appears as a straight line (**c**). The fluorescent signals arising from the membrane are summed up for every scan path (**d**) and autocorrelated. Reproduced from Petrásek, Z.; Ries, J.; Schwille, P. Scanning FCS for the characterization of protein dynamics in live cells. Meth. Enzymol. 2010, 472, 317–343

number of subsequent scans. The advantage of linear SFCS is that it uses the scanning unit of standard fluorescence microscopes and can be readily used on most commercial set-ups.

The photon arrival times are grouped according to the scan rate. For example, if the measurement time is 600 s and the scan rate 1 kHz, the data is grouped in 1 ms segments, each according to one scan. Furthermore, if the data is binned in pixel times of $2\ \mu s$, the result consists of 3×10^8 points in 6×10^6 scans. The fluorescence intensity is then represented in a pseudo-image where the horizontal axis denotes the position along the scan and the vertical axis denotes the number of the scan (Fig. 2b). The region where the detection volume passes through the membrane is clearly visible in the pseudo-image as a trace of high intensity values. Due to thermal membrane undulations and drifts, this trace is usually not straight vertical but the membrane appears at a different position from scan to scan. However, this effect can be corrected by aligning the scans with respect to the position of maximum of the membrane contributions (Fig. 2c). By summing up the membrane contributions for each scan (Fig. 2d), one creates a discrete fluorescence trace that can be auto-correlated as in confocal FCS (Fig. 2e). The theoretical model function for the auto-correlation curve in case of Brownian diffusion within a Gaussian detection volume is given by [25]

$$G(\tau) = \frac{1}{N}\left(1 + \frac{\tau}{\tau_D}\right)^{-1}\left(1 + \frac{\tau}{\tau_D S^2}\right)^{-\frac{1}{2}}. \qquad (6)$$

Table 2
Diffusion coefficients of fluorophores commonly used in FCS calibration experiments

Fluorophore	Diffusion coefficient ($\mu m^2/s$)	References
Rhodamine 6D	426, 414 ± 1*	[47, 48]
Alexa488	435	[48]
eGFP	95	[48]
Alexa543	341	[48]
Atto655	426 ± 8*	[49]

Diffusion coefficients obtained in water solutions at 22.5 °C, except those marked with
*which were obtained at 25 °C

The only additional parameter here is the structure parameter S and denotes the aspect ratio of the focal volume: $S = w_z/w_0$. As discussed above, w_0 and w_z are the extensions of the focal volume in radial and axial directions. To obtain S, a calibration experiment with a well-characterized dye is required. If the diffusion coefficient of the dye is known, S can be calculated from the diffusion time. Table 2 gives an overview over some dyes which have been characterized in other studies. However, one should take into account that calibration is usually performed in solution and does not correspond perfectly to measurements in membranes.

Some advanced FCS techniques may be easily combined with SFCS and allow for studying interactions of two differently labeled species (two-color fluorescence cross correlation spectroscopy, two-color FCCS) or enable calibration-free measurements (two-focus FCCS) [25–27]. Two-color FCCS is suitable for the study of molecule binding by determining the concentration of complexes formed by two species. In this set-up, two laser lines excite the fluorophores in the same focal volume and the fluorescence traces of both lines are recorded. To avoid cross talk, the excitation can be performed alternately such that at any time point only one laser excites the fluorophores. In two-focus FCCS, the detection volume is scanned along two parallel paths; the possibility of obtaining the distance between the paths with high precision spares the determination of the shape of the focal volume. It can be easily realized with one single laser if the same focal volume alternately scans along two parallel paths. Both techniques can be combined for calibration-free binding and diffusion measurements [28]. In both methods, two different fluorescence traces are cross-correlated following the expression

$$G_x(\tau) = \frac{\langle \delta F_1(t) \cdot \delta F_2(t+\tau) \rangle}{\langle F_1(t) \rangle \cdot \langle F_2(t) \rangle}. \tag{7}$$

Similar to other FCS variants, theoretical model functions are fitted to the measured cross-correlation curves to obtain physical parameters. In case of two-focus FCCS, the theoretical cross-correlation function is given by

$$G_x(\tau) = G_1(\tau) \cdot G_2(\tau) \cdot exp\left(-\frac{d^2}{4D\tau + w_0^2}\right) \qquad (8)$$

With $G_1(\tau)$ and $G_2(\tau)$ being the autocorrelation curves of the individual traces and d the distance between the scan paths.

1.3 Model Membranes

Dynamic living systems such as cells are highly complex and introduce sample-to-sample variability when studied in experiments. To overcome this problem and to exclude unwanted effects due to poor chemical control, using membrane model systems is an important approach. Giant Unilamellar Vesicles (GUVs) are widely used to model membranes since they consist of free-standing bilayers and have a similar size to eukaryotic cells. This makes them accessible to light microscopy on the one hand, and on the other hand they show virtually no curvature in the membrane area contained in the focal plane of a confocal microscope setup. GUVs can be formed easily via electroformation [29]. In this setup, a dried lipid film or proteo-lipid mixture is used on an electrode surface (usually Pt). FCS studies on GUVs have been successfully carried out and provided the characterization of lipid structures [30, 31], protein-lipid interactions with SNARE proteins [32], the mechanosensitive channel MscL [33], Bcl-2 proteins [34], bacteriorhodopsin [35], and fibroblast growth factor 2 (FGF2) [36]. In the following section, we give a detailed protocol for the application of SFCS on GUVs and also list all required materials for such experiments.

2 Materials

2.1 Confocal Microscopy Setup

Currently, a series of commercial FCS setups are available on the market. Originally, SFCS has been implemented in Zeiss microscopes but nowadays other microscopes can in principle be used, for example those from PicoQuant or Leica Microsystems. Usually, these setups combine a confocal microscopy system with FCS capabilities and contain user-optimized platforms. Several lasers are included for excitation and several detection channels with fiber-coupled avalanche photo diodes (APDs) are implemented to detect the photons. Generally, an FCS system consists of an inverted microscope together with a high-numerical aperture objective like the 40× NA 1.2 UV-VIS-IR C-Apochromat water-immersion objective from Zeiss (Jena, Germany). A dichroic mirror reflects the laser light which is then focused by the microscope objective to a spot of 0.3–0.5 μm diameter. The sample emits fluorescence light which passes the dichroic and the emission filter to remove residual Rayleigh and Raman scattered light. Additionally, a pinhole is inserted into the image plane to enforce axial resolution. An optical fiber of the same diameter can be used instead. APDs with single-photon sensitivity are usually used for detection. To correlate the fluorescent trace, a hardware or software correlator can be used. For Scanning FCS, it is essential to record photon arrival times with sufficient time resolution. If your set-up does not allow it, you can use for example the photon mode of the hardware correlator Flex 02-01 D (Bridgewater, NJ, USA, www.correlator.com).

2.2 Giant Unilamellar Vesicles

Model membranes are very helpful for the study of the principles governing cell membrane organization. Excellent model systems which are highly suitable for FCS measurements are giant unilamellar vesicles (GUVs). GUVs are free-standing lipid bilayers with diameters in the range of 10–100 μm and thus visible in optical microscopy. They provide a flat, single-spherical and closed bilayer. In the following sections, we explain the preparation of GUVs and their application in scanning FCS perpendicular to the membrane. For their preparation, following material is needed:

1. Lipid mixtures from stock solutions, prepared by dissolving lipids of interest in chloroform (lipids can be purchased for example from Avanti Polar Lipids, Alabaster, AL, USA). Lipids can be used in form of purified lyophilized lipids or extracted lipid mixtures (*see* **Note 1** for proper lipid handling). To label the membrane for FCS measurements, add around 0.01 % of a lipophilic fluorescent dye (*see* **Note 2** for information about lipophilic fluorescent dyes and **Note 3** for information on protein dyes). The stocks of your desired lipid mixtures at 1 mg/ml should be stored in a freezer.

2. Sucrose solution of 300 mM: dilute 10.26 g of sucrose in 100 mL distilled water. Store at 4 °C.

3. Blocking solution: 2 mg/mL bovine serum albumin (BSA) or casein; dissolve 100 mg of BSA or casein in 50 mL of water. Store at 4 °C.

4. Phosphate-buffered saline (1× PBS Buffer): 2.7 mM KCl, 1.5 mM KH_2PO_4, 8 mM Na_2HPO_4, and 137 mM NaCl (pH = 7.2); weigh 0.2237 g KCl, 0.2041 g KH_2PO_4, 1.1357 g Na_2HPO_4 and 8.0065 g NaCl in a beaker. Fill to around 800 mL with water. Adjust the pH by adding NaOH. Add enough water to arrive at 1 L solution. Store at room temperature. A different aqueous solution isoosmolar with the sucrose solution can be used (*see* **Note 4**).

5. Homemade Teflon-and-Pt chambers (Fig. 3a).

6. Home-made Al heating block (Fig. 3b).

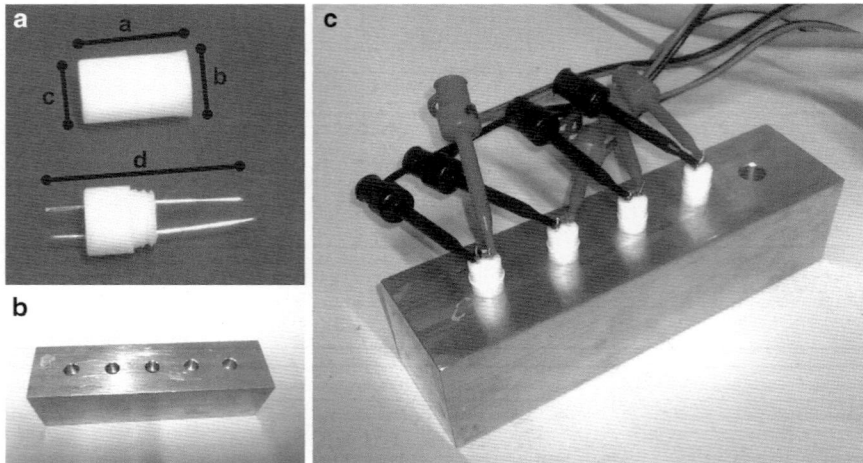

Fig. 3 The Pt-wire chamber for electroformation. (**a**) A homemade Pt-and-Teflon chamber with the following dimensions is shown: (*a*) 15 mm, (*b*) 8 mm, (*c*) 4 mm, (*d*) 30 mm. The distance between the electrodes is around 4 mm. (**b**) Homemade aluminum block is shown. (**c**) The connection of the chambers to the function generator and the whole setup is shown

7. Electric heating plate.

8. Function generator, for example TTi TG315 from Turlby-Tandar Instruments Ltd, Huntingdon, Cambs, England.

9. Connecting clamps and cables, for example BNC Male Minigrabber Clip Cable from Pomona Electronics (sold by Farnell, www.farnell.de).

10. Observation chambers, for example Lab-Tek chambered cover glasses from Nalge Nunc Intl. (Rochester, New York, USA, www.nuncbrand.com).

3 Methods

3.1 Preparation of Giant Unilamellar Vesicles

1. Clean the chamber for electroformation with ethanol and water and let it dry.

2. Take the lipid mixture out of the freezer and wait until it reaches room temperature before opening the vial.

3. Spread 2.5 µl of the lipid solution on each of the Pt wires on the side that screws inside the electroformation chamber. Wait until the chloroform is completely evaporated.

4. If necessary, warm up adequate amounts of 300 mM sucrose to above the transition temperature of the lipid mixture. Fill the electroformation chamber with 300 µL of the sucrose solution and close it by screwing the cap with the Pt wires.

5. Place the chamber in an Al heating block and connect the cables of the function generator to each of the two Pt wires

(Fig. 3c). Avoid contact between them! Here, the heating block is used rather for carrying the Pt-wire chambers than for heating purposes.

6. Switch on the function generator with a sinus wave at a voltage of 1.5 V (RMS) and a frequency of 10 Hz (*see* **Note 5**). The electroformation should be processed under these conditions for around 30 min to 2 h, depending on the lipid composition. Then, switch the frequency to 2 Hz and wait 30 min to let the GUVs gently detach from the Pt wires. It is important that the electroformation process is carried out above the transition temperature of the lipid mixture. Turn off the function generator and disconnect the cables. For salt-containing buffers the electroformation has to be carried out in three steps at high frequency (~500 Hz) and low voltage (~1 V) as described in [37].

7. Fill the wells of an observation chamber with 700 µl of a blocking solution of the BSA blocking solution and incubate for around 1 h. After blocking, rinse the wells of the observation chamber extensively with water.

8. Add 200–500 µl 1× PBS. Proteins or other molecules of interest can also be added to the well.

9. Cut the end of a micropipette tip to widen the opening and gently take the GUV suspension from the electroformation chamber (usually 50–100 µL). Add it carefully on top of the PBS on the observation chamber (*see* **Note 6** for proper GUV handling).

10. Wait 5 min until the GUVs sediment to the bottom of the chamber.

3.2 Microscope Alignment and Calibration

Before data acquisition, you have to align properly the microscope setup and perform a calibration experiment. The calibration is necessary to determine the shape of the detection volume (w_0 and S). Use a dye with similar spectral characteristics as the dye in your measurements. For example, free Alexa-488 is a well-known dye in the green spectrum. Some other dyes are listed in Table 2.

1. Switch on the excitation laser around 60–90 min before starting the measurements to guarantee stabilization of the light source.

2. Put 200 µL of 10 nM dye solution in an observation chamber similar to the one that will be used.

3. Make sure the FCS configuration of the microscope fits the spectral properties under study.

4. Adjust the laser power to give maximum counts per molecule without significant photobleaching during the acquisition time. This is usually 15–25 µW, depending on the microscope and sample.

5. Position the focal volume inside the solution at least 100 µm above the coverslip surface (*see* **Note 7**).

6. Set the size of the pinhole to 40 μm or 1 airy unit and align it with respect to the beam path. For this purpose, select the X and Y positions of the pinhole that produce the highest fluorescence count rate.

7. If necessary in your set-up, adjust the position of the collimator to achieve maximum count rate.

8. Adjust the correction collar of the objective to correct for glass thickness of the observation chamber by maximizing the count rate.

9. Perform a first FCS measurement on the calibration dye solution. For instance, perform three acquisitions of 20 s each.

10. Save the obtained data for further analysis. Depending on the software you use for data analysis, it might be necessary to export the data in ASCII format.

11. Fit an appropriate model function to the measured autocorrelation curve and estimate the diffusion time and the structure parameter (*see* **Note 8**).

3.3 Data Acquisition Using Scanning FCS on GUVs

1. Set the laser power to 15–25 μW.

2. Place the observation chamber containing the GUVs on the sample holder of the microscope.

3. In the microscope's imaging mode, set up the light path to use APDs for detection. If not possible, perform GUV selection and positioning of the focal volume in the normal imaging mode and change to APD detection for the FCS measurement.

4. Focus the sample and look for a uniform GUV with a size of around 50 μm (*see* **Note 9**).

5. Place the GUV on the center of your imaging field and zoom in. Find the focal plane of the equator of your GUV where it appears largest (*see* **Note 10**).

6. Align the scanning path of the microscope so that the GUV membrane is at the center and perpendicular to the scanned line. We generally acquire 32 pixel in x and 1 or 2 pixel in y (one focus or two focus SFCS, respectively) with a zoom of 12 to have a pixel size of 0.03 μm. In the case of two focus SFCS, we measure with an angle of 45° for practical reasons (in our microscope these conditions provide the closest to two parallel lines). We recommend checking this out for your microscope.

7. Select the scanning rate of your microscope at maximum speed (unidirectional, no averaging).

8. If saving the photon arrival times is possible with the microscope software, start a time-lapse experiment.

9. If the microscope setup does not allow saving photon arrival times, you can use the photon mode of the hardware correlator mentioned in Subheading 2.1 (*see* **Note 11**).

10. For statistical reasons, the measuring time has to be of the order of magnitude of 10,000 times longer than the diffusion time of the fluorophore under study. For instance, the measuring time has to be set to 30 s if the diffusion time in the membrane is around 3 ms (*see* **Note 12**).

11. Take several curves per measurement point.

3.4 Autocorrelation of SFCS Data

To process the row data in form of photon arrival times and also to perform the autocorrelation, we use a self-written Matlab program. Here, we shortly present the algorithm [25].

1. Bin the photon arrival times in bins of 100 ns to 5 μs depending on the scan rate. In theory, the bin size is determined by the photon count rate, but should be in any case much smaller compared to the scan rate.

2. Arrange the photon beam according to the scan rate to a pseudo-image such that every row corresponds to one line scan.

3. To smooth out small intensity irregularities along the fluorescence trace, convolute every line scan with an averaging filter.

4. Every line scan has a maximum corresponding to the position of the membrane. Shift all line scans such that the maxima are positioned on the same column. This corrects for membrane movement during the measurement.

5. Fit a Gaussian to each line scan.

6. For each line scan, sum up the values between -2.5σ and 2.5σ to produce one intensity value. Doing this for all line scans creates a discrete fluorescence intensity trace over time.

7. Compute the auto- and cross-correlation curves of the resulting intensity trace with a multiple tau correlation algorithm [38].

3.5 Fitting a Model Function to the Autocorrelation Curve

To obtain physical parameters, a theoretical model has to be fitted to the experimental autocorrelation curve. You can use a program with mathematical fitting options, such as Origin or Matlab. For instance, we use a self-written Matlab program to fit model functions to FCS autocorrelation curves.

1. Open the obtained autocorrelation curve with an appropriate fitting program.

2. Plot the autocorrelation curves.

3. For qualitative analysis, the shape of the curves is related to the process causing the fluorescence fluctuations. Among others, the decay of the autocorrelation curve depends on the type of particle motion, such as transport or diffusion. In addition, the amplitude of the autocorrelation curve is a function of the particle area concentration (*see* **Note 13**).

4. Discard the curves with distorted shapes since they usually result from major instabilities.

5. For quantitative analysis, fit the curves with an appropriate model function using a nonlinear least-squares fitting algorithm (*see* Table 1). A plot of the fitting residuals provides information about the goodness of the fit. Successful fitting of the data will provide an estimation of the diffusion time τ_D and number of particles N in the focal area. The diffusion coefficient as well as the area concentration may then be derived using the information about the shape of the focal volume as obtained in the calibration experiment.

6. In the case of two-focus SFCCS, absolute measurements of fluorophore concentration and diffusion coefficient are realized by fitting the data with Eq. 8.

4   Notes

1. Do not open container vials of your lipid stocks right after taking them out of the freezer. Condensed water can degrade phospholipids. Instead, allow to equilibrate at room temperature. Also, avoid exposure to oxygen since this can also lead to lipid degradation. After usage, before closing the vials, flux an inert gas like N_2 or Argon into the vial and seal it with parafilm.

2. FCS measurements on membranes are usually performed with an area concentration of fluorescent dye between 1 and 1,000 molecules/μm^2 [18]. We therefore suggest a lipidic dye concentration of 0.01–0.05 mol %. The widely used family of long-chain dialkylcarbocyanines, like DiD, DiI and DiO (Invitrogen) cover a wide range of wavelengths and have been successfully used to determine the lateral organization of lipids in model membranes. Alternatively, fluorescently labeled lipids like rhodamine-phosphatidylethanolamine (Avanti Polar Lipids) or BODIPY-cholesterol (Invitrogen) can be used. Note that fluorophore bound to the lipid may change the lipid behavior and functioning [39].

3. When membrane proteins are studied in FCS, label them with fluorescent probe. If possible, proteins can be chemically labeled with reactive dyes specific for cysteins or amine groups. These dyes include the Alexa (Invitrogen) and Atto (Atto-TEC) family, which span a broad range of wavelengths. FCS measurements of proteins fused to fluorescent proteins are also possible. Again, take into account that labeling can affect protein behavior. A review that contains further information about protein labeling can be found in [40, 41].

4. The vesicles can shrink or swell or even burst under osmotic pressure. Therefore, sugar solutions and buffers in and outside the GUVs should be isoosmolar. Usually, a concentration between 10 and 300 mM is appropriate. The outside buffer may

be substituted by glucose solution with the same osmolarity (you can use 300 mM glucose solution instead of PBS buffer).

5. The conditions under which the electroformation is performed determine the vesicle characteristics [29, 42–44]. To support lipid swelling and liposome formation, it is advantageous to use AC currents [43]. The relationship between the AC-field strength and frequency is described in a recent study [45].

6. GUVs are very sensitive, so always handle them very gently, especially while pipetting and mixing. Cut off pipette tips to reduce shear that can destroy the vesicles.

7. FCS measurements are sensitive to distortions in the detection volume. These distortions arise from incorrect axial positioning, a varying refractive index along the beam path, coverslip thickness and other effects [18, 46]. Make sure that the calibration measurement is performed away from the coverslip surface.

8. The calibration measurement is required to measure the diffusion time τ_D of a dye with well-known characteristics and the structure parameter S of the focal volume. This is done by fitting an appropriate model function to the measured autocorrelation curve. Since the diffusion coefficient of the dye is known, one obtains the waist w_0 using (Eq. 5). Both parameters have to be known to perform measurements on particles with unknown characteristics. Since these parameters are characteristic for a microscope, we recommend keeping a record of these calibration measurements to test for consistency over several months. Eventual problems in the microscope setup can be detected this way.

9. Depending on the lipid composition, the electroformation produces GUVs with sizes between 20 and 300 μm. However, larger vesicles are more suitable for FCS measurements because they show less curvature in the detection volume and cause less photobleaching problems. When using proteoliposomes, getting large enough vesicles might be difficult. Make sure to only select truly unilamellar, tensed, homogeneous and immobile vesicles.

10. The scanning mode of a typical LSM scans in the horizontal x–y plane. SFCS measurements have to be performed perpendicular to the ideally flat membrane. This coincides with the equator of the GUV, which is the focal plane where GUV appears largest. Otherwise, the scan path is not perpendicular to the membrane and the autocorrelation curve is heavily distorted.

11. If you use the hardware correlator, open the software for the photon mode and set up the analysis time, data storage, etc. In the image mode of your microscope, perform continuous scanning over the area of interest. Start acquisition on the hardware correlator and stop the experiment (acquisition and continuous scanning) once the desired time has elapsed.

12. For statistical reasons, FCS data has to be acquired with sufficient time resolution for the species diffusion [18, 23]. Normally, a measuring time of 10,000 times longer than the typical diffusion time is required.

13. The shape of the autocorrelation curve indicates the diffusion characteristics of the particles. For a given detection volume, the amplitude is related to the particle concentration; larger amplitude corresponds to lower concentration. The decay time estimates the type of diffusion; a steep decay is typical for transport phenomena, while slower decay corresponds to random Brownian or anomalous diffusion.

References

1. Rigler R, Elson E (2001) Fluorescence correlation spectroscopy: theory and applications. Springer, New York

2. Bacia K, Schwille P (2003) A dynamic view of cellular processes by in vivo fluorescence auto- and cross-correlation spectroscopy. Methods 29: 74–85

3. Magde D, Elson EL, Webb WW (1972) Thermodynamic fluctuations in a reacting system—measurement by fluorescence correlation spectroscopy. Phys Rev Lett 29:705–708

4. Aragon SR, Pecora R (1975) Fluorescence correlation spectroscopy and Brownian rotational diffusion. Biopolymers 14:119–137

5. Elson EL, Magde D (1974) Fluorescence correlation spectroscopy. I. Conceptual basis and theory. Biopolymers 13:1–27

6. Koppel D (1974) Statistical accuracy in fluorescence correlation spectroscopy. Phys Rev A 10:1938–1945

7. Koppel DE, Axelrod D, Schlessinger J et al (1976) Dynamics of fluorescence marker concentration as a probe of mobility. Biophys J 16:1315–1329

8. Magde D, Elson EL, Webb WW (1974) Fluorescence correlation spectroscopy. II. An experimental realization. Biopolymers 13:29–61

9. Magde D, Elson EL, Webb WW (1978) Fluorescence correlation spectroscopy. III. Uniform translation and lamellar flow. Biopolymers 17:361–376

10. Fahey PF, Koppel DE, Barak LS et al (1977) Lateral diffusion in planar lipid bilayers. Science 195:305–306

11. Schwille P, Korlach J, Webb WW (1999) Fluorescence correlation spectroscopy with single-molecule sensitivity on cell and model membranes. Cytometry 36:176–182

12. Korlach J, Schwille P, Webb WW et al (1999) Characterization of lipid bilayer phases by confocal microscopy and fluorescence correlation spectroscopy. Proc Natl Acad Sci 96:8461–8466

13. Bacia K, Kim SA, Schwille P (2006) Fluorescence cross-correlation spectroscopy in living cells. Nat Methods 3:83–89

14. Haustein E, Schwille P (2003) Ultrasensitive investigations of biological systems by fluorescence correlation spectroscopy. Methods 29: 153–166

15. Hess ST, Huang S, Heikal AA et al (2002) Biological and chemical applications of fluorescence correlation spectroscopy: a review. Biochemistry 41:697–705

16. Kim SA, Heinze KG, Schwille P (2007) Fluorescence correlation spectroscopy in living cells. Nat Methods 4:963–973

17. Petrov E, Schwille P (2008) State of the art and novel trends in fluorescence correlation spectroscopy. In: Resch-Genger U (ed) Standardization and quality assurance in fluorescence measurements II: Bioanalytical and biomedical applications. Springer, Berlin

18. Ries J, Schwille P (2008) New concepts for fluorescence correlation spectroscopy on membranes. Phys Chem Chem Phys 10:3487–3497

19. Schwille P (2001) Fluorescence correlation spectroscopy and its potential for intracellular applications. Cell Biochem Biophys 34: 383–408

20. Thompson NL, Lieto AM, Allen NW (2002) Recent advances in fluorescence correlation spectroscopy. Curr Opin Struct Biol 12: 634–641

21. Rigler R, Mets Ü, Widengren J (1993) Fluorescence correlation spectroscopy with high count rate and low background: analysis

of translational diffusion. Eur Biophys J 22: 169–175

22. Ries J, Weidemann T, Schwille P (2012) Fluorescence correlation spectroscopy. In: Egelman E (ed) Comprehensive biophysics. Academic, New York, pp 210–245

23. Tcherniak A, Reznik C, Link S et al (2009) Fluorescence correlation spectroscopy: criteria for analysis in complex systems. Anal Chem 81: 746–754

24. Ries J, Chiantia S, Schwille P (2009) Accurate determination of membrane dynamics with line-scan FCS. Biophys J 96:1999–2008

25. Ries J, Schwille P (2006) Studying slow membrane dynamics with continuous wave scanning fluorescence correlation spectroscopy. Biophys J 91:1915–1924

26. Bacia K, Schwille P (2007) Practical guidelines for dual-color fluorescence cross-correlation spectroscopy. Nat Protoc 2:2842–2856

27. Dertinger T, Loman A, Ewers B et al (2008) The optics and performance of dual-focus fluorescence correlation spectroscopy. Opt Express 16:14353–14368

28. Ries J, Petrásek Z, García-Sáez AJ et al (2010) A comprehensive framework for fluorescence cross-correlation spectroscopy. N J Phys 12

29. Angelova MI, Dimitrov DS (1986) Liposome electroformation. Faraday Discuss 81:303

30. Chiantia S, Schwille P, Klymchenko AS et al (2011) Asymmetric GUVs prepared by MbetaCD-mediated lipid exchange: an FCS study. Biophys J 100:L1–L3

31. Chiantia S, Ries J, Schwille P (2009) Fluorescence correlation spectroscopy in membrane structure elucidation. Biochim Biophys Acta 1788:225–233

32. Bacia K, Schuette CG, Kahya N et al (2004) SNAREs prefer liquid-disordered over "raft" (liquid-ordered) domains when reconstituted into giant unilamellar vesicles. J Biol Chem 279:37951–37955

33. Doeven MK, Folgering JH, Krasnikov V et al (2005) Distribution, lateral mobility and function of membrane proteins incorporated into giant unilamellar vesicles. Biophys J 88: 1134–1142

34. Garcia-Saez AJ, Ries J, Orzaez M et al (2009) Membrane promotes tBID interaction with BCL(XL). Nat Struct Mol Biol 16:1178–1185

35. Kahya N, Wiersma DA, Poolman B et al (2002) Spatial organization of bacteriorhodopsin in model membranes. Light-induced mobility changes. J Biol Chem 277: 39304–39311

36. Steringer JP, Bleicken S, Andreas H et al (2012) Phosphatidylinositol 4,5-bisphosphate (PI(4,5)P2)-dependent oligomerization of fibroblast growth factor 2 (FGF2) triggers the formation of a lipidic membrane pore implicated in unconventional secretion. J Biol Chem 287:27659–27669

37. Montes LR, Alonso A, Goni FM et al (2007) Giant unilamellar vesicles electroformed from native membranes and organic lipid mixtures under physiological conditions. Biophys J 93:3548–3554

38. Magatti D, Ferri F (2001) Fast multi-t real-time software correlator for dynamic light scattering. Appl Opt 40:4011–4021

39. Chiantia S, Kahya N, Schwille P (2007) Raft domain reorganization driven by short- and long-chain ceramide: a combined AFM and FCS study. Langmuir 23:7659–7665

40. Marks KM, Nolan GP (2006) Chemical labeling strategies for cell biology. Nat Methods 3:591–596

41. Prescher JA, Bertozzi CR (2005) Chemistry in living systems. Nat Chem Biol 1:13–21

42. Angelova MI, Dimitrov DS (1987) Swelling of charged lipids and formation of liposomes on electrode surfaces. Mol Cryst Liq Cryst 152: 89–104

43. Dimitrov DS, Angelova MI (1988) Lipid swelling and liposome formation mediated by electric fields. Bioelectrochem Bioenerg 19: 323–336

44. Angelova MI, Soléau S, Méléard P (1992) Preparation of giant vesicles by external AC fields. Kinetics and applications. Prog Coll Polym Sci 89:127–131

45. Politano TJ, Froude VE, Jing B et al (2010) AC-electric field dependent electroformation of giant lipid vesicles. Colloids Surf B Biointerfaces 79:75–82

46. Enderlein J, Gregor I, Patra D et al (2005) Performance of fluorescence correlation spectroscopy for measuring diffusion and concentration. Chemphyschem 6:2324–2336

47. Petrasek Z, Schwille P (2008) Precise measurement of diffusion coefficients using scanning fluorescence correlation spectroscopy. Biophys J 94:1437–1448

48. Culbertson CT, Jacobson SC, Michael Ramsey J (2002) Diffusion coefficient measurements in microfluidic devices. Talanta 56:365–373

49. Dertinger T, Pacheco V, von der Hocht I et al (2007) Two-focus fluorescence correlation spectroscopy: a new tool for accurate and absolute diffusion measurements. Chemphyschem 8:433–443

Chapter 16

X-Ray Diffraction of Lipid Model Membranes

Arwen I.I. Tyler, Robert V. Law, and John M. Seddon

Abstract

In this chapter the use of X-ray diffraction to study the structure of lyotropic phases and lipid model membranes is described. Determination of the phase symmetry and lattice parameters from small-angle X-ray scattering (SAXS), and of the nature of the hydrocarbon chain packing from wide-angle X-ray scattering (WAXS), are discussed. Methods by which the sign of the interfacial curvature of non-lamellar phases may be determined are then presented. Finally, the calculation of electron density profiles from the intensities of the observed Bragg peaks is described, for the lamellar phase and for the inverse hexagonal phase.

Key words X-ray, SAXS, WAXS, Phase

1 Introduction

Amphiphilic molecules such as surfactants, phospholipids, and glycolipids can spontaneously self-assemble into a wide range of ordered lyotropic liquid-crystalline phases, possessing long range periodicity and short range disorder [1]. These phases adopted may possess one-dimensional order (lamellar sheets, L), two-dimensional order (hexagonally arranged cylinders, H), or three-dimensional order (cubic phases, Q). For example, the fluid lamellar L_α phase consists of a 1D periodic stack of lipid bilayers alternating with water layers. X-ray scattering can reveal not just the structure, but also the elastic parameters of fluid bilayers [2, 3]. The structural ordering within such phases can extend from the nanoscale (2–3 nm), up to much larger values (>100 nm), although locally the molecules are liquid-like (or soft solids in the case of the low-temperature lamellar gel phases). Apart from the fluid bilayer L_α phase and the liquid-ordered L_o phase, all other fluid lyotropic mesophases are based upon ordered arrangements of *curved* interfaces, separating the amphiphiles from the water regions.

The long range order of the structure appears in the small-angle region of the diffraction pattern (SAXS) and its symmetry defines the phase of the lyotropic liquid crystal. The wide-angle

Dylan M. Owen (ed.), *Methods in Membrane Lipids*, Methods in Molecular Biology, vol. 1232,
DOI 10.1007/978-1-4939-1752-5_16, © Springer Science+Business Media New York 2015

X-ray scattering (WAXS) gives information about the short range order of the system, i.e., the packing of the lipid acyl chains [4], so it can to be used to differentiate between the various lamellar phases [lamellar crystal (L_c), lamellar gel (L_β), fluid lamellar (L_α and liquid-ordered L_o)] [5]. Non-bilayer phases are thought to be of considerable biological relevance. Whenever there is a topological change in the membrane, corresponding to events such as membrane fusion, fission, exocytosis or endocytosis, non-bilayer structures are assumed to be adopted locally in the membrane [6]. To further our knowledge of such events, a comprehensive understanding of the processes governing phase transitions, the type of intermediates formed and the mechanism by which transitions occur is vital [7]. X-ray diffraction of lipid mesophases is at present the only experimental technique which can give structural information on model membranes down to Angstrom length scales and hence is invaluable in studying the type of equilibrium or intermediate (dynamic) structures formed. X-ray scattering experiments are ideal for studying the structure and dynamics of lipid membranes under near-physiological conditions and allow us to inspect structural changes of model membranes by altering parameters such as temperature, pressure, pH, ionic strength, etc. Moreover, SAXS is one of the best methods for determining bilayer thickness of unsupported lipid membranes in near-native conditions, and also provides information on the hydrocarbon chain region thickness of $L\alpha$ and L_o regions of a membrane. Furthermore, X-ray scattering can be used to study the effect of additives such as drug molecules or peptides on the headgroup or chain region of the bilayer [8].

1.1 Small and Wide Angle X-Ray Scattering Theory

Only a short overview of the X-ray scattering technique is presented; the interested reader can refer to numerous excellent books on X-ray diffraction theory [for example refs. 9 and 10]. Lyotropic liquid crystals exhibit long range periodicity and short range disorder, resulting in a small number of sharp diffraction peaks from the sample. Theoretically, the diffracted beams should be as sharp as the incident beam, but in reality however the lines are broadened by a variety of factors such as crystallite size and disorder. In X-ray diffraction a parallel beam of X-rays with a wavelength comparable to atomic distances irradiates the sample and scatters in all directions from the electrons in the sample. When the X-ray beam is incident on a set of periodic points in a lattice, diffraction maxima occur in certain directions where the scattered X-rays are in phase. This holds for any periodic distribution of electron density, regardless of whether these points are atoms, molecules or a collection of molecules such as liquid crystals, colloids, or polymers. For centrosymmetric crystals the phase of the scattered peak can only be 0 or π with respect to the phase of the incident X-ray. Bragg visualized diffraction by imagining discrete, parallel planes in the crystal,

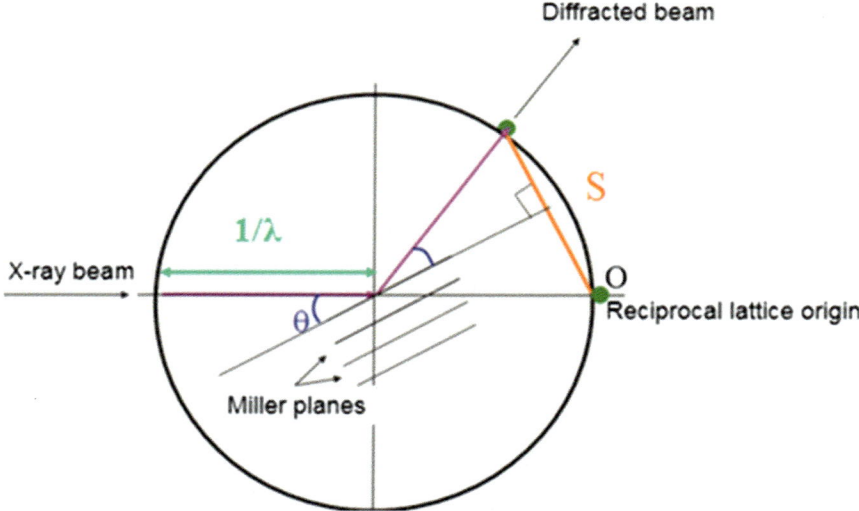

Fig. 1 Schematic of the Ewald sphere

separated by distance d, reflecting the X-ray beam. For the X-ray beams reflected from different planes to interfere constructively and give a diffraction peak, they must arrive at the detector in phase and hence the path difference between them must be an integer multiple, n, of the X-ray wavelength, λ, as described by the Bragg equation:

$$n\lambda = 2d_{hkl} \sin \theta$$

where n is the order of the reflection, λ is the wavelength of radiation, d_{hkl} is the distance between repeated planes in the lattice, and 2θ is the scattering angle.

The diffraction pattern is related to the reciprocal lattice described by the lattice planes (h, k. l). The observed Bragg peaks corresponding to specific lattice planes are due to reciprocal lattice points (defined by the crystal) intersecting the Ewald sphere (Fig. 1) which is defined by the X-ray beam. The Ewald sphere is drawn with radius $1/\lambda$ and consequently has the same properties as Bragg's law; the distance S (shown as an orange vector in Fig. 1) reduces to $S = 2 \sin\theta/\lambda$, where $S = 1/d_{hkl}$ or $q/2\pi$.

Single diffracted points will be observed from diffraction from a single crystal; however, in powder diffraction, the grains or the domains of the powder/liquid crystalline sample are randomly orientated and hence the reciprocal lattice is allowed to explore all possible orientations giving rise to diffracted rings rather than points.

2   Materials

2.1   Samples and Preparation Tools

1. Lipids and cholesterol either as a powder or a stock solution in organic solvent.
2. Silver Behenate ($CH_3(CH_2)_{20}COOAg$).
3. Deionized HPLC grade water.
4. A cylinder of nitrogen gas for solvent evaporation.
5. Liquid nitrogen and a hair dryer for freeze-thawing.
6. A vortexer.
7. Desktop centrifuge capable of spinning samples up to $4,000 \times g$.
8. 1.5 mm diameter glass X-ray capillary tubes.
9. Silicon sealant.
10. Teflon sheets 1.5 mm thick.
11. Mylar.
12. 8 and 3 mm punches (*see* **Note 1**).
13. Double sided adhesive tape.

2.2   Equipment and Data Analysis

The method described below assumes the user is familiar with the basic concepts of X-ray diffraction. SAXS and WAXS measurements can be collected using either an in-house X-ray beamline or a synchrotron source. In-house beamlines typically have a copper anode that produces $K\alpha$ (1.54 Å) and $K\beta$ (1.392) radiation. The $K\beta$ radiation is removed (for example by a nickel filter) and the $K\alpha$ beam is monochromated and focussed before hitting the sample. At a synchrotron source, X-rays are produced by electrons traveling at relativistic speeds inside the synchrotron ring. As for the in-house beamlines, the X-ray beam is passed through a monochromator and focussed before hitting the sample. The energy of the X-rays used at a synchrotron can be tuned to match the user's requirements. In our experience an energy of 17 keV (0.729 Å) is ideal for minimizing radiation damage to the sample. The resulting diffraction patterns are recorded on a detector such as a CCD camera. The sample environment should be connected to a temperature control unit. The sample to detector distance depends on the required q range of the experiment. Some synchrotron beamlines have two detectors which allow simultaneous recording of both SAXS and WAXS data. The WAXS detector sits above the sample chamber while the SAXS detector is found further downstream which allows a wide q range to be covered. For the in-house set-up in this case, the detector is positioned very close to the sample and offset so that half of the diffraction pattern appears on the detector. This allows the simultaneous recording of both the small and wide angle regions; however, the q range that can be accessed is limited to around 0.09–6.28 $Å^{-1}$.

Data processing and analysis can be carried out using packages such as FIT2D (www.esrf.eu/computing/scientific/FIT2D/). All diffraction images shown here were analyzed using the IDL-based AXcess software package, developed in-house by Dr. A. Heron [11] which is available upon request from the authors. Electron density reconstructions can be constructed using a programming language such as MATLAB (The Mathworks, Inc.), Mathematica (Wolfram Research Inc.) or FORTRAN. The CrystalDiffract software package (www.crystalmaker.com/crystaldiffract/index.html) is useful in simulating expected reflection positions and intensities for a particular spacegroup. The CrystalMaker (CrystalMaker Software Ltd, Oxford, England—www.crystalmaker.com) can be used to visualize 2D sections through the model structure of a specified spacegroup.

3 Methods

3.1 Sample Preparation

1. Individual lipid powders are lyophilised for at least 4 h to remove any water. For lipid mixtures, the desired amount of lyophilised powders (*see* **Note 2**) are then weighed and mixed together in a glass vial. The lipid mixture is dissolved in either cyclohexane with a few drops of methanol to aid dissolution, or chloroform (*see* **Note 3**).

2. The mixture is then frozen in liquid nitrogen and lyophilised for at least 24 h. Excess H_2O (<60 wt%) is then added to the samples. The samples are then subjected to a minimum of five freeze–thaw cycles in order to achieve reproducibility and sample homogeneity, and sometimes mechanical mixing may also be required to homogenize the samples.

3. The hydrated samples should also be shaken in a vortexer as well as centrifuged between each freeze-thaw cycle. When not used, the samples should be stored in a freezer.

4. Between 5 and 40 mg of sample is required for the X-ray experiments depending on whether a capillary or a Teflon spacer is used. A 3 mm diameter sample aperture in the Teflon spacer will hold 11 mm^3 of sample. To transfer the sample into a capillary, the sample is placed in the wide, open end of the capillary tube using a spatula and placed into a centrifuge so that the sample can be spun to the bottom. The capillary is then flame sealed and further sealed using silicone sealant (Dow Corning Corp.) in order to prevent any evaporation of water. If the sample is too viscous or sticky to travel down the capillary then the sample can be heated with a hairdryer to above its gel–fluid phase transition which will facilitate the transfer, or a Teflon spacer can be used instead.

5. To make the spacers, an 8 mm punch is used to cut the Teflon, followed by making a smaller 3 mm hole in the center. 3 mm

diameter holes (*see* **Note 1**) are punched into double sided adhesive tape which is then centered on the spacer and stuck on either side of it. An 8 mm diameter Mylar window is then fixed to the Teflon from one side using the double sided adhesive tape. The sample is then transferred into the spacer and the spacer is finally sealed using a Mylar window [12]. Further freeze–thaw cycles or annealing of the sample above its melting transition when in the sample container can also be applied to achieve homogeneity.

3.2 Data Acquisition and Processing

6. Exposure times vary depending on the instrument and concentration of the sample. Typical single exposures vary between 2 and 30 min for in-house beamlines and between 0.1 and 120 s at a synchrotron source. Care should be taken not to saturate the detector; if a sample is diffracting weakly then multiple images can be recorded and then added together in order to increase signal to noise. An empty sample container as well as a sample container filled with deionized HPLC H_2O can be used to correct for background. In order to be able to calculate d-spacings from samples, the sample to detector distance is required. This is determined by calibrating the system using silver behenate, which is a known calibrant with a characteristic d-spacing ($a = 58.38$ Å) [13]. The sample to detector distance is calculated using $x = \Delta y \, (d/\lambda)$ where Δy is the position of the first order peak in mm, d is the lattice parameter of AgBe and λ is the radiation wavelength. If the WAXS data are recorded on a different detector than the SAXS then NBS silicon powder can be used as the WAXS calibrant [http://www.nist.gov/mml/mmsd/functional_properties/diffraction-metrology.cfm].

7. In order to calculate d-spacings, each 2D diffraction image is opened in the software used for X-ray diffraction analysis and the center of the pattern is found by the program. The image is then radially integrated (butterfly integration) to generate a linear intensity plot (intensity against S). Each peak in the linear intensity plot is fitted, within prespecified constraints, to the desired functional form such as a modified Gaussian. The d-spacing, height, and half width of each peak can then be extracted. This process is shown in Fig. 2 using AgBe as an example. Once the calibration is finished then the process is repeated for each image in the data set. Some software packages, such as AXcess, are capable of batch processing.

3.3 Determining the Mesophase Adopted by Lipid Samples Using SAXS and Indexing

8. The long range order of the structure appears in the low angle region of the diffraction pattern (small angle X-ray scattering, SAXS) and is important in determining the mesophase of the liquid crystal. Each mesophase (lamellar, hexagonal, or cubic) possesses long range translational ordering, but with different

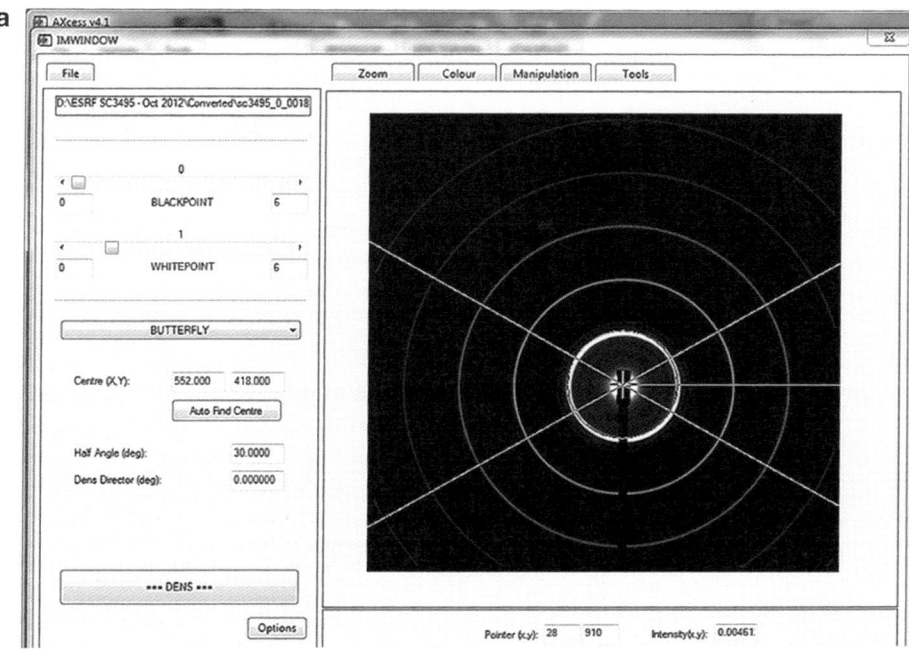

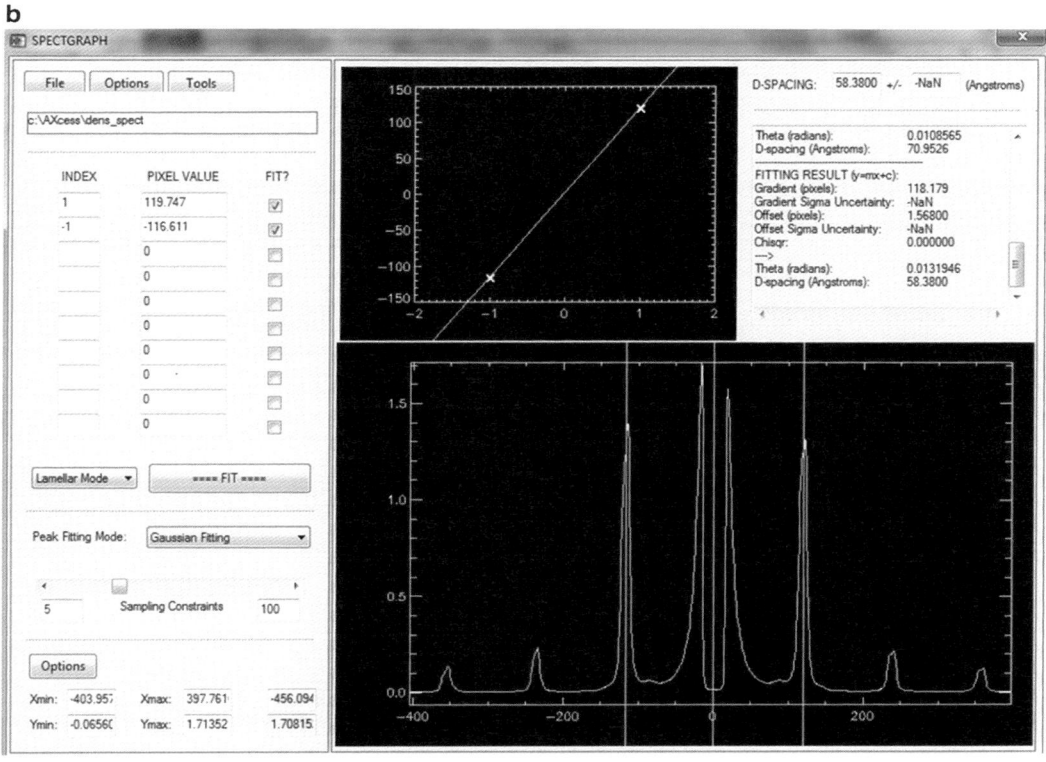

Fig. 2 (**a**) A SAXS pattern of silver behenate. The system is calibrated using silver behenate which is a known calibrant with a characteristic d-spacing 58.38 Å. The program finds the center of the pattern which is then radially integrated (*butterfly integration*) to (**b**) generate a linear intensity plot. Each peak in the linear intensity plot is fitted, within prespecified constraints, to the desired functional form, such as a modified Gaussian. The resulting peaks are indexed and fitted. The d-spacing, height and half width of each peak can then be extracted

1D, 2D, or 3D symmetry, and hence will give rise to a unique series of Bragg reflections which are characteristic of the symmetry of the phase.

9. Analysis of the small angle X-ray diffraction pattern will give the repeat spacing of the electron density profile (i.e., the repeat distance of the lattice), the d-spacing. For example, the repeat distance of a liquid crystal in a lamellar phase is the sum of the lipid bilayer thickness and the thickness of one adjacent interlamellar water layer. The d-spacing can be used to provide information about the thickness of the bilayer and hydration limits.

10. The spacegroup and lattice parameter of the phase adopted is determined by the position of the Bragg peaks, whose reciprocal spacings ($S_{hkl} = 1/d_{hkl}$) are in characteristic ratios (*see* Table 1). Some examples are described below. In the equations given below, S_{hkl} is the reciprocal spacing of the set of lattice planes (h, k, l) characterized by the Miller indices h, k, and l. Lamellar phases of lipids, which consist of alternating bilayers separated by water, give rise to diffraction patterns with Bragg peaks in the ratio of 1, 2, 3, 4…as the interplanar spacing equation is given by:

$$S_h = \frac{h}{d}$$

11. 2D hexagonal phases, which consist of monolayers of lipids arranged in cylinders which are then packed onto a 2D hexagonal lattice, give rise to diffraction patterns with Bragg peaks in the ratio of 1, $\sqrt{3}$, 2, $\sqrt{7}$, 3…as the interplanar spacing equation is given by:

$$S_{hk} = \frac{2\left(h^2 + k^2 - hk\right)}{\sqrt{3}a}$$

12. There are 36 possible 3D cubic phases [14]; however, only a handful of different space groups have so far been identified in lipid mixtures. For a simple cubic lattice, Bragg peaks appear in the ratio 1, $\sqrt{2}$, $\sqrt{3}$, 2, $\sqrt{5}$, $\sqrt{6}$, $\sqrt{8}$… as the interplanar spacing equation is given by:

$$S_{hkl} = \frac{\left(h^2 + k^2 - l^2\right)^{1/2}}{a}$$

13. There are 27 possible spacegroups with a 3D hexagonal lattice [14]; however, only one spacegroup belonging to this lattice type has been found in a lipid system [15]. The Bragg peak ratio for this lattice type is more complicated as h and k are

Table 1
Examples of mesophases adopted by lipids, along with the allowed Bragg reflections of the spacegroup and characteristic diffraction patterns

Phase	Ratio of peaks and corresponding Miller indices					Example diffraction patterns
Lamellar	1: (100)	2: (200)	3: (300)	4: (400)	5... (500)	
2D—Hexagonal (p6mm) Figure taken from [14]	1: (10)	√3: (11)	2: (20)	√7: (21)	3: (30)	√12: (22)	√13... (31)	

(continued)

Table 1
(continued)

Phase	Ratio of peaks and corresponding Miller indices							Example diffraction patterns
Im3m 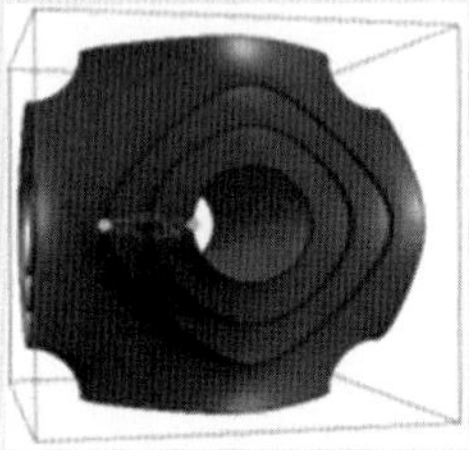 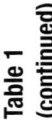	$\sqrt{2}$: (110)	$\sqrt{4}$: (200)	$\sqrt{6}$: (211)	$\sqrt{8}$: (220)	$\sqrt{10}$: (310)	$\sqrt{12}$: (222)	$\sqrt{14}$...: (321)	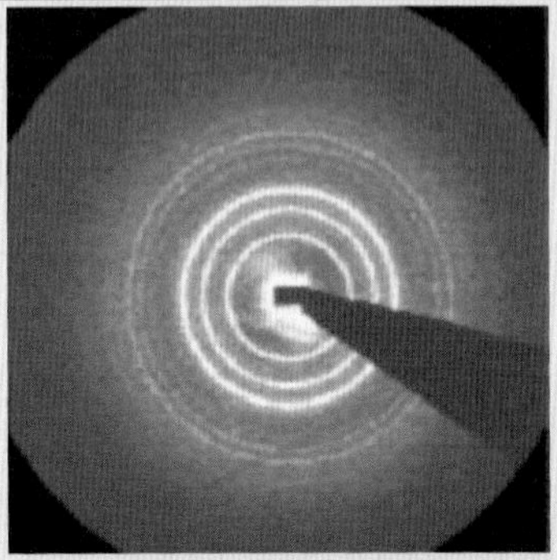
Pn3m 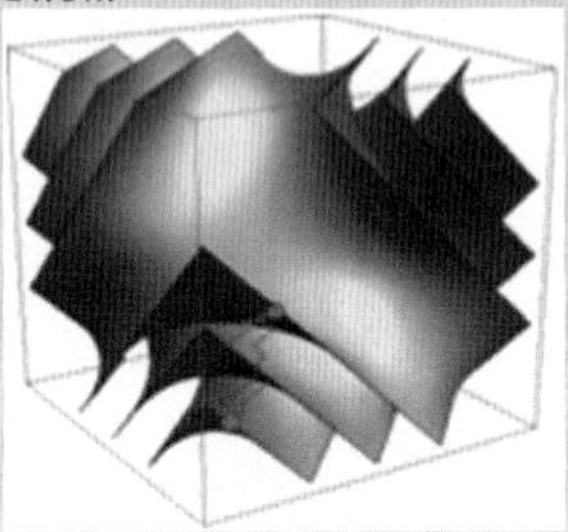	$\sqrt{2}$: (110)	$\sqrt{3}$: (111)	$\sqrt{4}$: (200)	$\sqrt{6}$: (211)	$\sqrt{8}$: (220)	$\sqrt{9}$: (221)	$\sqrt{10}$...: (310)	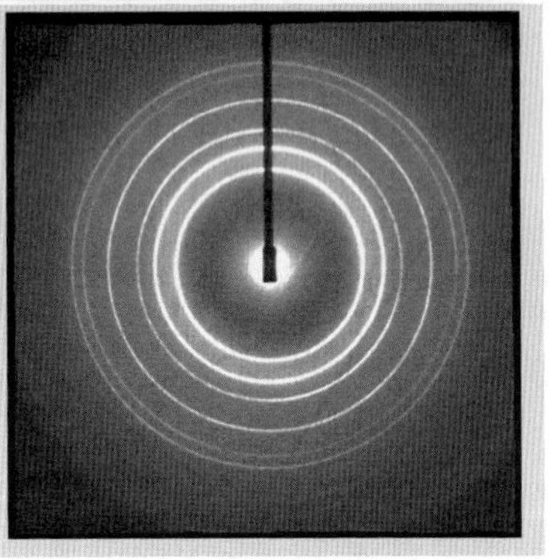

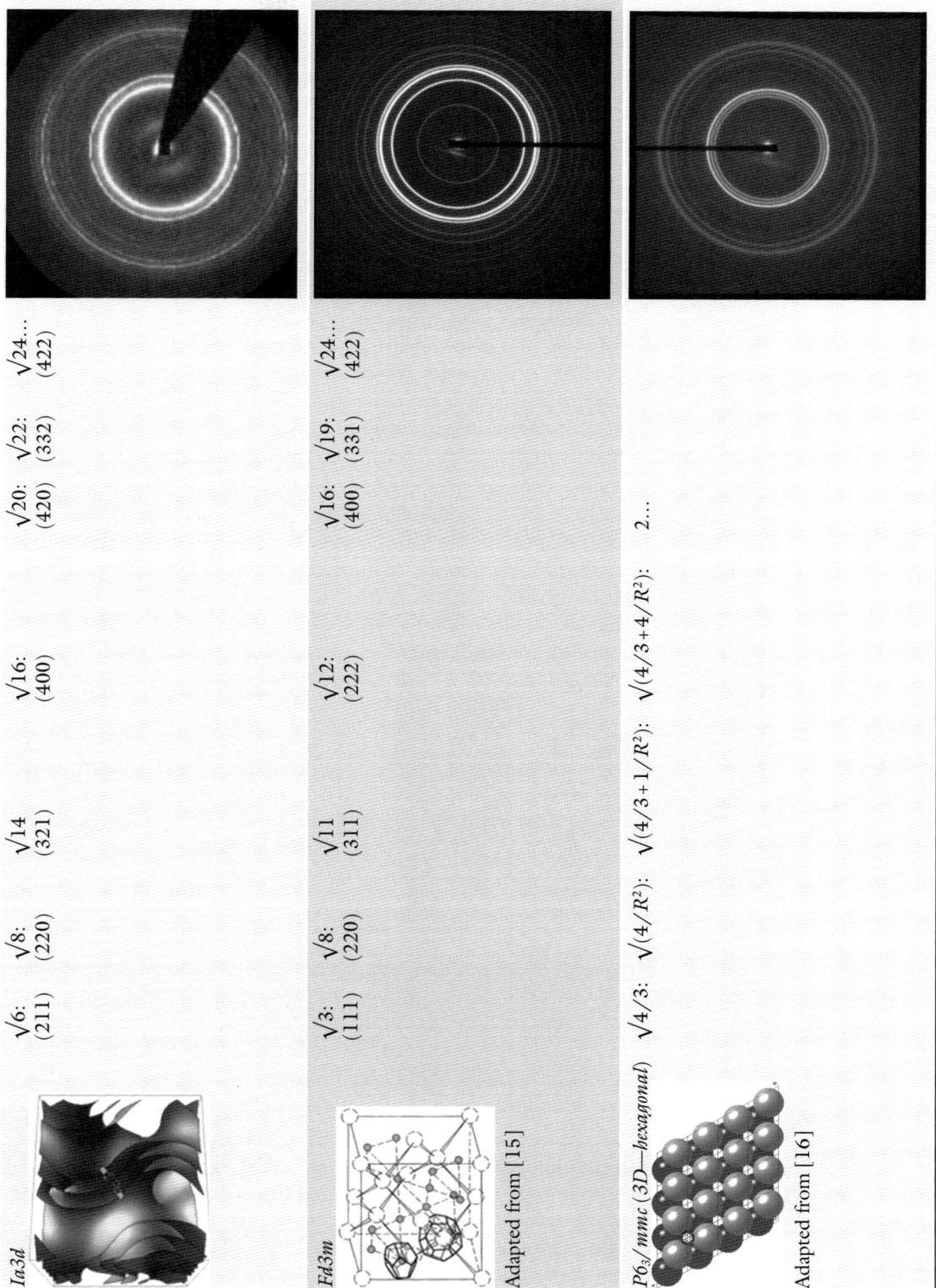

Ia3d

$\sqrt{6}$:
(211)

$\sqrt{8}$:
(220)

$\sqrt{14}$
(321)

$\sqrt{16}$:
(400)

$\sqrt{20}$:
(420)

$\sqrt{22}$:
(332)

$\sqrt{24}\ldots$
(422)

Fd3m

$\sqrt{3}$:
(111)

$\sqrt{8}$:
(220)

$\sqrt{11}$
(311)

$\sqrt{12}$:
(222)

$\sqrt{16}$:
(400)

$\sqrt{19}$:
(331)

$\sqrt{24}\ldots$
(422)

Adapted from [15]

P6_3/mmc (3D—hexagonal)

$\sqrt{4/3}$: $\sqrt{(4/R^2)}$: $\sqrt{(4/3+1/R^2)}$: $\sqrt{(4/3+4/R^2)}$: $2\ldots$

Adapted from [16]

permutable but *l* is not and there are two lattice parameters, *a* and *c*. The interplanar spacing equation takes the form:

$$S_{hkl} = \frac{4}{3}\left[\frac{\left(h^2 + k^2 + hk\right)}{a^2}\right] + \frac{l^2}{c^2}$$

14. Once the lattice type has been identified then the crystallographic spacegroup to which the phase belongs has to be identified. As different phases have different symmetry elements, this allows determination of the phase adopted provided that there are a sufficient number of Bragg reflections (*see* **Note 4**). This however is not a trivial task as due to the large thermal disorder in these systems only a few small angle peaks are normally observed. For unaligned samples the exact spacegroup cannot be absolutely distinguished experimentally without forming a monodomain sample and only the cubic aspect can be identified with certainty [16].

15. If a monodomain sample is available, this uncertainty can be resolved as depending on the spacegroup, *hkl* reflections might not fully permutable; the observed intensity of an *hkl* reflection might not be the same as the intensity of a *khl* reflection. Although the extinction symbol can be usually identified with certainty there will be a few spacegroups that fall under it. For example, for the extinction symbol P-c there are five possible spacegroups, three with a hexagonal and two with a trigonal symmetry, thus making them indistinguishable from systematic absences using powder diffraction data.

16. There are a few ways to help assign a spacegroup for lipid phases. It is common practice with lyotropic liquid crystal systems to assign the space group with the highest symmetry [17]. Moreover, the relative intensities for each peak can help elucidate the correct space group determination.

17. The expected reflection positions and intensities for a spacegroup can be simulated using the CrystalDiffract software package and see if they match to the ones obtained experimentally. Moreover, if a phase with the same spacegroup but of different sign of interfacial curvature (i.e., oil in water versus water in oil) has a known diffraction pattern, then the intensities of this phase could be used as a comparison, invoking Babinet's Principle.

18. Each Bragg reflection for a certain structure can be indexed using the interplanar spacing equation, and using the lattice parameter(s) value(s), all possible *hkl* combinations permitted by the structure factor can be determined. Although this is a valid method which is commonly used to index a pattern, it can sometimes be problematic as peaks can accidently be

Table 2
SAXS data for the sample used to index a Pn3m cubic phase, where *hkl* are Miller indices and θ is the experimentally determined scattering angle for each peak

d-spacing (Å)	$\sin^2\theta$ ($\times 10^{-5}$)	Normalize	Clear fractions	*hkl*?	$\frac{\sin^2\theta}{h^2 + k^2 + l^2}$
115.984	1.2697	1	2	110	6.3485E-06
94.5706	1.90978	1.504125	3	111	6.3659E-06
81.1688	2.59249	2.041822	4	200	6.4812E-06
66.6762	3.84197	3.025899	6	211	6.4033E-06
57.1755	5.22488	4.11506	8	220	6.5311E-06
54.824	5.6827	4.475636	9	221	6.3141E-06
51.8672	6.34908	5.000467	10	310	6.3491E-06

missed out if a reflection is too weak and hence leading to incorrect assignment of the *hkl* values.

19. In order to confirm that each Bragg reflection has been correctly assigned to its corresponding Miller plane, an analytical method can also be used. As an example, a cubic and 3D hexagonal lattice will be shown here.

20. Combining the cubic interplanar spacing equation with Bragg's Law yields:

$$\frac{\lambda^2}{4a^2} = \frac{\sin^2\theta}{\sqrt{h^2 + k^2 + l^2}}$$

The lattice parameter is constant as it is determined by the crystal, λ is a constant as it is fixed experimentally, consequently even though the angle θ varies with values for *hkl*, the ratio between the two varies so as to remain constant. Hence to index the sample, values of $\sin^2\theta$ for each Bragg peak are calculated and then normalized by dividing through with the first value ($\sin^2\theta_n/\sin^2\theta_1$). The fractions are then cleared by multiplying by a common number to generate whole numbers.

21. Finally the *hkl* values, which if expressed as $h^2 + k^2 + l^2$, would generate the whole numbers generated by clearing the fractions are deduced. This is shown in Table 2 for indexing a Pn3m cubic phase. By doing so, then systematic absences become evident which help identify the type of lattice and spacegroup. For example, for a primitive (P) cubic lattice all *hkl* values are allowed, for a body centered cubic (BCC) lattice the sum of $h + k + l$ must be an even number and for a face

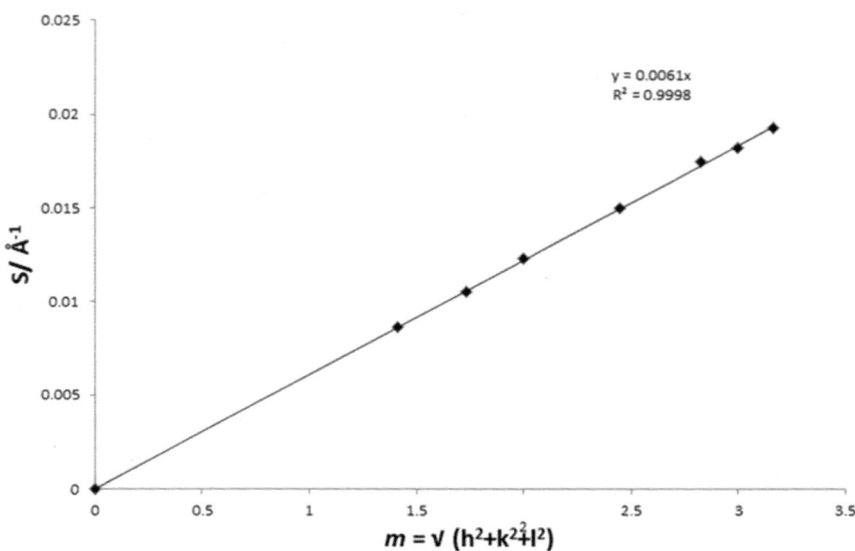

Fig. 3 The observed reciprocal spacings versus $\sqrt{(h^2 + k^2 + l^2)}$ for DOPC:DOG:CHOL 17:3:4 (60 wt% water) at 68 °C and 1 bar fit perfectly to a linear plot through the origin confirming the indexing as cubic spacegroup Pn3m

centered cubic lattice the values of h, k, and l must be either all even or all odd.

22. To further ascertain that the indexing is correct two very similar approaches can be made. (A) For each θ value, the value of $\sin^2\theta/(h^2 + k^2 + l^2)$ on the basis of the assumed hkl values is computed. The result should be a constant for all the values if the indexing is correct, with the constant equal to $\lambda^2/4a^2$. (B) Alternatively, the indexing of the diffraction data can be shown as a plot of the reciprocal d-spacings $(1/d_{hkl})$ versus $m = \sqrt{(h^2 + k^2 + l^2)}$. For a correctly indexed cubic phase, such a plot should be linear, with a gradient equal to the reciprocal of the lattice parameter of the phase, and should pass through the origin (*see* Fig. 3).

23. The second example is for a 3D hexagonal phase belonging to spacegroup P6₃/mmc. Combining the hexagonal interplanar spacing equation with Bragg's law yields:

$$\sin^2\theta = \frac{\lambda^2}{4a^2}\left[\frac{(h^2 + k^2 + hk)}{a^2} + \frac{l^2}{c^2}\right]$$

The a and c parameters are constant as they are determined by the crystal, λ is a constant as it is fixed experimentally, consequently even though the angle θ varies with values for hkl,

Table 3
SAXS data for the sample used to determine the quotient A, where hkl are Miller indices and θ is the experimentally determined scattering angle for each peak

Peak	$\theta/°$	$\sin\theta$	$\sin^2\theta$	$\dfrac{\sin^2\theta}{3}$	$\dfrac{\sin^2\theta}{4}$	$\dfrac{\sin^2\theta}{7}$	$\dfrac{\sin^2\theta}{9}$	hkl
1	0.348	0.006065	3.6784×10^{-05}	1.22613E-05	9.196E-06	5.25486E-06	4.08711E-06	100
2	0.370	0.006458	4.1702×10^{-05}	1.39005E-05	1.04254E-05	5.95736E-06	4.63351E-06	
3	0.393	0.006859	4.70471E-05	1.56824E-05	1.17618E-05	6.72102E-06	5.22746E-06	
4	0.507	0.008840	7.81452E-05	2.60484E-05	1.95363E-05	1.11636E-05	8.6828E-06	
5	0.602	0.010498	0.000110207	3.67357E-05	2.75518E-05	1.57439E-05	1.22452E-05	110
6	0.654	0.011414	0.000130284	4.3428E-05	3.2571E-05	1.8612E-05	1.4476E-05	
7	0.699	0.012208	0.000149042	4.96807E-05	3.72605E-05	2.12917E-05	1.65602E-05	020
8	0.707	0.012339	0.000152255	5.07517E-05	3.80637E-05	2.17507E-05	1.69172E-05	
9	0.7185	0.012540	0.000157248	5.2416E-05	3.9312E-05	2.2464E-05	1.7472E-05	
10	0.739	0.012889	0.000166124	5.53746E-05	4.15309E-05	2.3732E-05	1.84582E-05	

the ratio between the two varies as to remain constant. Hence we can write:

$$\sin^2\theta = A\left(h^2 + k^2 + hk\right) + Cl^2$$

where $A = \dfrac{\lambda^2}{3a^2}$ and $C = \dfrac{\lambda^2}{4c^2}$

Since $h^2 + k^2 + hk$ values adopt values such as 0, 1, 3, 4... and l^2 values adopt 0, 1, 4, 9... the observed reflections can be divided by these values and a common quotient can be determined.

24. The common quotient A, which is used to determine the $hk0$ type reflections, was found to be 3.67×10^{-5} (Table 3) and the value for C, used to determine the $00l$ type reflections, was found to be 1.04×10^{-5} (Table 4). Table 5 summarizes the combinations of A and C which were used to index the diffraction pattern.

Table 4
SAXS data for the sample used to determine the quotient C, where hkl are Miller indices and θ is the experimentally determined scattering angle for each peak

Peak	$\theta/°$	$\sin^2\theta$	$\sin^2\theta - A$	$\sin^2\theta - 3A$	hkl
1	0.348	3.6784×10^{-05}	–	–	
2	0.370	4.1702×10^{-05}	4.91754E-06	–	002
3	0.393	4.70471E-05	1.02631E-05	–	
4	0.507	7.81452E-05	**4.13612E-05**	–	012
5	0.602	0.000110207	7.34232E-05	–	
6	0.654	0.000130284	9.35001E-05	1.99321E-05	
7	0.699	0.000149042	0.000112258	3.869E-05	
8	0.707	0.000152255	0.000115471	**4.1903E-05**	112
9	0.7185	0.000157248	0.000120464	4.68961E-05	
10	0.739	**0.000166124**	0.00012934	5.57717E-05	004

Table 5
Indexing of the diffraction pattern using the quotients A and C, where θ is the experimentally determined scattering angle for each peak, $\sin^2\theta_{obs}$ and $\sin^2\theta_{calc}$ are the experimentally determined and calculated $\sin^2\theta$ values respectively and hkl are the Miller indices assigned

Peak	$\theta/°$	$\sin^2\theta_{obs}$	$A + C$	$\sin^2\theta_{calc}$	hkl
1	0.348	3.6784×10^{-05}	$1A + 0C$	3.6784E-05	100
2	0.370	4.1702×10^{-05}	$0A + 4C$	4.15889E-05	002
3	0.393	4.70471E-05	$1A + 1C$	4.71669E-05	011
4	0.507	7.81452E-05	$1A + 4C$	7.82996E-05	012
5	0.602	0.000110207	$3A + 0C$	0.000110391	110
6	0.654	0.000130284	$1A + 9C$	0.000130483	013
7	0.699	0.000149042	$4A + 0C$	0.000147131	020
8	0.707	0.000152255	$3A + 4C$	0.00015204	112
9	0.7185	0.000157248	$4A + 1C$	0.000157686	021
10	0.739	0.000166124	$0A + 16C$	0.000166574	004

3.4 Determining Interfacial Curvature

25. The resulting phase behavior of a system is a balance of a large number of forces that can cause the monolayer to bend around a pivotal surface, away or towards the water interface, to form Type I (normal, positive interfacial curvature) or Type II (inverse, negative interfacial curvature) phases respectively. Although most lipids rarely form Type I phases due to their large chain splay, some lipids such as lysophospholipids do form Type I phases.

26. Babinet's Principle states that the diffraction pattern from an opaque body is identical to that from a hole of the same size and shape except for the overall forward beam intensity hence it is not possible to distinguish the sign of the interfacial curvature of the phase from the scattering pattern. Although the sign of the interfacial curvature cannot be ascertained by the diffraction data, there are certain approaches that can be used to establish this with some certainty [16, 18]:

    (a) Observation of a non-lamellar phase in coexistence with excess water is a strong indication of a Type II structure. Addition of water to a Type I phase will generally cause the system to break up into a micellar solution beyond a certain limiting water content.

    (b) If a phase of unknown type (I or II) is neighboring a non-lamellar phase of known type in the phase diagram then it is highly likely that the unknown phase will also be of the same type (I or II).

    (c) By calculating the interfacial area per molecule (which can be extracted from the diffraction data) and plotting it against water concentration, it should increase or stay constant with increasing water content, but never decrease if the assumed type is correct. Such plots stay relatively constant for Type I but increase substantially for Type II phases. A Type I phase can swell at constant surface area; however, a Type II phase must increase its interfacial area per molecule as water penetrates the hydrophilic cores of the phase. Favorable interactions between the lipid head-groups and the water molecules typically cause the head-groups to expand laterally, hence the interfacial area per molecule can never decrease with increasing water content.

    (d) The value of the interfacial area per molecule at the lipid–water interface is usually lower for a Type II non-lamellar phase than a neighboring lamellar phase in the phase diagram either by driving the transition between the two phases by varying the composition or varying a thermodynamic parameter such as temperature. Type I phases show the opposite trend.

    (e) Comparing the intensities of the Bragg peaks of the phase for a range of water contents with ones calculated from models of the structure should provide clear evidence for whether the phase is Type I or Type II. This is done by plotting the experimentally observed intensities of the Bragg peaks against q and seeing how they fit to the calculated underlying transforms using parameters derived from the known dimensions of the phase at each water content.

(f) Lamellar phases usually appear at higher water contents than non-lamellar phases for a Type II system and vice versa for Type I systems. However, it should be noted that this rule does not always hold and should be applied with caution. Monoacylglycerols for example form Type II non-lamellar phases at a higher water content than the lamellar phase [19].

3.5 Determining the Type of Lamellar Phase by Wide Angle X-Ray Scattering

27. In the wide-angle X-ray scattering (WAXS) region, the short range order of the system, i.e., the packing of the lipid acyl chains, is probed. WAXS can differentiate between the various lamellar phases, having different chain packings (L_c, $L\beta$, $L\alpha$, and L_o).

28. The lamellar crystalline phase (L_c) can be identified by the presence of a number of sharp wide-angle diffraction peak(s). In the lamellar gel phase ($L\beta$) phase, the hydrocarbon chains are essentially fully extended and adopt a hexagonal packing, which gives rise to a relatively sharp and symmetrical diffraction peak around 4.2 Å. The tilted gel phase ($L\beta'$) gives rise to a sharp reflection, with a broad shoulder to higher q. As the fluid lamellar phase ($L\alpha$) is disordered and the acyl chains undergo rapid *trans-gauche* isomerisation, the liquid-like chain packing results in a broad diffraction peak centered around 4.6 Å. It is difficult to identify the liquid-ordered L_o phase from WAXS due to its similarity to the $L\alpha$ phase: the L_o phase is also characterized by fast lateral diffusion and disordered chain packing, resulting in a broad diffraction pattern very similar to that of the $L\alpha$ phase. Although it is difficult to distinguish this phase, it usually shows a broad peak between 4.2 and 4.6 Å and the peak usually has a larger temperature dependence compared to the $L\alpha$ phase.

3.6 Electron Density Reconstructions of Lamellar and Inverted Hexagonal Phases

29. The intensities of the Bragg peaks are determined by the distribution of electron density in the unit cell which is constrained by the symmetry of the spacegroup (*see* **Note 4**). From the peak intensities one can deduce the structure factors, and low resolution electron density maps can be reconstructed by Fourier transformation.

30. The intensity of a diffracted Bragg peak has two distinct contributions: the first is the contribution from the individual atoms in the unit cell (atomic form factor, f_n) and the second is

from the arrangement of these atoms in the molecule. The intensity contribution from the atom positions is known as the structure factor and is a function of all the atomic coordinates within the unit cell. The structure factor is the Fourier transform of the electron density within a unit cell. The structure factor is given by:

$$F(hkl) = \Sigma_n f_n \exp 2\pi i (hx_n + ky_n + lz_n)$$

31. It should be noted that the structure factor $F(hkl)$ has both a resultant total amplitude and a resultant total phase:

$$F(hkl) = |F(hkl)| \exp i\varphi(hkl)$$

If all parts of the crystal scatter independently, the intensity of any wave is proportional to the square of its amplitude. It is important to note that for a point, $\mathbf{x}$, in a reciprocal lattice which gives rise to a reflection there will be a reflection of equal magnitude (the Friedel mate) for point $-\mathbf{x}$ and these will have the same amplitude but will be π out of phase. This is known as Friedel's law which states that the squared amplitude is centrosymmetric; however, the phase is antisymmetric. As only the intensity is observable, the phase is lost and cannot be determined experimentally, so the original structure cannot be directly reconstructed purely from a single scattering experiment.

$$I(hkl) = I(\bar{h}\ \ \bar{k}\ \ \bar{l}) \alpha\ |F(hkl)|^2$$

The structure factor reduces to:

$$F(hkl) = \Sigma 2 f_n \cos 2\pi (hx_n + ky_n + lz_n)$$

and since this is a real quantity, the phase of the reflection, φ_{hkl}, is 0 or π for a centrosymmetric crystal and the phase problem condenses to ascertaining the signs of the structure factors:

$$F(hkl) = \pm|F(hkl)|$$

32. The structure factor equation given above assumes point scatterers in the unit cell. However, the structure factor can also be calculated by considering the unit cell being occupied by a smoothly varying electron density and taking into account scattering of all volume elements. Assuming the structure is infinitely periodic in all three dimensions, the electron density can be represented by a three-dimensional Fourier series.

$$\rho(xyz) = \Sigma_h \Sigma_k \Sigma_l C(h'k'l') \exp[2\pi i(h'x + k'y + l'z)]$$

where $C(h'k'l')$ are the Fourier coefficients.

For a centrosymmetric crystal it can be shown that this becomes:

$$\rho(xyz) = 1 / V_{-\infty} \Sigma\Sigma\Sigma^{\infty} F(hkl) \cos\left[2\pi(hx + ky + lz) + \varphi_{hkl}\right]$$

where the phase angle, φ_{hkl}, of the Bragg reflection is either 0 or π.

33. Although the amplitude of the Bragg reflections can be measured, the phases cannot. Determining these phases is essential to structure reconstruction but is not a trivial task. The number of phase combinations (φ-sets) is 2^{n-2}, where n is the number of the individual Bragg reflections obtained in the experiment. For example, if 10 Bragg peaks are observed, the number of possible phase combinations becomes 256.

34. There are a number of methods that have been developed to deduce the phasing of the structure factors (*see* **Note 5**). In some cases, and in particular lamellar mesophases, water swelling experiments can be used to deduce the phasing, and this approach has been extensively used for such systems [20]. In this method structures at different hydrations are compared, as the lattice parameter will change at different water concentrations. By doing so, structure factors along the reciprocal space axis can be obtained and used to trace out the continuous Fourier transform of the bilayer.

35. It is assumed that the integrated difference of the square of the electron densities should be at a minimum if the correct phasing has been chosen, as a small change in hydration should only slightly shift the electron density. For each possible phase combination, $\pm|F(hkl)|$, continuous transforms are calculated using the sampling theorem [21] and the transform that best matches the structure factors is deemed the correct combination [22].

36. For 3D structures, such as the bicontinuous cubic mesophases, the most successful approach has been using the $\langle\rho_e^4\rangle$ technique, where ρ_e is the electron density. In this method, two structures of the same chemical composition are compared and the value of $\langle\rho_e^4\rangle$ should be the same for both structures. Hence, the structure of known phasing can be used to determine the phasing of the other [23].

37. Before taking the square root of the intensities of the Bragg peaks in order to convert them into structure factor moduli ($|F(hkl)|$), some corrections to the intensity of each Bragg peak have to be applied first. Theoretically, the diffracted beams should be infinitely sharp delta functions; in reality however, the lines are broadened by a variety of processes such as crystal size and disorder.

38. The integrated intensities must be corrected for certain factors. Factors that need to be considered when correcting the

integrated intensity of the Bragg peaks are the Lorentz, multiplicity, absorption, extinction, and polarization corrections. Absorption and extinction can usually be ignored for lipid systems due to the geometry of their adopted phases and the fact that they scatter weakly. Both can be minimized by using a thin sample. Polarization can affect integrated intensities. Usually at synchrotron sources, although the X-ray beam is polarized, polarization occurs perpendicular to the plane of scattering and since the monochromator and beam function in the vertical plane, the polarization is unchanged from source to detector and no intensity correction is required.

39. In-house beamlines however produce an unpolarized beam which becomes partially polarized when it hits the monochromator. In this case, a term which combines polarization in-plane and perpendicular to the plane of scattering is applied to the intensity and is given by:

$$\frac{1 + \cos^2(2\theta)}{2}$$

For small angle scattering, however, the polarization factor varies from unity (polarization perpendicular to the plane of scattering, no loss in intensity) by less than 1 % and hence can be ignored.

40. The two most common corrections applied to powder small angle diffraction data of lipids are the Lorentz and multiplicity corrections. The relative intensities are corrected by a factor of q^2 (Lorentz correction) [24], to take account of the SAXS geometry: the reciprocal lattice points in a powder sample are smeared over a spherical surface with area proportional to q^2; however, this only intersects the Ewald sphere, and hence only diffracts along a ring with length proportional to q, giving rise to a correction factor of q. Furthermore, full power diffraction ring patterns are recorded on a two-dimensional detector but conventionally during radial integration to give a one-dimensional diffraction plot, data are normalized to q, so results are comparable with those obtained using a linear detector, which gives a further correction factor of q. Combining these two considerations gives the full Lorentz correction.

41. Finally, the intensity has to be corrected for multiplicity (m_{hkl}). Single diffracted points will be observed from diffraction from a single crystal; however, in powder diffraction, the grains or domains of the powder/liquid crystalline sample are randomly orientated and hence the reciprocal lattice is allowed to explore all possible orientations giving rise to diffracted rings rather than points. Consequently, the d-spacings for related reflections are often equivalent. For example, in a simple cubic

lattice the following planes have the same d-spacing even though they have different orientations:

$$100, 010, 001, \bar{1}00, 0\bar{1}0, 00\bar{1}$$

These sets of planes will add to the intensity of the diffraction ring and hence this reflection is said to have a multiplicity of 6, and hence the intensity has to be divided by this number. The multiplicity factor is assessed by finding the number of variations in positions and sign which can be given to $\pm h$, k, and l that have planes with the same d-spacing. As a further example, the 110 peak in a simple cubic lattice has a multiplicity of 12 as $\pm h$, k, and l are fully permutable. This will depend on the spacegroup as $\pm h$, k, and l may not be fully permutable; for example for an Ia3d bicontinuous cubic phase $F_{hkl} = F_{hlk}$ only if $h + k + l = 4n$ and $F_{hkl} = -F_{hlk}$ only if $h + k + l = 4n + 2$.

3.7 Electron Density Reconstruction of the Lamellar Phase

42. The intensities of the peaks from a lamellar phase are Lorentz and multiplicity corrected (for a lamellar phase the latter correction is unity), normalized to the first order amplitude, and converted to structure factor moduli. For a lamellar phase the electron density is given by [25]:

$$\rho(x) = \left(\frac{2}{d}\right) \sum_1^h \varphi(h) |F(h)| \cos\left(\frac{2\pi hx}{d}\right)$$

where x is the distance from the center of the bilayer, d the lamellar repeat spacing, $\varphi(h)$ is the phase for the h-th order, and can be $+1$ or -1, and $|F(h)|$ is the structure factor amplitude.

43. The electron density reconstructions are based on an arbitrary electron density scale. Based on the structure of the lamellar phase, the most electron rich region is the phosphorus headgroup and hence these regions should correspond to maxima. The terminal methyl groups from the hydrocarbon chains should have the lowest electron density ($0.16e/\text{Å}^3$) and hence appear as minima in the electron density profile. The maxima and minima should be bound together with an intermediate electron density due to the hydrocarbon chain regions. Finally, there should be a decrease in electron density outside the headgroups as that represents the water region ($0.33e/\text{Å}^3$).

44. An example of various phase combinations, taken from [26], for the Lα phase of 1,2-di-O-[trans-9,10-octadecenoyl ("elaidoyl")]-3-O-phosphatidylethanolamine-sn-glycerol (18:1$t\Delta$9-PE) is shown in Fig. 4. As noted earlier, the maxima (arising from the phospholipid headgroups) should be followed by a decrease in the electron density in the water region hence this leaves just two phasing combinations (– – + – and – + + –) as the correct ones. The correct phasing combination

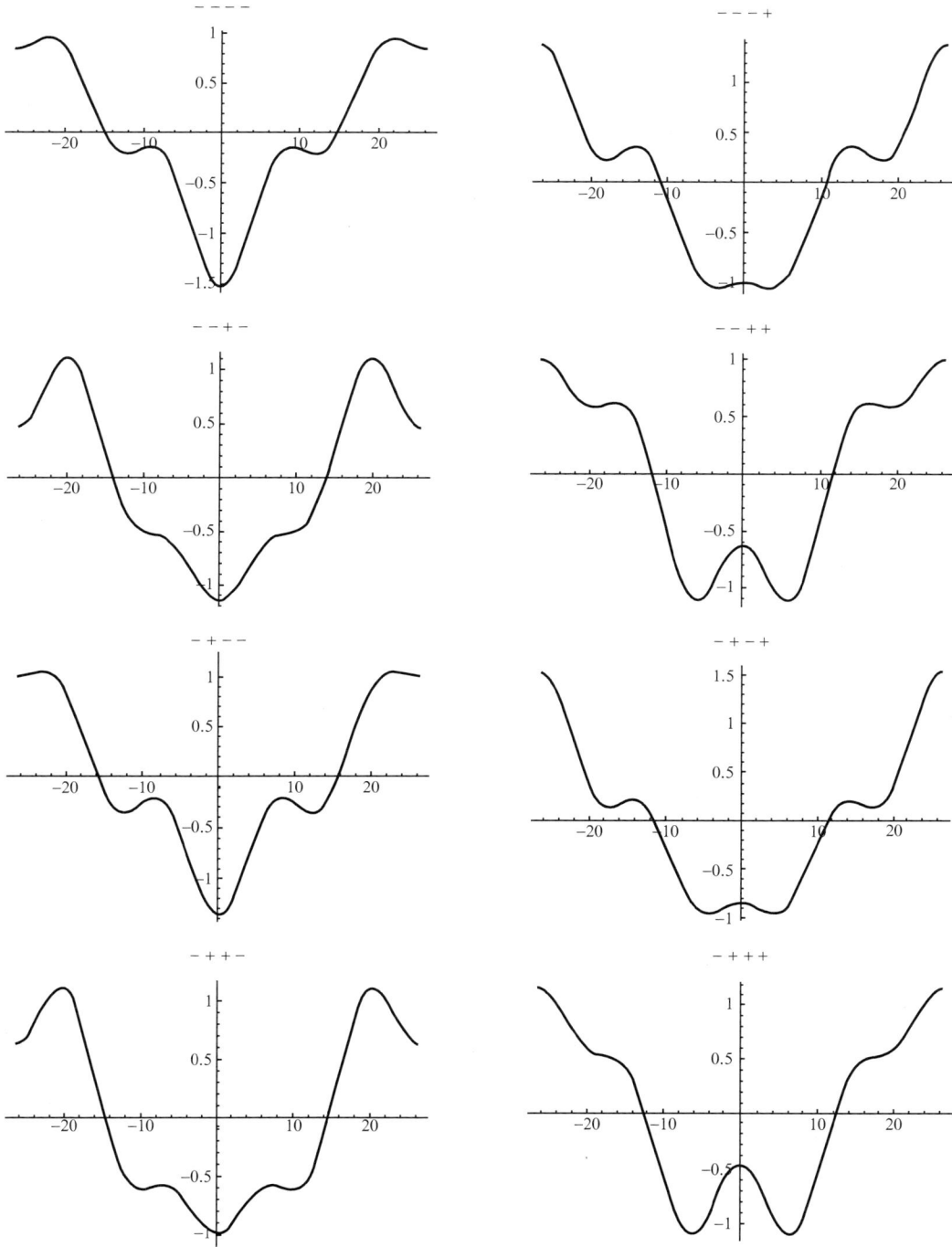

Fig. 4 Using some of the possible phase combinations to generate electron density reconstructions of the $L\alpha$ phase of (18:1$t\Delta$9-PE). The y-axis represents electron density and has arbitrary units and the x-axis is in Angstroms. The correct phase choice was judged to be $- - + -$. Taken from ref. 26

was judged to be − − + − as it has a deeper dip towards the water region and also because a very structurally similar lipid is known to have the same phase combination.

45. Electron density reconstructions of lamellar phases can be very useful in determining the structural parameters of a bilayer. They can be used for example to determine the thickness of unfixed bilayers in phase separated systems showing liquid ordered (L$_o$) and fluid lamellar (Lα) phase coexistence.

3.8 Electron Density Reconstruction of the Inverse Hexagonal (H$_{II}$) Phase

46. The intensities of the peaks from an H$_{II}$ phase are Lorentz and multiplicity corrected, normalized to the first order amplitude, and converted to structure factor moduli. For an H$_{II}$ phase, the peaks (21), (31), (32), and (41) have a multiplicity of 12 and the remaining peaks have a multiplicity of 6. For a 2D-hexagonal phase (planegroup p6mm) the electron density is given by [27, 28]:

$$\rho = \frac{1}{A_c}\left\{ F(00) + 2\left\{ \sum_1^\infty hF(h0)\cos 2\pi hX + \sum_1^\infty kF(0k)\cos 2\pi kY \right\} \right.$$

$$+ 2\left\{ \sum_1^\infty h\sum_1^\infty kF(hk)\cos 2\pi(hX + kY) \right.$$

$$\left. \left. + \sum_1^\infty h\sum_1^\infty kF(-hk)\cos 2\pi(hX - kY) \right\}\right\}$$

where A_c is the volume of the unit cell.

47. For the hexagonal phase we have to convert from hexagonal to Cartesian coordinates, and hence the following adjustment has to be made to the X and Y axis and use X' and Y' in the electron density equation instead:

$$X' = \frac{3}{2}X - \frac{\sqrt{3}}{2}Y$$

$$Y' = \sqrt{3}Y$$

48. The 3D electron density reconstruction and 2D contour plot showing the hexagonal unit cell of the inverse hexagonal (H$_{II}$) phase using Lorentz and multiplicity corrected amplitudes and phasing calculated by Harper et al. [26] for an unsaturated phosphatidylethanolamine (18:1tΔ9-PE) at 85 °C is shown in Fig. 5.

4 Notes

1. Punches are used to make Teflon spacers for the samples. The outer diameter will depend on the size of the Teflon spacer holder available, which in this case is 8 mm. The central sample

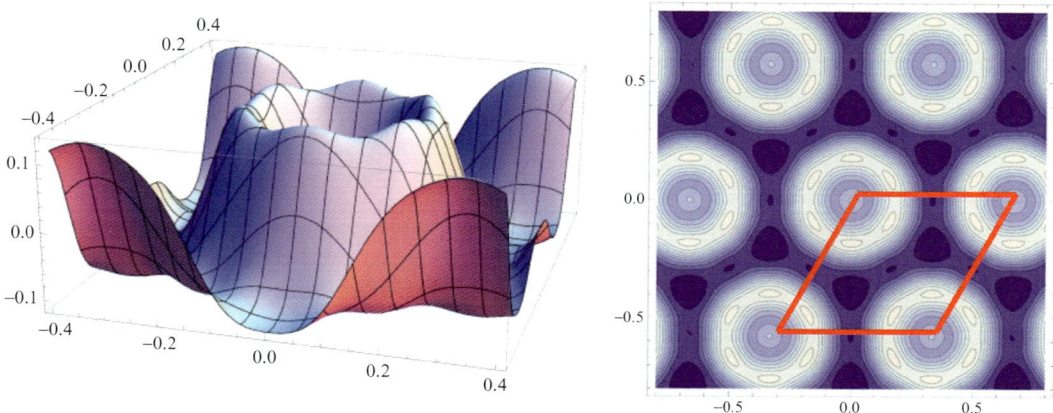

Fig. 5 *Left*: A 3D view of the electron density reconstruction. *Right*: A 2D contour plot of the electron density. The *red rhombus* shows the hexagonal unit cell

aperture (inner diameter of the Teflon spacer) can be adjusted to the user's requirements; however, below 1.5 mm sample, alignment in the X-ray beam becomes difficult.

2. For very accurate lipid concentration calculations the lipid crystal structures should be checked to see if any water molecules in the headgroup are present in the crystal structure. For example phosphatidylcholines and sphingomyelins tend to have two water molecules attached to their headgroup and hence the mass of two water molecules should be added to the molecular weight of the lipid when calculating mole ratios. Cholesterol is assumed to be anhydrous.

3. Most lyophilizers cannot cool down below –20 °C and hence any solvent such as chloroform with a freezing point below –20 °C should first be evaporated with a stream in nitrogen before putting the sample in the lyophiliser.

4. Care should be taken in phase identification, as it is possible for a symmetry-allowed reflection to have zero intensity because the unit cell transform passes through a minimum at that particular diffraction angle. For example, if the $\sqrt{3}$ and $\sqrt{7}$ peaks of an H_{II} phase are too weak to be observed then the pattern may be incorrectly assigned to that of a lamellar phase. One should explore carefully around a particular point in the phase diagram, checking that the data index consistently as the composition or a thermodynamic parameter is altered within the single phase region.

5. Standard methods such as isomorphous replacement (for example brominating one component) tend not to be suitable for lipid systems as it may disrupt the packing of the system.

References

1. Luzzati V (1968) X-ray diffraction studies of lipid-water systems. In: Chapman D (ed) Biological membranes. Academic, New York, pp 71–123

2. Tristram-Nagle S, Nagle JF (2004) Lipid bilayers: thermodynamics, structure, fluctuations and interactions. Chem Phys Lipids 127:3–14

3. Pabst G, Kucerka N, Nieh M-P, Rheinstaedter MC, Katsaras J (2010) Applications of neutron and X-ray scattering to the study of biologically relevant model membranes. Chem Phys Lipids 163:460–479

4. Tardieu A, Luzzati V, Reman FC (1973) Structure and polymorphism of the hydrocarbon chains of lipids: a study of lecithin-water phases. J Mol Biol 75:711–733

5. Mills TT, Tristram-Nagle S, Heberle FA, Morales NF, Zhao J, Wu J, Toombes GES, Nagle JF, Feigenson GW (2008) Liquid-liquid domains in bilayers detected by wide angle x-ray scattering. Biophys J 95:682–690

6. Luzzati V, Vargas R, Mariani P, Gulik A, Delacroix H (1993) Cubic phases of lipid-containing systems. Elements of a theory and biological connotations. J Mol Biol 229:540–551

7. Aeffner S, Reusch T, Weinhausen B, Salditt T (2009) Membrane fusion intermediates and the effect of cholesterol: an in-house x-ray scattering study. Eur Phys J E Soft Matter 30:205–214

8. Pabst B, Heberle FA, Katsaras J (2013) X-ray scattering of lipid membranes. In: Roberts GCR (ed) Encyclopedia of biophysics. Springer, Berlin, pp 2785–2791

9. Wolfson MM (1997) An introduction to X-ray crystallography, 2nd edn. Cambridge University Press, Cambridge

10. Pecharsky VK, Zavalij PY (2003) Fundamentals of powder diffraction and structural characterisation of materials. Springer, Berlin

11. Seddon JM, Squires AM, Conn CE, Ces O, Heron AJ, Mulet X, Shearman GC, Templer RH (2006) Pressure-jump X-ray studies of liquid crystal transitions in lipids. Philos Trans A Math Phys Eng Sci 364:2635–2655

12. Brooks NJ, Gauthe LLE, Terrill NJ, Rogers SE, Templer RH, Ces O, Seddon JM (2010) Automated high pressure cell for pressure-jump x-ray diffraction. Rev Sci Instrum 81:064103

13. Huang TC, Toraya H, Blanton TN, Wu Y (1993) X-ray-powder diffraction analysis of silver behenate, a possible low-angle diffraction standard. J Appl Cryst 26:180–184

14. International Tables for Crystallography (1983) In Th Hahn (ed) Volume A: space-group symmetry

15. Shearman GC, Tyler AII, Brooks NJ, Templer RH, Ces O, Law RV, Seddon JM (2009) A 3-D hexagonal inverse micellar lyotropic phase. J Am Chem Soc 131:1678–1679

16. Seddon JM, Templer RH (1995) Polymorphism of lipid-water systems. In: Lipowsky R, Sackmann E (eds) Handbook of biological physics vol. 1, structure and dynamics of membranes. Elsevier SPC, Amsterdam, pp 97–160

17. Mariani P, Luzzati V, Delacroix H (1988) Cubic phases of lipid-containing systems—structure-analysis and biological implications. J Mol Biol 204:165–189

18. Seddon JM (1990) Structure of the inverted hexagonal (HII) phase and non-lamellar phase transitions of lipids. Biochim Biophys Acta 1031:1–69

19. Kulkarni CV, Tang T-Y, Seddon AM, Seddon JM, Ces O, Templer RH (2010) Engineering bicontinuous cubic structures at the nanoscale-the role of chain splay. Soft Matter 6:3191–3194

20. Franks NP, Levine YK (1981) Low-angle X-ray diffraction. In: Grell E (ed) Membrane spectroscopy. Springer, Berlin, pp 437–487

21. Shannon CE (1949) Communication in the presence of noise. Proc Inst Radio Eng 37:10–21

22. McIntosh TJ, Simon SA, Ellington JC, Porter NA (1984) A new structural model for mixed-chain phosphatidylcholine bilayers. Biochemistry 23:4038–4044

23. Luzzati V, Vargas R, Gulik A, Mariani P, Seddon JM, Rivas E (1992) Lipid polymorphism: a correction. The structure of the cubic phase of extinction symbol Fd– consists of two types of disjointed reverse micelles embedded in a 3D hydrocarbon matrix. Biochemistry 31:279–285

24. Janssen T, Janner A, Looijenga-Vos A, de Wolff PM (1995) International tables for crystallography. Kluwer, London

25. Blaurock AE, Worthing CR (1966) Treatment of low angle x-ray data from planar and concentric multilayered structures. Biophys J 6:305–312

26. Harper PE, Mannock DA, Lewis NAH, McElhaney RN, Gruner SM (2001) X-ray

diffraction structures of some phosphatidyl-ethanolamine lamellar and inverted hexagonal phases. Biophys J 81:2693–2706

27. Lonsdale K (1936) Simplified structure factor and electron density formulae for the 230 space groups of mathematical crystallography. G. Bell & Sons, London

28. International Tables for X-ray Crystallography. Vol I. In NFM Henry, K Lonsdale (eds), Kynoch Press, Birmingham (1969)

29. Seddon JM, Robins J, Gulik-Krzywicki T, Delacroix H (2000) Inverse micellar phases of phospholipids and glycolipids. Invited lecture. Phys Chem Chem Phys 2:4485–4493

Chapter 17

Solid State NMR of Lipid Model Membranes

Arwen I.I. Tyler, James A. Clarke, John M. Seddon, and Robert V. Law

Abstract

In this chapter we describe the use of solid state nuclear magnetic spectroscopy to study the structure of lyotropic phases and lipid model membranes and show its ability to probe, site specifically, at a sub-Ångstrom resolution. Here, we demonstrate the immense versatility of the technique and its ability to provide information on the different liquid crystalline phases present. A multinuclear for example ^{31}P, ^{1}H, and ^{13}C approach is able to elucidate both the structure and dynamics over a wide variety of timescales. This coupled with a non-perturbing label ^{2}H is able to provide information such as the order parameters for a wide variety of different liquid phases.

Key words NMR, Nuclear magnetic resonance, Phase, Order

1 Introduction

Phospholipids represent a fundamental component of biological membranes [1–4]. They are a major component of the plasma membrane, which is amongst the most fundamental barriers of a cell, which demarcates the spatial domain of a cell from the outside world. They can be viewed as liquid crystalline materials which have all the concomitant properties of these materials. Due to their amphiphilic nature, lipids spontaneously form extended structures in aqueous environments, so as to protect their hydrophobic chains, driven by the hydrophobic effect. These extended structures are lyotropic liquid crystal phases, possessing long range periodicity and short range disorder. These phases adopted may possess one-dimensional order (lamellar sheets, L), two-dimensional order (hexagonally arranged cylinders, H), or three-dimensional order (cubic phases, Q). The geometry of the phase adopted by the lipid aggregates and the dynamics of the molecules within it are inherently related to the structural properties of each lipid molecule.

Increasingly, phospholipids have been understood, not just a simple barrier or the framework for ion channel or other membrane proteins, important though this is, but that phospholipids

Dylan M. Owen (ed.), *Methods in Membrane Lipids*, Methods in Molecular Biology, vol. 1232,
DOI 10.1007/978-1-4939-1752-5_17, © Springer Science+Business Media New York 2015

can acts as cell signalling molecules in their own right. Examples of this are the diacylglycerols, phosphoinositol, lysophosphocholine, cardiolipin, fatty acids, and the sphingomyelin-ceramide signalling system amongst many others.

Furthermore, there has been substantial controversy over the role of amongst the most commonly occurring steroid molecule in the plasma membrane prevalent in mammalian cell walls: cholesterol. These interesting micro-domains or "lipid rafts" have reinvigorated the use of the techniques to examine the physical properties of the plasma membrane and plasma membranes themselves [5–7].

Phospholipids have a rich number of nuclei which make them particularly amenable to solid state NMR spectroscopy. These include ^{31}P, ^{13}C, ^{1}H, and with the incorporation of the stable isotope ^{2}H NMR. There have also been an increasing number of reports on ^{14}N of the headgroup for phosphocholines [8–10]. These biological liquid crystals have a number of important physiological liquid crystalline phases which are readily observed by solid state NMR spectroscopy.

2   Materials

2.1   Samples and Preparation Tools

1. Lipids and cholesterol either as a powder or a stock solution in organic solvent.

2. Deuterated water (D_2O) with a purity >98 % is required for ^{31}P, ^{13}C, and ^{1}H experiments. High purity deuterium depleted water is required for the ^{2}H experiments.

3. A cylinder of nitrogen gas for solvent evaporation.

4. Liquid nitrogen and a hair dryer for freeze-thawing.

5. A vortexer.

6. Desktop centrifuge capable of spinning samples up to $4,000 \times g$.

7. Zirconia (4 mm diameter) MAS rotors.

8. Glass slides.

2.2   NMR Instrumentation and Data Analysis

9. Solid state NMR spectrometer such as a Bruker Avance III 600 spectrometer operating at 14.09 T or the equivalent.

10. For the ^{31}P, ^{1}H, and ^{13}C NMR experiments a double resonance ^{1}H/X 4 or 2.5 mm cross-polarization magic angle spinning (CP MAS) probe is required. For these experiments a ^{2}H lock is unnecessary. For the ^{2}H NMR experiments a static NMR probe with a singly tuned horizontal 5 mm solenoid sample coil is required. All probes should be equipped with a thermocouple and heater for temperature control.

11. A Bruker VT (variable temperature) and BCU (air cooling) units such as a Bruker VT3000 and Bruker BCU05 units respectively

for temperature control. This is achieved by the VT unit, the gas required for MAS is passed through the cooling unit which cools it to ~–5 °C and passed to the heating coil in the NMR probe before transferring to the sample in the probe head. The error in temperature should be less than ±0.5 °C. A Bruker pneumatic unit is required for sample spinning.

12. A PC with relevant software required for experimental set-up and data acquisition, such as Bruker's TopSpin 3.5 software or later. Standard pulse programs provided by Bruker should be used.

13. For data analysis, a PC with TopSpin 3.5 installed for data processing, spectral subtraction and spectrum plotting (XWINPLOT) is required. For dePaking spectra Bruker's Amix 3.6 software is required.

3 Methods

3.1 Sample Preparation

1. Individual lipid powders are lyophilized for at least 4 h to remove any water, as lipid behavior is directly related to the water content. For lipid mixtures, the desired amount of lyophilized powders (see **Note 1**) are then weighed and mixed together in a glass vial. The lipid mixture is dissolved in either cyclohexane with a few drops of methanol to aid dissolution, or chloroform (see **Note 2**).

2. The mixture is then frozen in liquid nitrogen and lyophilized for at least 24 h. Excess H_2O or D_2O (<60 wt%) is then added to the samples or a known quantity H_2O or D_2O when samples need to be examined "dehydrated" or limited hydration conditions [11].

3. For the ^{31}P, ^{13}C, and ^1H NMR experiments D_2O should be added to the samples, whereas for the ^2H NMR experiments deuterium depleted H_2O should be added to the samples instead to avoid sharp isotropic peaks which promote Fourier artifacts in the NMR spectrum.

4. The samples are then subjected to a minimum of five freeze–thaw cycles in order to achieve reproducibility and sample homogeneity and sometimes mechanical mixing may also be required to homogenize the samples. The hydrated samples should also be shaken in a vortexer as well as centrifuged (see **Note 3**) in-between each freeze–thaw cycle. When not used, the samples should be stored in a freezer at ~–5 °C.

5. The sample preparation described above should yield multilamellar vesicles (MLVs) in the micron range; however, it should be noted that very small vesicles (<150 nm), due to their small radii of curvature and adequate diffusion rates for the lipids to experience all angles to the applied magnetic field on the NMR

timescale, will result in an averaging out of quadrupolar splittings or the chemical shift anisotropy (CSA) giving a single peak for each isotropic chemical shift [12]. This is also true for other phases such as the cubic phases and micellar phases.

6. Between 5 and 50 mg of sample is required for the NMR experiments depending on the nucleus studied. More sensitive nuclei such as ^1H require less sample mass and shorter acquisition times compared to the less sensitive ^{13}C due to shorter ^1H T_1's and ~100 % isotopic abundance. The acquisition time increases quadratically with decreasing sample size therefore much longer acquisition times are required to give a good signal to noise for smaller sample sizes.

7. The sample is transferred into an NMR rotor appropriate for either MAS or static experiments (e.g., 4 mm ZrO_2 rotor). A convenient way of doing this is by taking a pipette tip, cutting the thin end of it and inserting it into the rotor.

8. The pipette tip and rotor can be secured together with some Parafilm®, the sample is then placed on the wide end of the pipette tip using a spatula and the whole assembly is placed into a centrifuge so that the sample can be transferred into the rotor by low speed spinning. If the sample is too viscous then the sample can be gently heated with a hairdryer to above its gel—fluid phase transition temperature which will facilitate the transfer. Further freeze–thaw cycles or annealing of the sample above its melting transition when in the rotor can also be applied to achieve homogeneity.

9. For both ^{31}P and ^2H NMR aligned samples can also be prepared [13, 14, 15]. Here, the bilayers are supported on glass slides which are aligned in the magnetic field at a known angle. These have many advantages over the above method including higher sensitivity, the ability to carefully control the angle of the bilayer with respect to the magnetic field and also it makes the dePaking procedure unnecessary (Subheading 3.2.5). They have been particularly useful for peptide-phospholipid studies. However, it can be slow to prepare samples using this method.

10. To prepare aligned samples, glass slides of the correct dimension need to be cleaned with fuming nitric acid or sulphuric acid/hydrogen peroxide for 2–4 h and washed with excess distilled water to remove any acid traces. These are stored in distilled water or dry in a desiccator until required. The thin (70–80 µm) glass slides or cover slips (Marienfeld Laboratory Glassware, Lauda-Koenigshofen, Germany; Whale Apparatus, Hellertown, PA, USA) should be of a dimension to fit into the thin-walled glass tube or cuvette which is then inserted, after sealing, into the horizontal NMR coil. Exact dimensions of the slides vary depending upon the size and coil design.

11. A suitable phospholipid solution is made up of the desired concentrations (5–20 mg/mL) in organic solvent is placed on the flat slides with a coverage of 0.1–0.2 mg/cm². They are then dried carefully with a stream of dry nitrogen until they reach a constant mass, further drying can be achieved by use of low vacuum to ensure complete removal of any organic solvent. The slides are then placed in a hydration chamber with a salt solution of the appropriate (e.g., K_2SO_4) to produce the desired hydration level, e.g., 50–100 % relative humidity. The rehydration solutions which are used to control the humidity need to be made from deuterium depleted water for ²H NMR spectroscopy. The slides are then inserted into the thin-walled glass tube (square or round, depending upon the coil geometry). Multiple slides (ca. 15–60) can be stacked on top of each other into the NMR coil to improve signal-to-noise ratio. The samples are then sealed to prevent moisture loss or gain. The angle of the sample slides with respect to the magnetic field can be carefully controlled. For samples that are prone to oxidation a similar procedure can be carried out inside a nitrogen flushed glove box.

12. This method has been found to be particularly suitable for the studying of peptide-lipid interactions [13, 14] where understanding the peptide orientation with respect to the bilayer is of great value. However, care has to be taken that any buffer has sufficient water present to allow the full buffering capacity if salt/pH buffer is used. Furthermore, caution must be used to determine that delamination does not occur from the surface of the glass slide, which allows the formation of isotropic phases, e.g., micelles, as some phospholipids are more susceptible to this than others, e.g., phosphoethanolamine. Finally there can be subtle differences between 100 % hydration and the presence of excess water which might affect the phase behavior of some phospholipid systems [16].

13. In addition there should be a brief mention of an alternative method for the study of protein/peptide-phospholipid interactions. These are disk-shaped bilayers or bicelles, which composed of short phospholipid at the edge of the disk and longer one in the center. These will naturally align, with their long axis with the magnetic field, due to magnetic susceptibility effects. They have been employed with great success in order to understand the protein/peptide-phospholipid interactions [17–19].

3.2 Multinuclear Solid State NMR

3.2.1 ³¹P Static CSA NMR

Data Acquisition and Processing

1. Set up the NMR spectrometer for detection of the ³¹P frequency and decoupling on the ¹H channel (*see* **Note 4**). However, as the ³¹P-¹H dipolar coupling is very weak (~10–100 Hz) and the contribution to the linewidth of the static ³¹P NMR is small, it is possible to carry out this without decoupling. If decoupling is used the ¹H decoupling frequency should be placed in the center of the ¹H spectrum.

Table 1

Typical experimental parameters used for ^{31}P, ^{1}H and ^{13}C solid state NMR. Though these parameters can be dependent upon the amplifier type, spectrometer field strength and manufacturer

	^{31}P CSA	^{31}P MAS	^{1}H MAS	^{13}C MAS
Pulse program	SPE	SPE	SPE	NOE
Time domain	2,048	3,018	4,096	4,096
No. scans	2,048	128	16	2,048
Acquisition time	21 ms	31 ms	51 ms	51 µs
Sweep width	48 kHz	48 kHz	40 kHz	40 kHz
Relaxation delay	3 s	3 s	2 s	3 s
Lorentzian broadening	100 Hz	10 Hz	1 Hz	10 Hz
X-pulse length	2 µs	2 µs	N/A	4 µs
^{1}H-pulse length	3.6 µs	3.6 µs	4 µs	3.6 µs

2. Place the probe in the spectrometer, tune and match both the ^{31}P and ^{1}H channels. A Bruker Avanace III 600 spectrometer operating at 14.09 T has a ^{1}H resonance of 600.01 MHz and a ^{31}P resonance of 242.9 MHz. Set a variable temperature (VT) experiment and allow the sample to equilibrate for a minimum of 15 min before the time dependent NMR signal (FID—free induction decay) is acquired.

3. Acquire the FID using a single pulse excitation (SPE) or Bloch decay sequence. This involves a 90° or $\pi/2$ pulse to rotate the magnetization to the x, y-plane, the signal is acquired followed by a delay time to allow the system to return to equilibrium before the sequence is repeated. If, however, there is reason to believe there is distortion of the broad signal then a Hahn echo can be used instead of a Bloch decay. Typical acquisition parameters on a Bruker Avance III 600 MHz spectrometer are summarized in Table 1.

4. In order to reduce noise the FID is multiplied with an exponential window function characterized by the Lorentzian line broadening factor (typically between 10 and 100 Hz) followed by a Fourier transformation to get the spectrum. The phase of the spectrum is modified by the software to correct for instrumental time lags in acquisition and a flat baseline is obtained.

5. A baseline correction can then be made, either linear or nonlinear; however, caution has to be used because of the possible distortion introduced by this process. The same processing should be applied to all spectra that are compared, i.e., the same

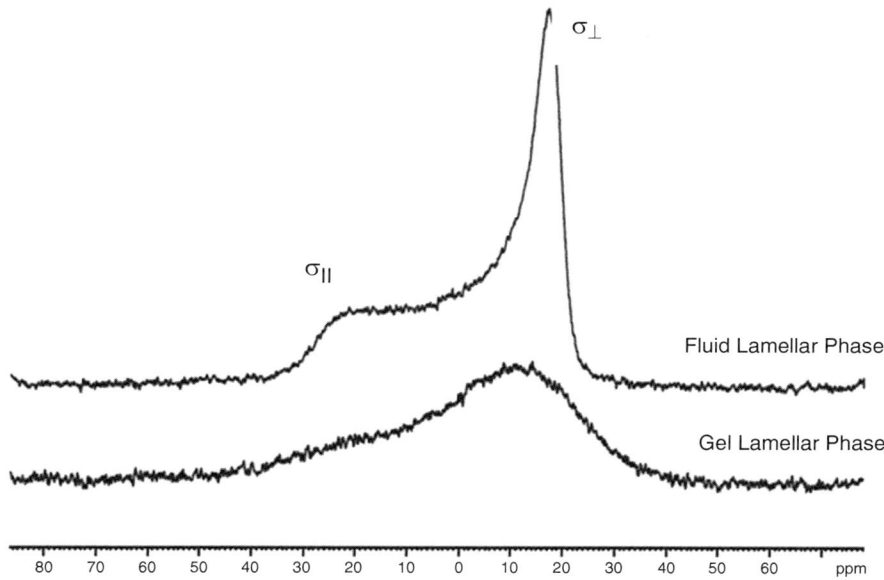

Fig. 1 Characteristic ^{31}P CSA powder patterns of MLVs of a phospholipid gel (L_β) and fluid (L_α) lamellar phase

line broadening and baseline correction routine. The chemical shift scale of ^{31}P NMR spectra is reported by making external reference to 85 % H_3PO_4 (0 ppm). The convention regarding the measurement for anisotropy of the ^{31}P static spectrum is given in ref. [20].

Interpretation of Spectra

1. Non-spherical electron distribution around a nucleus in a molecule can cause chemical shift anisotropy [20–23]. In an anisotropic system, e.g., liquid crystalline or solid, the isotropic chemical shift of the nucleus is now dependent upon its orientation to the magnetic field, $\mathbf{B}_0$. Consequently for a vesicle system, a spread of chemical shifts will result in with the two extremes of chemical shift being with the molecule orientated parallel or perpendicular to the magnetic field. This results in a powder pattern which is a statistical distribution of all the possible orientations of the molecules in the sample with respect to the magnetic field.

2. The CSA of the phosphorus signal is defined in a coordinate system by three orthogonal shielding tensors [20, 24]. In a liquid crystalline lamellar phase these are reduced to two effective tensors, $\sigma_\parallel$ (downfield, high frequency) and $\sigma_\perp$ (upfield, low frequency) (Fig. 1), which arise from parts of the chemical shift which are parallel and perpendicular to the long axis of molecular rotation. In a lamellar vesicle there is small population of phospholipids parallel to the magnetic field (at the top and bottom of the vesicle) and a larger population around the

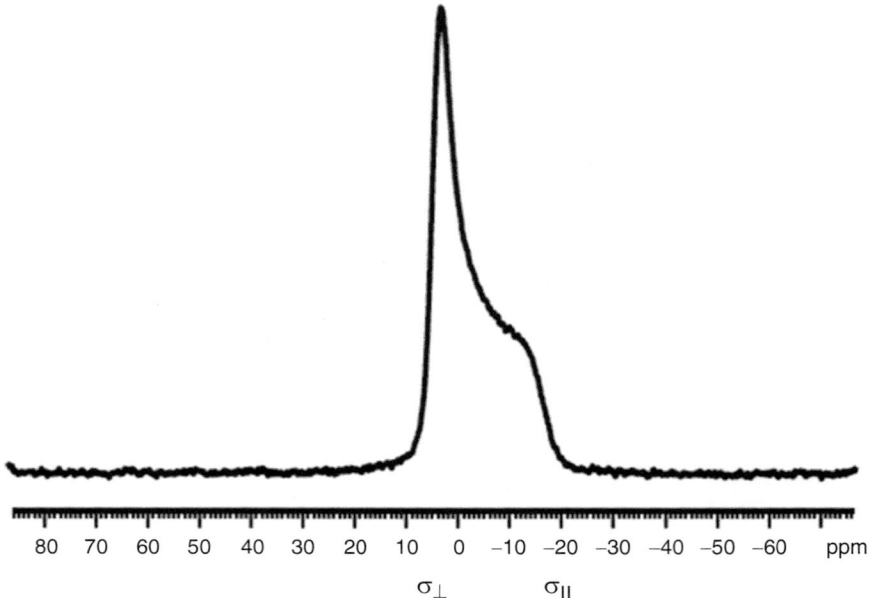

Fig. 2 Characteristic ^{31}P CSA powder pattern of a lipid in a hexagonal phase. Note the reversal of the tensors due to the geometry of the phase

equator of the vesicle which are perpendicular to the field. It is this distribution that gives the characteristic symmetry of the ^{31}P powder pattern. The CSA tensor ($\Delta\sigma$) is defined as: $\Delta\sigma = \sigma_{\parallel} - \sigma_{\perp}$. The edges off the powder spectrum (Fig. 1) obtained from MLVs (*see* **Note 5**) are $\sigma_{\parallel}$ (low intensity) and $\sigma_{\perp}$ (high intensity).

3. The powder pattern of bilayer gel phase is very broad and has a large $\Delta\sigma$ value. Upon melting to a fluid phase the powder pattern becomes much sharper and narrower, with typical $\Delta\sigma$ values for phospholipids in the region of 45 ± 5 ppm (Fig. 1). Increasing the temperature has little effect on $\Delta\sigma$ for a fluid lamellar phase. By setting a temperature scan one can determine the gel to fluid transition temperature (T_m) for a sample based on the resulting spectra; however, care should be taken in approximating the T_m from ^{31}P CSA experiments as they report on the conformational freedom of the headgroup which may not reflect the state of the hydrocarbon chains (*see* **Note 6**).

4. The hexagonal (H_{II}) phase can be distinguished from the lamellar phase as the phospholipids in the H_{II} phase possess additional averaging. The diffusion of a lipid around a cylinder in the H_{II} phase is faster than the NMR timescale hence the resulting in a reduction of the width of the CSA pattern. H_{II} powder patterns are half the magnitude of the lamellar phase ones and of opposite sign (Fig. 2).

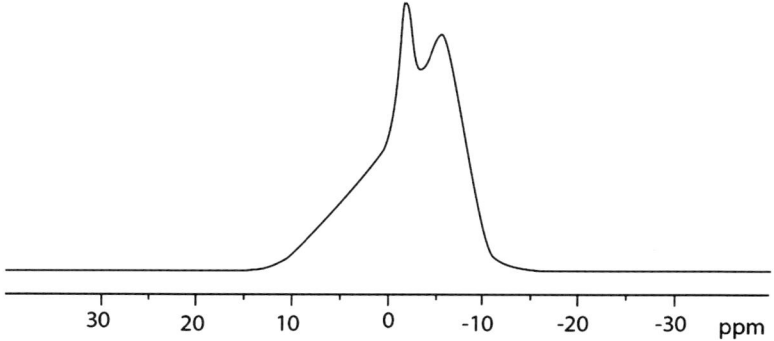

Fig. 3 Simulated ^{31}P CSA patterns of a lamellar–isotropic phase coexistence

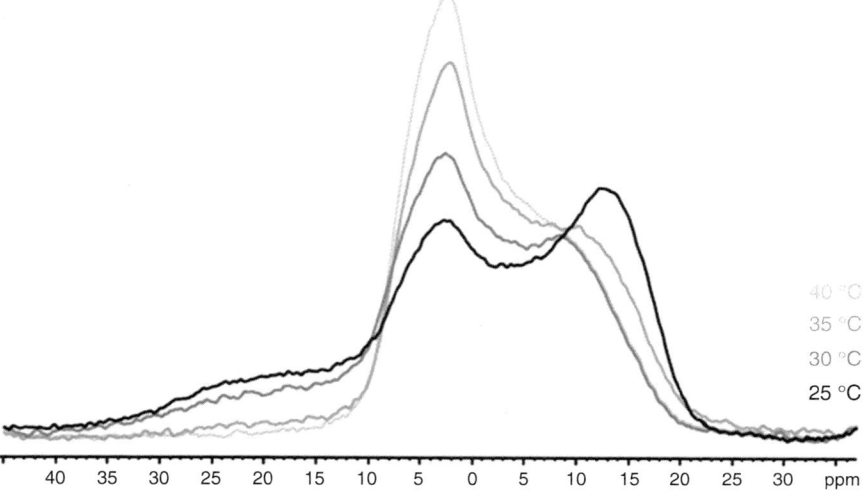

40 °C
35 °C
30 °C
25 °C

Fig. 4 ^{31}P CSA patterns of a lamellar (L_α)–hexagonal (H_{II}) phase coexistence showing a decrease in the population of the lamellar phase with temperature until a pure H_{II} phase is seen at 40 °C

5. The lipids in cubic and micellar phases undergo fast molecular averaging in all dimensions resulting in an isotropic ^{31}P NMR signal. An example of a lamellar phase coexisting with an isotropic peak is shown in Fig. 3.

6. ^{31}P CSA can also be used to quantitatively measure phase coexistence. A change in the population of superimposed peaks from the coexisting phases can be followed as a function of temperature (Fig. 4).

3.2.2 ^{31}P MAS NMR

Data Acquisition
and Processing

1. The experiment is set up as in subheading "Data Acquisition and Processing" with the addition that the sample is spun at a MAS frequency of around 3–10 kHz and the sample temperature is adjusted to reflect the desired temperature in the rotor (*see* **Note 7**).

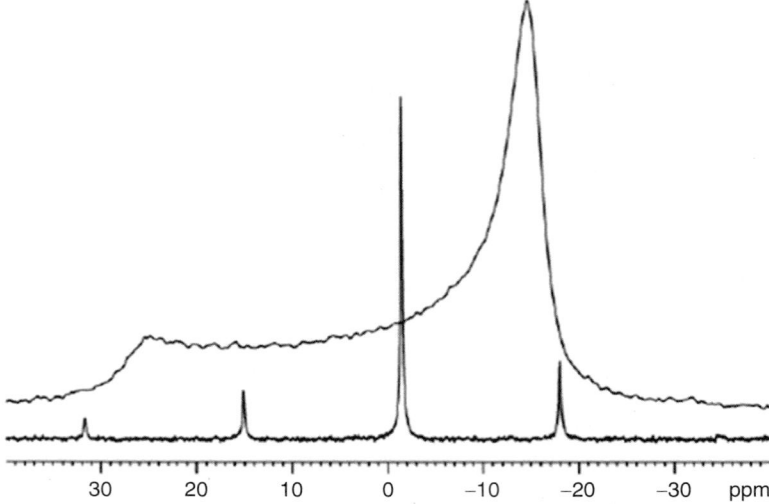

Fig. 5 ^{31}P static (*top*) and MAS (*bottom*) spectra of a phospholipid in the L$_\alpha$ phase. The MAS spectrum shows the removal of line broadening of the powder pattern to leave an isotropic peak with spinning sidebands. Note distortion in the line-shape are due to magnetic susceptibility effects of the vesicle [24]

2. Typical acquisition parameters on a Bruker Avance III 600 MHz spectrometer are summarized in Table 1. The FID is multiplied with an exponential window function corresponding to a line broadening factor of around 10 Hz followed by a Fourier transformation to obtain the spectrum.

Interpretation of Spectra

1. Fast spin rates are required to completely remove the CSA and dipolar coupling contribution to the phosphorus lineshape to yield a spectrum comparable to a solution state spectrum; however, this causes frictional heating and sometimes leads to sample dehydration (*see* **Note 7**). Consequently, intermediate spin rates are used to partially remove the broadening and give an isotropic peak (relative to 85 % H$_3$PO$_4$) with spinning sidebands at intervals equal to the spinning frequency (Fig. 5). This is particularly useful in distinguishing between individual lipids or looking at the effect of additives to the phosphate headgroup.

2. ^{31}P MAS can be used to distinguish between lipids in the L$_\alpha$ and L$_\beta$ phases. The L$_\alpha$ phase gives rise to a single sharp peak at the isotropic chemical shift with a half-height peak width of around 50 Hz due to the large degree of conformational freedom around the phosphate headgroup. The L$_\beta$ phase has a larger peak width as slow exchange of conformations leads to a distribution of chemical shifts. An example of ^{31}P MAS spectra obtained from a binary system of two phase separated phospholipids (in the L$_\alpha$ and L$_\beta$ phases) at low temperature and the

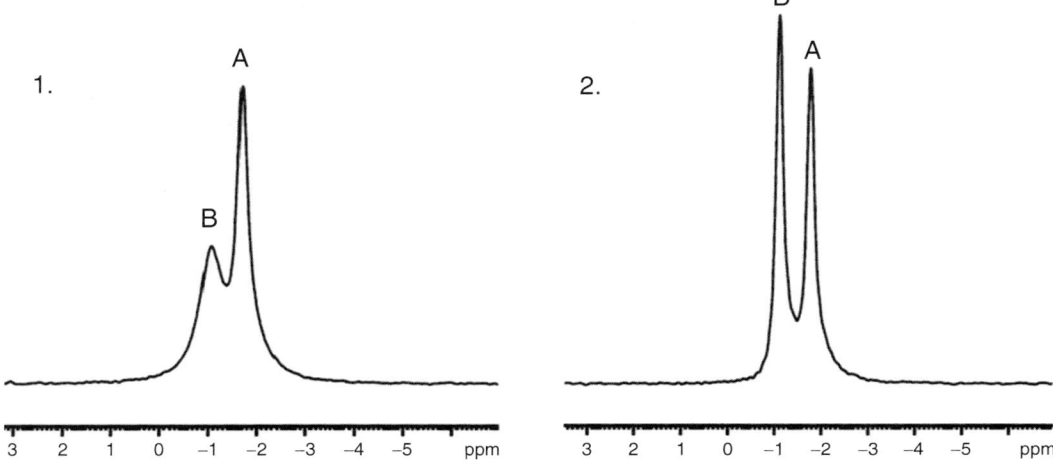

Fig. 6 ^{31}P MAS spectra of a binary system of phospholipids. (*1*) At low temperatures phospholipid A is in the L_α phase whereas phospholipid B is in the L_β phase. (*2*) Heating the sample above the T_m of phospholipid B yields a spectrum where both lipids are in the L_α phase

transformation of the phospholipid in the L_β phase at a higher temperature is shown in Fig. 6 [25]. It should be noted that phospholipids with identical headgroups will give rise to a peak at the same isotropic chemical shift, whereas phospholipids with different headgroups, e.g., phosphoethanolamine, will give rise to a peak at different isotropic chemical shifts under MAS.

3. The peak width and normalized intensity of the ^{31}P signal can be used as an indication of the lipid phase. For example, a mixture of DOPC–DPPC–CHOL 1:1:1 shows a broad peak at lower temperatures corresponding to a coexisting DPPC–CHOL rich L_β phase and a DOPC rich L_α phase. Increasing the temperature causes the gel phase to transform into a liquid ordered (L_o) phase, coexisting with a DOPC-rich L_α phase [26]. As there is rapid exchange both in the L_α and L_o phase a sharp peak is observed in the ^{31}P MAS spectrum (Fig. 7).

4. Care should be taken in approximating the T_m from ^{31}P MAS experiments as they report on the conformational freedom of the headgroup which may not reflect the state of the hydrocarbon chains (*see* **Note 6**). Moreover, the L_α and L_o phases cannot be distinguished from ^{31}P experiments so further experiments (such as ^{2}H static NMR) are required to ascertain the phase adopted by the system.

3.2.3 ^{1}H MAS NMR

Data Acquisition and Processing

1. Set up the NMR spectrometer for detection of the ^{1}H frequency. Place the probe in the spectrometer, tune and match the ^{1}H channels. An Avance III 600 spectrometer operating at 14.09 T has a ^{1}H resonance of 600.01 MHz. Due to the large

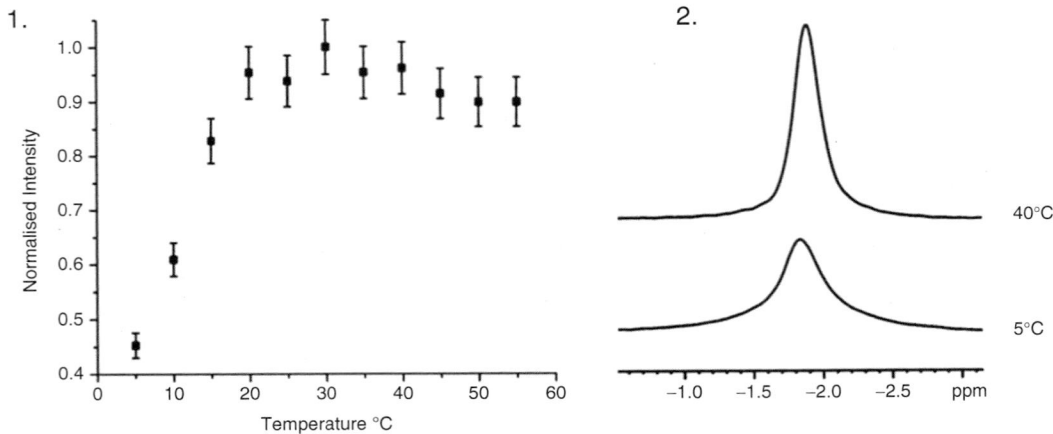

Fig. 7 (*1*) Normalized ³¹P MAS intensities for DOPC–DPPC–CHOL 1:1:1 as a function of temperature. The data reaches a plateau at around 20–25 °C when the system transforms from a coexisting L_β and L_α phase to a coexisting L_α and L_o phase. (*2*) ³¹P MAS spectra of DOPC–DPPC–CHOL 1:1:1 at 5 and 40 °C [26]

inherent sensitivity and high natural abundance of the ¹H nucleus, small quantities of sample are required (ca. 5 mg). A typical spectrum takes only a few minutes to acquire. The sample is spun at a MAS frequency of around 3–10 kHz and the sample temperature is adjusted to reflect the desired temperature in the rotor (*see* **Note 7**). Set a VT experiment and allow the sample to equilibrate for a minimum of 15 min before the FID is acquired.

2. Acquire the FID using a single pulse excitation (SPE) or Hahn echo sequence. Typical acquisition parameters on a Bruker Avance III 600 MHz spectrometer are summarized in Table 1. The FID is multiplied with an exponential window function characterized by the Lorentzian line broadening factor (typically between 0.3 and 10 Hz) followed by a Fourier transformation to get the spectrum. The phase of the spectrum is modified to correct for time delays in acquisition and a flat baseline is obtained.

Interpretation of Spectra

1. The inherent sensitivity of the ¹H nucleus results in spectra with remarkably broad peaks as the ¹H nucleus suffers from residual dipolar coupling and magnetic susceptibility effects [27, 28]. As for the ³¹P nucleus, these can be removed by fast spin rates; however, this causes frictional heating and sometimes leads to sample dehydration (*see* **Note 7**) hence slower MAS rates (3–5 kHz) are employed to partially remove the chemical shift anisotropy and give rise to a series of spinning sidebands.

2. A typical static ¹H spectrum of DPPC–CHOL 60:40 at 40 °C as well as one acquired using MAS is shown in Fig. 8. It is seen that

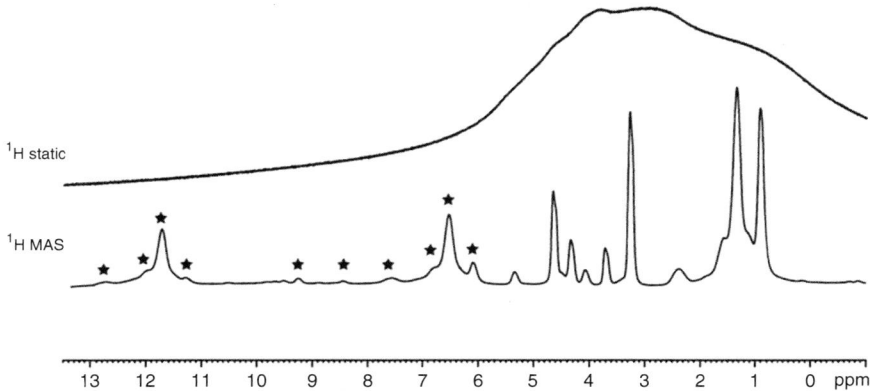

Fig. 8 ¹H NMR spectra of a phospholipid (DPPC) acquired without spinning (*top*) and with MAS (*bottom*). The *asterisks* point out to spinning sidebands [25]

Table 2
¹H chemical shift assignment for DPPC in a DPPC–CHOL 60:40 mixture

Chemical shift/ppm	Assignment
0.87	CH_3
1.31	$(CH_2)_n$
2.3	$CH_2–CO$
3.23	$N–(CH_3)_3$
3.66	$CH_2–N$
4.01	$CH_2–OP$ (glycerol)
4.42	HDO
4.28	$PO–CH_2$ (choline)
4.28	$OCO–CH_2$ (glycerol)
5.28	$OCO–CH$ (glycerol)

the CSA and dipolar coupling are partially removed by MAS to produce a spectrum similar to a solution state spectrum [29]. The peak assignments for the DPPC molecules in the mixture are shown in Table 2. For complete structure elucidation as well as confirmation of correct peak assignment, 2-D NMR experiments (TOCSY, COSY, NOESY) are required [30, 31].

3. It should be noted that in the region of 0.0–2.5 ppm the signal is dominated by the lipid chains; however, there is also contribution from the cholesterol molecules. Perdeuterated DPPC can be used in order to resolve the cholesterol peaks as by removing the signal of the hydrocarbon chains.

4. The distance between nuclear spins affects the strength of dipolar coupling and additionally it is averaged by fast molecular motion. Intermolecular interactions are averaged by fast lateral diffusion of lipids within the lipid phase and intramolecular interactions are reduced with increased long axis rotation. Consequently, ^{1}H spectra serve as a useful indicator of the structure and dynamics of lipid interactions.

5. As the lipids in an L_β phase are tightly packed (*see* **Note 8**) neighboring dipoles are closer to each other and coupled with the fact that slow lateral diffusion and long axis rotation takes place in this phase, the spectrum is dominated by a much greater extent of dipolar coupling compared to a fluid phase, hence the spectrum obtained from an L_β phase is very broad (sometimes the only resolvable peak is from the headgroup which experiences higher degrees of motion). In contrast, the fast lateral diffusion and long axis rotation as well as the large inter-lipid spacing in the L_α phase result in a spectrum with sharp resolvable peaks.

6. ^{1}H MAS spectra can be used to monitor the L_β–L_α phase transition by monitoring the narrowing of the linewidth of the chain methylenes' peak. This transition is often broad but it is always centered around the T_m of the system.

7. Distinguishing an L_o phase is slightly problematic using ^{1}H MAS NMR due to its similarities with both the L_β and L_α phases. ^{2}H static NMR experiments are required to properly distinguish the L_o phase; however, it has been shown [28] that the peak width of the first order spinning sideband as well as the spinning sideband/main peak intensity ratio of the methylene peaks can act as an indicator of the formation of an L_o phase (*see* **Note 9**).

8. Ordered phases have a higher spinning sideband intensity and so the L_β phase can be distinguished from the L_o phase by its larger spinning sideband—main peak intensity ratio. Moreover, the linewidth of the methylene peak increases in the order $L_\alpha < L_o < L_\beta$.

9. Discontinuities in plots of intensity or first order spinning sideband intensity against temperature can be used as an indicator of phase transitions. It should be noted however that it can be difficult to distinguish phase coexistence using ^{1}H MAS NMR. Although an L_α–L_β or L_o coexistence can sometimes be distinguished by the intensity of the methylenes peak being closer to an L_α phase but with the spinning sideband intensity closer to that of the ordered phase, the linewidth of both the methylene peak and the first order spinning sideband of a single L_o phase at low temperatures can have a similar peak width to an L_β or L_o coexistence region.

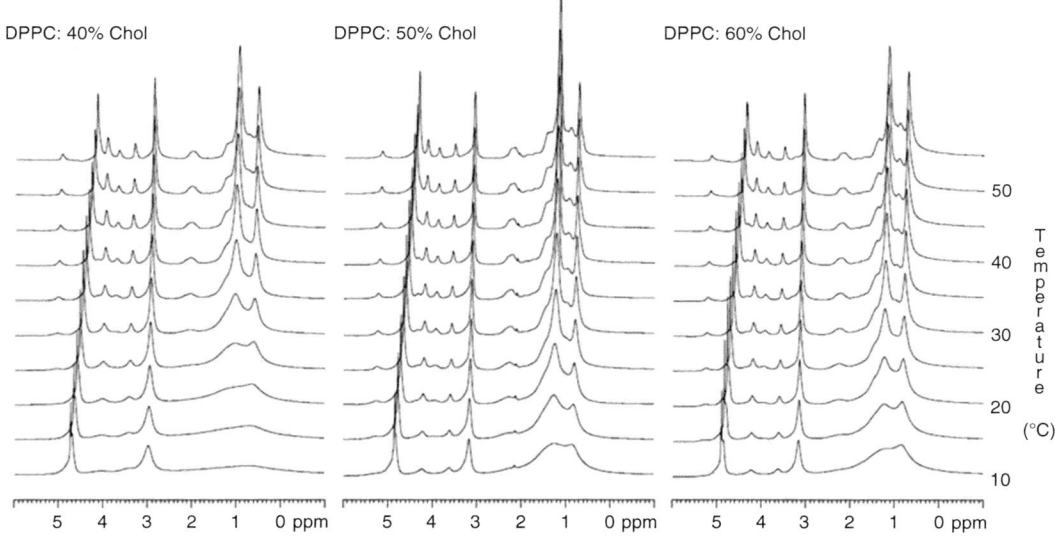

DPPC: 40% Chol   DPPC: 50% Chol   DPPC: 60% Chol

Temperature (°C)

Fig. 9 ¹H NMR spectra of DPPC mixtures with 40, 50, and 60 mol% cholesterol between 10 and 55 °C [32]

10. Spectra from other nuclei, such as ^{13}C MAS, show superposition of peaks from coexisting phases which cannot be seen in ^{1}H spectra (apart from a decrease in the linewidth of the peaks as the phases become more fluid). This can be explained by spin diffusion which is due to the strong ^{1}H-^{1}H interactions (*see* **Note 8**).

11. The effect of cholesterol on phospholipids can easily be seen in ^{1}H MAS spectra (Fig. 9) [32]. Increasing the cholesterol content from 40 to 60 % at 5 °C causes the methylene peaks to sharpen. This reduction in dipolar coupling is due to an increase in the distance between coupled protons as well as increased lipid motion due to the fluidizing effect of cholesterol. The methylene peaks also sharpen as temperature is increased from 5 to 60 °C as the system transforms from a coexisting L_β and L_o phase to a pure L_o phase. It should be noted however that further experiments (such as ^{2}H NMR) are required to correctly identify the phases adopted by the system. The glycerol peaks also sharpen as the temperature is increased which is a result of either residual dipolar coupling or a distribution of chemical shifts from additional conformational freedom around the interfacial region.

3.2.4 ^{13}C MAS NMR

Data Acquisition and Processing

1. Set up the NMR spectrometer for detection of both the ^{13}C frequency and decoupling on the ^{1}H channel, with the decoupling frequency set to the middle of the ^{1}H spectrum (*see* **Note 4**). Place the probe in the spectrometer, tune and match both the ^{1}H and ^{13}C channels. A Bruker Avance III 600 spectrometer

operating at 14.09 T has a ^{1}H resonance of 600.01 MHz and a ^{13}C resonance of 150.9 MHz. The sample is spun at a MAS frequency of around 3–10 kHz and the sample temperature is adjusted to reflect the desired temperature in the rotor (*see* **Note 7**). Set a VT experiment and allow the sample to equilibrate for a minimum of 15 min before the FID is acquired.

2. Acquire the FID using a single pulse excitation (SPE), or heteronuclear Overhauser (nOe) effect enhanced to increase sensitivity. Cross-polarization can also be employed; however, care has to be taken with heating effects. Typical acquisition parameters on a Bruker Avance 600 MHz spectrometer are summarized in Table 1. The FID is multiplied with an exponential window function characterized by the Lorentzian line broadening factor (typically between 5 and 15 Hz) followed by a Fourier transformation to get the spectrum. The phase of the spectrum is modified to correct for time delays in acquisition and a flat baseline is obtained.

Interpretation of Spectra

1. The sensitivity of the approximately 1 % naturally abundant ^{13}C nucleus can be expanded to provide information about the system of interest without the need for isotopic labelling. As with other nuclei, the use of MAS is used to remove the CSA to produce a spectrum similar to a solution state spectrum which can then be manipulated using a variety of standard pulse sequence programs to explore the structure and dynamics of a lipid system [32].

2. Using single pulse experiments one can extract information on a lipid system by observing changes in the chemical shift—a distribution of chemical shifts is seen as molecular segments of the lipid system can adopt a large number of conformations. ^{13}C NMR can be employed to look at the phase behavior of a system as well as look at the backbone of all the components in a lipid mixture and compliment results from other techniques, such as ^{2}H NMR spectra. The carbon chemical shifts of lipids acquired using solution state NMR serve as a guide for peak assignment.

3. Typical spectra from an L_β and L_α phase are shown in Fig. 10. The rate of exchange between each conformation in an L_β phase is slower and there is a larger distribution of chemical shifts resulting in peak broadening [32, 26]. The peaks in the region between 55 and 70 ppm are from the headgroup region and as this undergoes fast motion on the NMR timescale the peaks are sharp. The signals from the interfacial region appear in the region between 65–75 and 170–180 ppm and are broadened due to the large distribution of conformations. Peak broadening is also seen in the hydrocarbon chain region (between 10 and 40 ppm) and the peaks appear at a higher

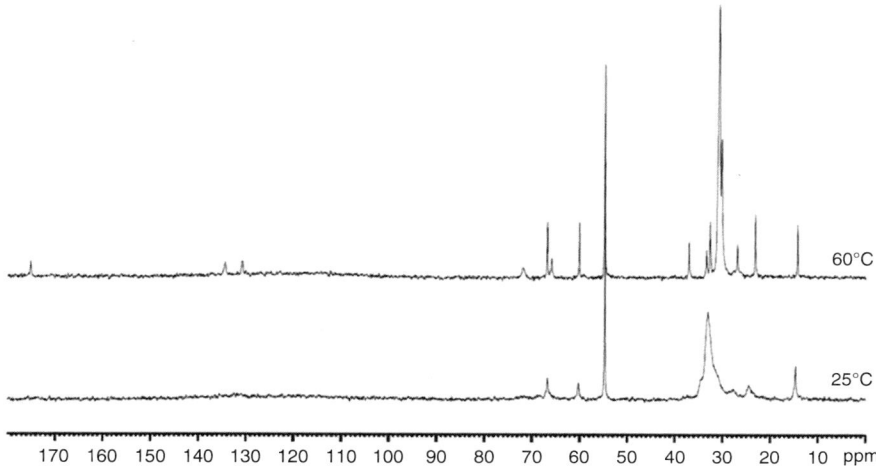

60°C

25°C

170 160 150 140 130 120 110 100 90 80 70 60 50 40 30 20 10 ppm

Fig. 10 ^{13}C MAS NMR spectra from a sphingophospholipid at 25 and 60 °C in the L_β and L_α phase respectively [25]

chemical shift compared to what would be anticipated from a solution state spectrum. In fluid phases, such as the L_α, H_{II} or cubic phases, the rate of exchange between each conformation is much faster, resulting in a sharp peak that is an average of all the possible conformations that are adopted. Both the interfacial and hydrocarbon chain regions give rise to resolvable peaks and due to the fast long axis rotation of the chains in these phases different chain segments can be identified.

4. The chains in an L_o phase have a higher degree of order compared to a fluid phase and are mostly in an *all-trans* conformation resulting in the chain peaks having similar chemical shifts to the L_β phase. Lipids in an L_o phase however undergo fast long axis rotation hence give rise to much sharper peaks compared to the L_β phase. It should be noted that it can become difficult to distinguish between the L_β and L_o phase, especially when the system is in a phase coexistence region or close to a phase boundary and further experiments (such as ^{2}H) are required for correct phase assignment. However, phase coexistence between a gel and fluid phase can often be seen in the peaks arising from the chain region which appear superimposed.

5. The peaks from the chains in an L_β phase appear at a higher chemical shift compared to those in a fluid phase (*see* below). It should be noted however that intermolecular packing, hydrogen bonding as well as distribution of *trans-gauche* states are likely to affect the spectra making them differ from a solution state spectrum. These interactions are also temperature dependent.

6. ^{13}C MAS NMR spectroscopy can also be used to determine the proportion of *gauche* states in the system (associated with the

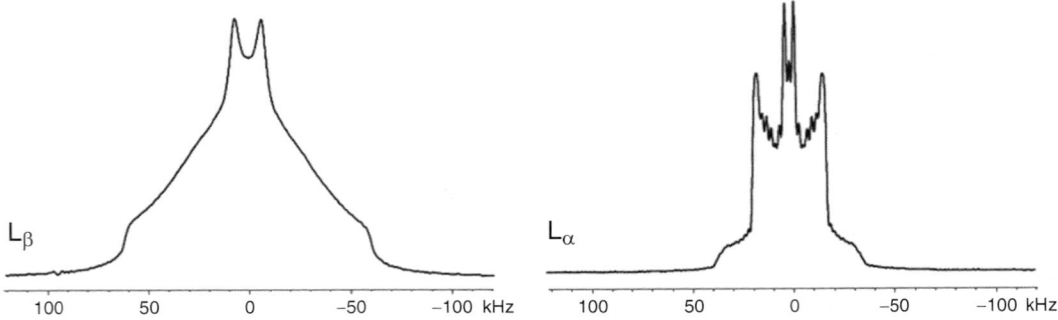

Fig. 11 Typical ^{2}H powder patterns from a perdeuterated phospholipid in the L$_\beta$ (*left*) and L$_\alpha$ (*right*) phases

L$_\alpha$ phase) by examining the chemical shifts of the acyl chains. An upfield shift (towards 0 ppm) of the methylene peaks indicates an increase in chain disorder as the number of *gauche* states. This is due to the γ- and δ-*gauche* effects on the chemical shift observed along hydrocarbon chains increases. [33, 34] The effect of increasing the temperature from 5 to 60 °C on the methylenes peak of DPPC in a mixture of DPPC–CHOL 40:60 is shown in Fig. 11 [32].

7. Increasing the temperature causes an upfield shift of the peak, as the system transforms form an L$_\beta$ to an L$_o$ phase, which is due to differences between *trans* and *gauche* states of the chains. Increasing the temperature causes an increase the interlipid spacing allowing for an increased number of *gauche* states. Within the L$_o$ phase region this upfield shift is seen to continue; however, the peak is seen to continuously broaden. This could possibly be explained by a change in the distribution of *gauche* states, presumably with the bilayer having more *gauche* states towards its center.

3.2.5 ^{2}H NMR

Data Acquisition
and Processing

1. Set up the NMR spectrometer for detection of the ^{2}H frequency. Place the probe in the spectrometer, tune and match the ^{2}H channel. A Bruker Avance 600 spectrometer operating at 14.09 T has a ^{2}H resonance of 92.1 MHz. If running experiments on a system that has a selectively deuterated lipid segment surrounded by neighboring protonated groups then proton decoupling may be beneficial. Set a VT experiment and allow the sample to equilibrate for a minimum of 15 min before the FID is acquired. The sensitivity of ^{2}H NMR experiments is low hence large sample quantities and long acquisition times are required.

2. Acquire the FID using a standard Bruker phase-cycled quadrupole or solid echo pulse sequence. This involves an initial delay time followed by a 90° pulse which rotates the spins about the *x*-axis. Magnetization is spread in the plane from the interaction

Table 3
Typical experimental parameters used for ^2H solid state NMR

	^2H
Pulse program	Quadecho
Time domain	1,024
No. scans	4,096
Acquisition time	1.3 ms
Sweep width	0.4 MHz
Relaxation delay	1 s
Lorentzian broadening	100 Hz
Echo delay	0.02 ms
X-pulse Length	3 μs
X-power Level	–6 dB
^1H-pulse length	N/A
^1H-power Level	N/A

the electric field gradients of the C–^2H bonds and the electric quadrupole moment of the ^2H nucleus. Due to the different precession rates the spins become out of phase as relaxation occurs. After a delay time, a 90° pulse with rotation about the y-axis is applied which refocuses the signal to form an echo [35, 36]. The FID acquisition has to start before the echo maximum is reached. Typical acquisition parameters on a Bruker Avance III 600 MHz spectrometer are summarized in Table 3.

3. Left shift the data points in the FID so that the first data point matches the echo maximum. If this is not the case then intensity distortion will result in the spectrum. The FID is multiplied with an exponential window function characterized by the Lorentzian line broadening factor (typically between 50 and 200 Hz) followed by a Fourier transformation to get the spectrum. The phase of the spectrum is modified to correct for time delays in acquisition and a flat baseline is obtained.

Interpretation of Spectra

4. Due to its low Larmor frequency and natural abundance, lipids have to be labelled with the ^2H isotope [1, 37–40]. The ^2H nucleus is dominated by electronic quadrupolar interactions which can be used to probe molecular motion in a system. The ^2H NMR resonance is a doublet, known as a Pake doublet, and the magnitude of the quadrupolar splitting (ν_Q) is determined by the average orientation of the C–^2H bond with respect to the angle of the applied magnetic field.

5. The fewer orientations the C–²H bond is able to sample, the larger the splitting thus the more ordered the chains exhibit the largest the splittings. For the lamellar phases, the splittings follow the trend $L_\beta > L_o > L_\alpha$ [41, 42]. The chain order of a system in a hexagonal phase can also be probed by ²H NMR. The powder pattern of a hexagonal phase will be half the magnitude of the lamellar one, as in the ³¹P case. The angular dependence of ν_Q is given by:

$$\Delta \nu Q = \frac{3}{2}\left(\frac{e^2 qQ}{h}\right)\left(\frac{3\cos^2\theta - 1}{2}\right)$$

where e is the electronic charge, q is the electric field gradient and Q is the nuclear quadrupolar moment. $\frac{e^2 qQ}{h} = 167\,\text{kHz}$ so the maximum quadrupolar splitting for a methylene group is 250 kHz; however, due to motional averaging, this is rarely seen. A typical linewidth value for a methylene in an L_β or an L_α phase is ca. 50 kHz and ca. 25 kHz respectively.

6. The differences in chain motion of the various phases produce characteristic powder patterns for each phase. A typical spectrum from an L_β phase of a perdeuterated phospholipid is shown in Fig. 11. The chains in this phase are undergoing slow motion relative to the NMR timescale resulting in spectrum that consists of a featureless peak dominated by the central doublet which arises from the terminal methyl group that has substantial motion even in the gel phase [43, 44].

7. In the L_α phase, the fast long axis rotation as well as the large conformational disorder results in a spectrum results in a much narrower spectrum which has fine structure. The shape of the spectrum is characteristic of a fluid phase with axial symmetry (Fig. 11). In theory there should be a doublet from each C–²H group on the hydrocarbon chain as well as one from the terminal methyl group; however, these doublets are not always fully resolved and some signal are superimposed.

8. The methyl groups at the end of the chains cause the sharp, narrow doublet at the center of the ²H NMR powder pattern. The C–²H methylene groups resolve into doublets with larger splittings compared to the methyls, with the least mobile methylene groups causing the widest splitting. These methylene groups are the ones next the interfacial region (glycerol) and selective deuteration of specific segments has been used to confirm this. The chains in an L_o phase are ordered and mostly in an all *trans* configuration; however, they still undergo fast long axis rotation. Consequently the spectrum from an L_o phase (Figs. 12 and 13) reveals axial symmetry and shows fine structure even though it is considerably broader in linewidth than that of an L_α phase.

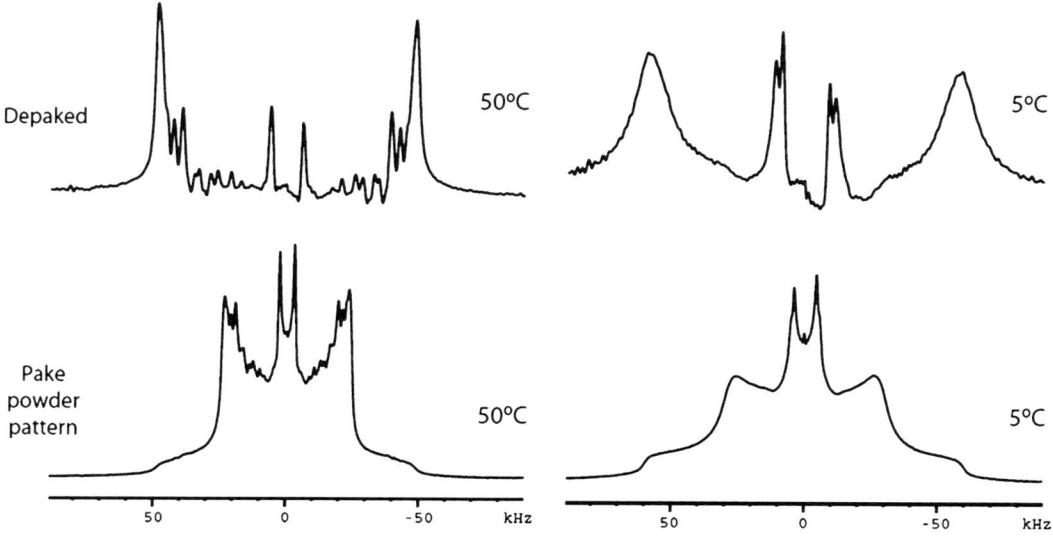

Fig. 12 ^{2}H powder patterns (*below*) and corresponding dePaked spectra (*above*) for DPPC–CHOL 60:40 in the L$_\beta$ (5 °C) and L$_o$ (50 °C) phases

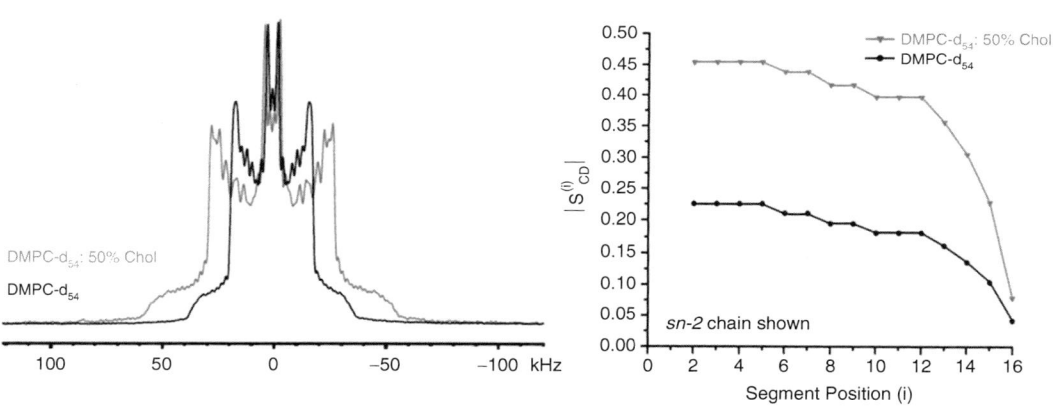

Fig. 13 ^{2}H powder patterns (*left*) for pure DMPC in the L$_\alpha$ phase as well as one obtained for a DMPC–CHOL 1:1 mixture in the L$_o$ phase and the corresponding order parameters (*right*) for the *sn*-2 chain [25, 58]

9. It is more appropriate to express the quadrupolar splitting as an order parameter (S_{CD}) which can be extracted from the quadrupolar splitting when the C–^{2}H is parallel to the applied magnetic field. Order parameters are a time and space average of a molecule or group with respect to the bilayer normal and are a result conformational disorder and bilayer fluctuation with respect to the applied magnetic field. As mentioned above, signals from doublets are often superimposed so a numerical method called dePaking [45–47] is required which calculates the quadrupolar splitting from the powder pattern (*see* **Note 10**).

10. DePaking numerically removes the orientational dependence of the quadrupolar splittings and hence allows calculation of the order parameter of each individual group in the hydrocarbon chain. Each splitting can be assigned to a position in either the sn-1 or sn-2 chain [39, 48, 49]. Various software are available to dePake spectra such as the Amix 3.6 package by Bruker that dePakes spectra through numerical iteration. ^{2}H powder patterns and corresponding dePaked spectra for a mixture of DPPC–CHOL 60:40 in the L_β and L_o phase are shown in Fig. 12 [45, 46, 50]. For an axially symmetric bond, the quadrupolar splitting of each group parallel to the applied magnetic field is related to the order parameter (S_{CD}) by:

$$\left| v_Q^{(i)} \right| = \frac{3}{2} x_Q \left| S_{CD}^{(i)} \right| P_2 (\cos\theta)$$

where $v_Q^{(i)}$ is the dePaked quadrupolar splitting for the ith C–^{2}H group, $x_Q = eq^2 Q/h = 167$ kHz which is the static quadrupolar constant, S_{CD}^i is the order parameter for the ith C–^{2}H group, P_2 is the second Legendre polynomial, θ is the angle between the bilayer normal and the applied magnetic field, $P_2(\cos\theta) = (3\cos^2\theta - 1)/2$ and as the dePaked spectra correspond to the $\theta = 0°$ orientation (bilayer normal to the applied magnetic field) this term becomes unity.

11. Plotting order parameters against hydrocarbon chain position can give information about the local chain dynamics. Example order parameter profiles for pure DMPC [51] in the L_α phase as well as one obtained for a DMPC–CHOL 1:1 mixture in the L_o phase are shown in Fig. 13. In the L_α phase, the order profile shows a plateau region (usually between carbon numbers 2 and 5) which corresponds to C–^{2}H bonds close to the interfacial region and have typical order parameter values of around 0.25–0.20. Progressing down the chain towards the terminal methyl group there is increased conformational disorder in the chain resulting in a decrease in the order parameter. The order profile from the L_o phase follows the same trend; however, it has elevated order parameter values due to an increase in chain order and number of *trans* states.

12. By looking at the lineshape, quadrupolar splittings, order parameters as well as calculated first spectral moments plotted against temperature, ^{2}H NMR can be used to probe phase transitions. It should be noted however that although deuterium labels do not affect lipid packing or conformation, they have a slight effect on hydrogen bond strength and phase transition temperatures of deuterated lipids are generally lowered by ca. 2–3 °C [35].

13. ^{2}H NMR spectroscopy is rich in dynamic information and this has been exploited by many authors [52–56]. An example of

this is the average acyl chain length and cross-sectional area which can be calculated using order parameters. A variety of models can be used to do this, for example the model shown below is based on a continuum distribution of segmental orientations [57]:

$$D_i = \frac{D_m}{2}\left(1 + \sqrt{\frac{8\left|S_{CD}^i\right| - 1}{3}}\right)$$

where i is the chain segment, $\langle D_i \rangle$ is the projection along the membrane normal, D_m is the maximum possible extension and equals 2.54 Å.

4 Notes

1. For very accurate lipid concentration calculations then the lipid crystal structures should be checked to see if any water molecules on the headgroup are present in the crystal structure. For example phosphatidylcholines and sphingomyelins tend to have two water molecules attached to their headgroup and hence the mass of two water molecules should be added to the molecular weight of the lipid when calculating mole ratios. Cholesterol is assumed to be anhydrous.

2. Most lyophilizers cannot cool down below −20 °C and hence any solvent such as chloroform with a freezing point below −20 °C should first be evaporated with a stream in nitrogen before putting the sample in the lyophiliser.

3. Some lipids are less dense than D_2O and hence centrifugation may lead to the excess D_2O accumulating at the bottom of the vial or the NMR rotor. If too much excess D_2O accumulates at the bottom of the NMR rotor, preventing sufficient sample material from being transferred, then it can be removed with a syringe.

4. Spin coupling leads to line broadening hence ^{31}P experiments are often ^{1}H decoupled in order to achieve narrower linewidths. Decoupling is achieved by applying a saturating pulse at the frequency of the nucleus to be decoupled whilst acquiring the spectrum from the nucleus of interest.

5. Aligned bilayer stacks on a glass plate show a single sharp peak as the phosphate is axially symmetric around the bilayer normal. The director can then be rotated about the magnetic field and the chemical shift recorded in order to get values of $\sigma_\parallel$ and $\sigma_\perp$.

6. The L_α–L_β transition of the hydrocarbon chains is a sharp first order transition; however, the narrowing of the CSA occurs progressively with temperature.

7. To completely remove the CSA and dipolar coupling contribution to the phosphorus lineshape to yield a spectrum comparable to a solution state spectrum, fast MAS rates are required (several tens of kHz). Although this can be achieved, fast MAS rates lead to frictional heating and sample dehydration so frequencies of 3–10 kHz are used to avoid this [59]. These intermediate spin rates partially remove the broadening giving an isotropic peak with spinning sidebands at intervals equal to the spinning frequency. Spinning frequencies of 5 kHz or less result in the rotor temperature being a couple of degrees lower than the bearing gas whereas faster spin rates raise the temperature of the rotor in a nonlinear manner compared to the bearing gas due to frictional heating at the bearings. For each spinning frequency, the temperature inside the spinning rotor can be calibrated against the ^{13}C MAS choline peak which is very sensitive to temperature changes and not affected by dipolar coupling. The temperature inside a rotor can be confirmed by carrying out a VT experiment on a phospholipid with known L_β–L_α transition temperature, such as DPPC. Temperature gradients can arise in the rotor as it is heated on either end but cooled at the center by the gas. To the authors' knowledge spinning rates between 3 and 5 kHz have minor temperature gradients across the sample as well as provide sufficient resolution of the peaks.

8. As many protons are in the hydrocarbon chains are very close to each other there is a possibility of spin diffusion taking place. This is known to take place in the L_β and L_o phases but has been debated for the L_α phase [29] and it is thought to be weak for this phase. Spin diffusion is related to internuclear distances and for strongly dipolar coupled nuclei, such as ^{1}H, this process has a considerable contribution to the peak linewidth. For a strongly coupled system, the spin diffusion coefficient between ^{1}H nuclei was measured to be 8×10^{-12} m^2/s [60] which is likely the value in the L_β phase due to the slow lateral diffusion and long axis rotation as well as the tight lipid packing in the phase. The fast lateral diffusion and long axis rotation as well as the large inter-lipid spacing in the L_α phase will lower the spin diffusion coefficient and it has also been suggested that molecular diffusion in this phase is faster than interlipid spin diffusion [31, 27, 28, 30]. Spin diffusion between lipid phases will also produce a single peak hence detecting phase coexistence is only possible in strongly dipolar coupled systems that have higher lateral diffusion rates for each phase compared to intermolecular speed diffusion rates.

9. At spinning speeds of 5 kHz the spinning sideband intensity is dominated by the methylenes. The methyl groups of a phospholipid or cholesterol experience smaller dipolar coupling and is therefore removed at these spin rates.

10. A number of numerical methods are available for dePaking based on full analytical solutions, fast Fourier or iterative methods [45]. DePaking can also be applied to spectra that are dominated by CSA. Phase coexistence can be detected if applied to ^{31}P CSA spectra as it will give a single peak for each coexisting phase. Experimentally, one can avoid dePaking by aligning of the sample on glass plates which are rotated about the applied magnetic field; however, this method requires extensive sample preparation and is carried out under conditions of relative humidity which limits the use of excess water.

References

1. Merz KM, Roux B (1996) Biological membranes: a molecular perspective from computation and experiment. Birkhäuser, Boston
2. Yeagle PL (2011) The structure of biological membranes, 3rd edn. CRC, Boca Raton
3. Stillwell W (2013) An introduction to biological membranes: from bilayers to rafts. Elsevier Science, London
4. Auger M (2000) Biological membrane structure by solid-state NMR. Curr Issues Mol Biol 2:119–124
5. Simons K, Ikonen E (1997) Functional rafts in cell membranes. Nature 387:569–572
6. Marsh D (2010) Liquid-ordered phases induced by cholesterol: a compendium of binary phase diagrams. Biochim Biophys Acta 1798:688–699
7. Ohvo-Rekila H, Ramstedt B, Leppimaki P, Slotte JP (2002) Cholesterol interactions with phospholipids in membranes. Prog Lipid Res 41:66–97
8. Ramamoorthy A, Lee D-K, Santos JS, Henzler-Wildman KA (2008) Nitrogen-14 solid-state NMR spectroscopy of aligned phospholipid bilayers to probe peptide-lipid interaction and oligomerization of membrane associated peptides. J Am Chem Soc 130:11023–11029
9. Lindstrom F, Williamson PTF, Grobner G (2005) Molecular insight into the electrostatic membrane surface potential by N-14/P-31 MAS NMR spectroscopy: nociceptin-lipid association. J Am Chem Soc 127:6610–6616
10. Rothgeb TM, Oldfield E (1981) N-14 nuclear magnetic-resonance spectroscopy as a probe of lipid bilayer headgroup structure. J Biol Chem 256:6004–6009
11. Faure C, Bonakdar L, Dufourc EJ (1997) Determination of DMPC hydration in the L(alpha) and L(beta′) phases by H-2 solid state NMR of D2O. FEBS Lett 405:263–266
12. Burnell EE, Cullis PR, Dekruijff B (1980) Effects of tumbling and lateral diffusion on phosphatidylcholine model membrane P-31-NMR lineshapes. Biochim Biophys Acta 603:63–69
13. Ando I, Naito A, Saito H (2006) Solid state NMR spectroscopy for biopolymers: principles and applications. Springer, Dordrecht
14. Ramamoorthy A (2005) NMR spectroscopy of biological solids. CRC, New YOrk
15. Das N, Murray DT, Cross TA (2013) Lipid bilayer preparations of membrane proteins for oriented and magic-angle spinning solid-state NMR samples. Nat Protocols 8:2256–2270
16. Katsaras J (1997) Highly aligned lipid membrane systems in the physiologically relevant "excess water" condition. Biophys J 73:2924–2929
17. Diller A, Loudet C, Aussenac F, Raffard G, Fournier S, Laguerre M et al (2009) Bicelles: a natural 'molecular goniometer' for structural, dynamical and topological studies of molecules in membranes. Biochimie 91:744–751
18. Park SH, Loudet C, Marassi FM, Dufourc EJ, Opella SJ (2008) Solid-state NMR spectroscopy of a membrane protein in biphenyl phospholipid bicelles with the bilayer normal parallel to the magnetic field. J Magn Reson 193:133–138
19. Cardon TB, Dave PC, Lorigan GA (2005) Magnetically aligned phospholipid bilayers with large q ratios stabilize magnetic alignment with high order in the gel and L-alpha phases. Langmuir 21:4291–4298
20. Seelig J (1978) P-31 nuclear magnetic-resonance and head group structure of phospholipids in membranes. Biochim Biophys Acta 515:105–140
21. Gorenstein DG (ed) (1984) Phosphorus-31 NMR principles and applications. Academic, Orlando
22. Dufourc EJ, Mayer C, Stohrer J, Althoff G, Kothe G (1992) Dynamics of phosphate head

groups in biomembranes—comprehensive analysis using P-31 nuclear-magnetic-resonance lineshape and relaxation-time measurements. Biophys J 61:42–57

23. Dufourc EJ, Mayer C, Stohrer J, Kothe G (1992) P-31 and H-1-nmr pulse sequences to measure lineshapes, T1z and T2e relaxation-times in biological-membranes. J Chim Phys Phys Chim Biol 89:243–252

24. Picard F, Paquet MJ, Levesque J, Belanger A, Auger M (1999) P-31 NMR first spectral moment study of the partial magnetic orientation of phospholipid membranes. Biophys J 77:888–902

25. Clarke JA (2005) Solid state NMR investigations of phospholipid/cholesterol interactions in model membranes. Thesis, Imperial College London, London

26. Clarke JA, Seddon JM, Law RV (2009) Cholesterol containing model membranes studied by multinuclear solid state NMR spectroscopy. Soft Matter 5:369–378

27. Polozov IV, Gawrisch K (2004) Domains in binary SOPC/POPE lipid mixtures studied by pulsed field gradient H-1 MAS NMR. Biophys J 87:1741–1751

28. Polozov IV, Gawrisch K (2006) Characterization of the liquid-ordered state by proton MAS NMR. Biophys J 90:2051–2061

29. Forbes J, Husted C, Oldfield E (1988) High-field, high-resolution proton magic-angle sample-spinning nuclear magnetic-resonance spectroscopic studies of gel and liquid-crystalline lipid bilayers and the effects of cholesterol. J Am Chem Soc 110:1059–1065

30. Chen ZJ, Stark RE (1996) Evaluating spin diffusion in MAS-NOESY spectra of phospholipid multibilayers. Solid State Nucl Magn Reson 7:239–246

31. Feller SE, Huster D, Gawrisch K (1999) Interpretation of NOESY cross-relaxation rates from molecular dynamics simulation of a lipid bilayer. J Am Chem Soc 121:8963–8964

32. Clarke JA, Heron AJ, Seddon JM, Law RV (2006) The diversity of the liquid ordered (L-o) phase of phosphatidylcholine/cholesterol membranes: a variable temperature multinuclear solid-state NMR and X-ray diffraction study. Biophys J 90:2383–2393

33. Tonelli AE, Schilling FC (1981) C-13 Nmr chemical-shifts and the microstructure of polymers. Acc Chem Res 14:233–238

34. Ishikawa S, Ando I (1992) Structural studies of dimyristoylphosphatidylcholine and distearoylphosphatidylcholine in the crystalline and liquid-crystalline states by variable-temperature solid-state high-resolution C-13 Nmr-spectroscopy. J Mol Struct 271:57–73

35. Davis JH (1979) Deuterium magnetic-resonance study of the gel and liquid-crystalline phases of dipalmitoyl phosphatidylcholine. Biophys J 27:339–358

36. Davis JH (1983) The description of membrane lipid conformation, order and dynamics by H-2-Nmr. Biochim Biophys Acta 737:117–171

37. Seelig J (1977) Deuterium magnetic-resonance—theory and application to lipid-membranes. Q Rev Biophys 10:353–418

38. Bechinger B, Seelig J (1991) Conformational-changes of the phosphatidylcholine headgroup due to membrane dehydration—a H-2-Nmr study. Chem Phys Lipids 58:1–5

39. Brown MF, Nevzorov AA (1999) H-2-NMR in liquid crystals and membranes. Coll Surf A Physicochem Eng Aspects 158:281–298

40. Grelard A, Loudet C, Diller A, Dufourc EJ (2010) NMR spectroscopy of lipid bilayers. Methods Mol Biol (Clifton, NJ) 654:341–59

41. Vist MR, Davis JH (1990) Phase-equilibria of cholesterol dipalmitoylphosphatidylcholine mixtures—H-2 nuclear magnetic-resonance and differential scanning calorimetry. Biochemistry 29:451–464

42. Veatch SL, Polozov IV, Gawrisch K, Keller SL (2004) Liquid domains in vesicles investigated by NMR and fluorescence microscopy. Biophys J 86:2910–2922

43. Hubner W, Blume A (1990) H-2 Nmr-spectroscopy of oriented phospholipid-bilayers in the gel phase. J Phys Chem 94:7726–7730

44. Smith ICP, Butler KW, Tulloch AP, Davis JH, Bloom M (1979) Properties of gel state lipid in membranes of acholeplasma-laidlawii as observed by H-2 Nmr. FEBS Lett 100:57–61

45. Bloom M, Davis JH, Mackay AL (1981) Direct determination of the oriented sample nmr-spectrum from the powder spectrum for systems with local axial symmetry. Chem Phys Lett 80:198–202

46. Sternin E, Bloom M, Mackay AL (1983) De-Pake-Ing of Nmr-spectra. J Magn Reson 55:274–282

47. Schafer H, Madler B, Sternin E (1998) Determination of orientational order parameters from H-2 NMR spectra of magnetically partially oriented lipid bilayers. Biophys J 74:1007–1014

48. Jansson M, Thurmond RL, Trouard TP, Brown MF (1990) Magnetic alignment and orientational order of dipalmitoylphosphatidylcholine bilayers containing palmitoyllysophosphatidylcholine. Chem Phys Lipids 54:157–170

49. Nevzorov AA, Brown MF (1997) Dynamics of lipid bilayers from comparative analysis of H-2 and C-13 nuclear magnetic resonance

relaxation data as a function of frequency and temperature. J Chem Phys 107:10288–10310

50. Sani M-A, Weber DK, Delaglio F, Separovic F, Gehman JD (2013) A practical implementation of de-Pake-ing via weighted Fourier transformation. Peer J 1:e30-e

51. Vermeer LS, de Groot BL, Reat V, Milon A, Czaplicki J (2007) Acyl chain order parameter profiles in phospholipid bilayers: computation from molecular dynamics simulations and comparison with H-2 NMR experiments. Eur Biophys J Biophys Lett 36:919–931

52. Heaton N, Reimer D, Kothe G (1992) Transverse deuteron spin relaxation in thermotropic liquid-crystal polymers—determination of viscoelastic properties. Chem Phys Lett 195:448–454

53. Jarrell HC, Smith ICP, Jovall PA, Mantsch HH, Siminovitch DJ (1988) Angular-dependence of H-2 Nmr relaxation rates in lipid bilayers. J Chem Phys 88:1260–1263

54. Jarrell HC, Jovall PA, Giziewicz JB, Turner LA, Smith ICP (1987) Determination of conformational properties of glycolipid head groups by H-2 Nmr of oriented multilayers. Biochemistry 26:1805–1811

55. Stohrer J, Grobner G, Reimer D, Weisz K, Mayer C, Kothe G (1991) Collective lipid motions in bilayer-membranes studied by transverse deuteron spin relaxation. J Chem Phys 95:672–678

56. Nagle JF (1993) Area lipid of bilayers from Nmr. Biophys J 64:1476–1481

57. Petrache HI, Brown MF (2007) X-ray scattering and solid-state deuterium nuclear magnetic resonance probes of structural fluctuations in lipid membranes. In: Dopico AM (ed) Methods in molecular biology, vol 400. Humana, Totowa, pp 341–53

58. Sternin E, Schafer H, Polozov IV, Gawrisch K (2001) Simultaneous determination of orientational and order parameter distributions from NMR spectra of partially oriented model membranes. J Magn Reson 149:110–113

59. Brus J (2000) Heating of samples induced by fast magic-angle spinning. Solid State Nucl Magn Reson 16:151–160

60. Mellinger F, Wilhelm M, Spiess HW (1999) Calibration of H-1 NMR spin diffusion coefficients for mobile polymers through transverse relaxation measurements. Macromolecules 32: 4686–4691

Chapter 18

Fluorescence Recovery After Photobleaching (FRAP): Acquisition, Analysis, and Applications

Michael Carnell, Alex Macmillan, and Renee Whan

Abstract

A significant number of biological processes occur at, or involve cellular membranes, including; cell adhesion, migration, endocytosis, signal transduction, and many biochemical reactions involving membrane anchored scaffolds. Each process involves a complex arrangement of interacting molecules whose location in space and time influence the outcome of the event.

In this protocol we discuss the application of fluorescence recovery after photobleaching (FRAP) to study the dynamics of membrane associated molecules. We discuss the principles, acquisition and the analysis of FRAP data and address issues surrounding its interpretation.

Key words Fluorescence recovery after photobleaching, Diffusion, Membranes, Dynamics, Binding kinetics, Live cell imaging, Quantitation, Analysis

1 Introduction

A central property underpinning many biological processes is the ability of proteins to translocate and interact with binding partners. Being able to measure and quantify these properties can be indispensable in elucidating the molecular mechanisms driving these events. There is a panoply of techniques available to measure these properties, including fluorescence recovery after photobleaching (FRAP) [1], fluorescence correlation spectroscopy (FCS) [2, 3], and particle tracking [4, 5].

FRAP was originally described over 35 years ago [1], but has since undergone a resurgence due to the development of confocal microscopes, the improvement and lower cost of lasers, and the increase in availability of fluorescent proteins [6]. FRAP provides a method for quantifying the two-dimensional lateral diffusion within membranes of living cells, and unlike FCS and particle tracking works at normal physiological protein concentrations. It can reveal information on the rate of diffusion [1], molecular binding kinetics [7], compartmentalization [8], and active transport [9, 10].

Dylan M. Owen (ed.), *Methods in Membrane Lipids*, Methods in Molecular Biology, vol. 1232,
DOI 10.1007/978-1-4939-1752-5_18, © Springer Science+Business Media New York 2015

The technique involves rapidly and irreversibly photobleaching a population of fluorescently labeled molecules in a small region of interest within the sample/cell, followed by observing and characterizing the recovery of fluorescence in that region over time. The underlying principle is that prior to bleaching the system under study is already at equilibrium. In this state the labeled molecules may be stationary with very little or no movement, freely diffusing within their environment, undergoing active transport, transiently binding to other molecules, or a complex mixture of all. Due to high molecular concentrations within living cells, and the diffraction limit of light, these conditions are typically indistinguishable from one another under a light microscope, appearing as a static unvarying signal. However, following a spatially restricted bleach, properties about the mobility of the labeled molecules are revealed in the kinetics of any fluorescent recovery. As bleached molecules move out of the region and fluorescing molecules move in from the surrounding area, the recovery profile over time is determined by diffusion, active transport and molecular interactions [1, 11]. For instance, the proportion of molecules that are static (immobile fraction) can be ascertained by observing what proportion of the bleached signal does not return (Fig. 1a, b). The rate of recovery and its relationship with the shape and size of the bleach area can reveal information about diffusion and binding coefficients of both single and complex populations (Fig. 1c).

The analysis of FRAP recovery curves remains an active area of research, striving to become a more quantitative tool. Mathematical models are continuously being developed that simulate the rate of recovery due to environmental factors. By comparing the models with experimental data, more accurate 2D and 3D diffusion and binding coefficients can be estimated [1, 11–14]. However, these models require a priori knowledge of the system under investigation as they make certain assumptions that if not true will result in substantial errors. The most prevalent assumptions being the bleaching profile of the laser, the homogeneity of the region, and that the parameter which you are measuring (diffusion or binding) is limiting fluorescence recovery. Although in some situations this may be true, it is likely that most recovery curves will contain both diffusion and binding information. Comprehensive models are being developed that account for both diffusion and binding, though these tend not to be robust. Indeed without good starting parameters the fitting process can fail, and the values generated are not necessarily unique [15, 16]. This means a fit can result from multiple combinations of different values resulting in lack of reproducibility and ambiguity as to which values are correct. This is not to say that these models are without use, but we do recommend that the researcher familiarizes themselves with the assumptions and limitations of any model that is

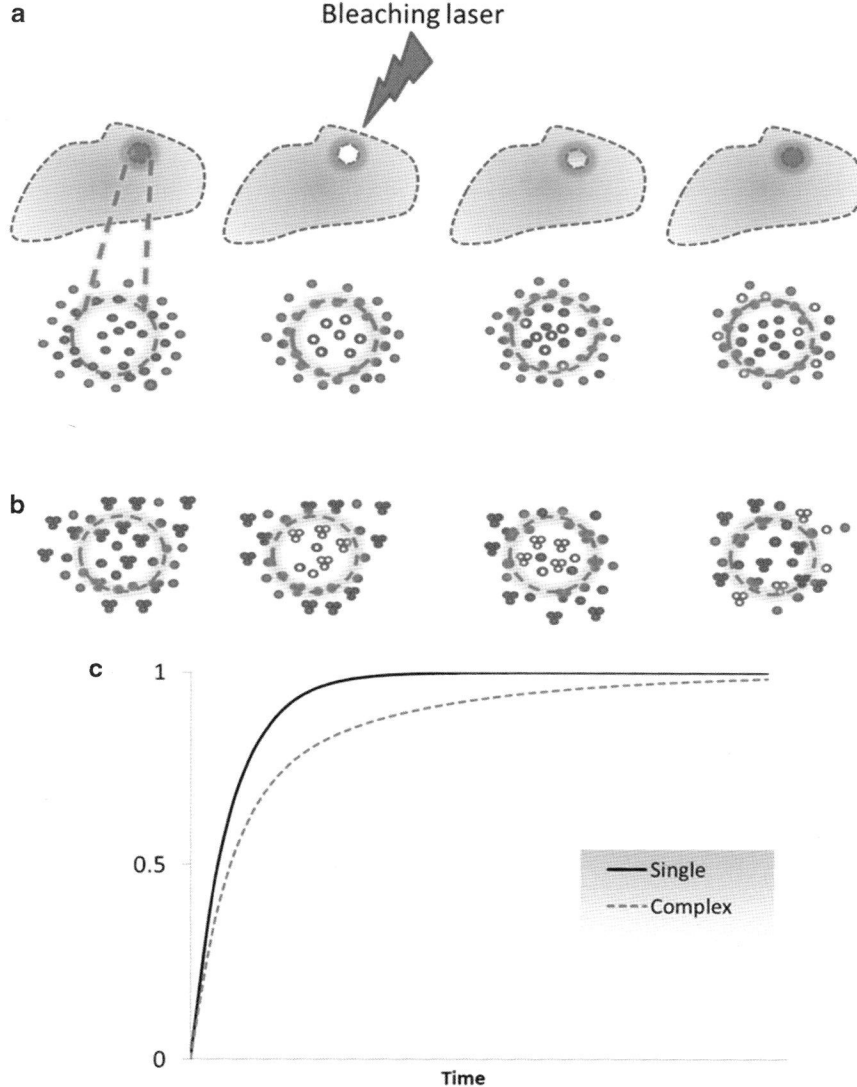

Fig. 1 Schematic of the principle underlying FRAP. Two cells, (**a**) a cell expressing a single monomeric species, and (**b**) expressing both monomeric and trimeric species. (**c**) Recovery curve for both single (**a**) and complex (**b**) curves. A baseline measurement is recorded prior to bleaching and the fluorescence recover monitored over time. Sample (**a**) illustrates a simple rapid recovery from fast diffusing monomers. Sample (**b**) illustration a complex recovery from rapidly diffusing monomers and more slowly diffusing trimers after bleaching, shown in the graph by a rapid initial recovery similar to (**a**) but this is attenuated over time

intended for implementation. In addition, the accuracy of a mixed-model can further be improved by empirically measuring one of the unknown coefficients through alternative techniques and entering it into the model as a constant.

2   Methods

2.1   FRAP Acquisition

The objective is to acquire a time-lapse image series containing a pre-bleach period suitable to characterise the initial intensity, followed by a rapid and spatially restricted bleaching phase, and ending with a post-bleached period that is adequate for characterizing the kinetics of fluorescent recovery. Suitable acquisition parameter will differ between systems, and will additionally depend on the brightness and dynamics of the sample under investigation. Here we describe general guidelines for acquiring data suitable for analysis.

Typically FRAP experiments are carried out on laser scanning confocal microscopes due to their inherent ability to perform restricted bleaching within a field of view. Customised wide-field systems can also be used if equipped with a suitable laser setup. Choosing microscope settings for fluorescent imaging is often a compromise between parameters that will affect signal-to-noise ratio, temporal, and spatial resolution. Typically for FRAP experiments the most important aspect of imaging is obtaining a high enough temporal resolution to qualitatively or quantitatively describe the characteristics of fluorescence recovery with minimal unintentional photobleaching during acquisition. With this in mind it is advisable to sacrifice spatial over temporal resolution while optimizing acquisition settings if needed.

Below we highlight some important factors that should be taken into consideration when setting up an acquisition:

- *Frame interval*: This will largely depend on the properties of the system you are investigating. If possible aim for a frame interval that will give you about ten frames by half recovery (*see* **Note 1**)

- *Bleaching during acquisition*: Unintentional bleaching during acquisition can hinder the interpretation of results. Although this bleaching can in part be corrected for, it is best to optimize your settings to limit this error during acquisition (*see* **Note 1**).

- *Pre-bleaching period*: The acquisition period prior to bleaching should provide an adequate measure of the pre-bleached intensity. This should cover 3–10 frames, however if your time interval is on the order of milliseconds, it is advisable to record more frames (~40–50) to allow any changes in the photophysics of the fluorophores to stabilize.

- *Bleaching duration*: Most models assume instantaneous bleaching which in practice is not possible. It is undesirable for bleached molecules to diffuse out of this region during bleaching, or for recovery to be occurring at the same time. Therefore it is imperative to use the shortest bleaching time possible, in practice it is advisable not to exceed a tenth of the time to half recovery.

- *Photo-damage*: The high laser power necessary for rapid photobleaching can also result in photo damage to the sample. This can alter cellular function as well as causing an artificial increase in immobile molecules (*see* controls). Aim for bleaching to cause a drop in intensity to 50–75 % of the pre-bleach value.

- *Bleaching shape*: A quantitative analysis requires fitting the recovery curve to a model. Determine what model you intend to use as it may be based on a specific shape of the bleached region. This chapter deals with circular bleach areas; however models do exist for analysing strip bleach patterns [17–19].

- *Post-bleaching period*: This should continue until the intensity within the bleach region plateaus, typically 10–50 times longer than the half time to full recovery (*see* **Notes 2** and **3**).

Controls

- *Fixed sample*: Quantitative models assume a certain bleach profile from the laser, e.g. Gaussian vs. Step function, and a model should be selected that is suitable for your system. Using a sample fixed in 4 % formaldehyde the bleach profile can be measured. Furthermore, it is advisable to measure a recovery curve on this sample to ensure that all bleaching is irreversible and there is not an increase in intensity after bleaching due to a long-lived dark state [19].

- *Acquisition image bleaching*: Bleaching during acquisition must be accounted for during analysis. In some cases a suitable reference region may not be available. If this is the case, it is possible to characterize the rate of unintentional bleaching on a separate sample and later use this to correct relevant measurements (*see* Subheading 2.2.2).

- *Photo-damage and photo-induced immobile fraction*: It is possible that the high laser power needed for rapid bleaching can cause photo-damage that results in perturbations to the system or causes a population of otherwise mobile molecules to become immobile. This can be assessed by re-bleaching the same region. Obtaining different half times or coefficient values with sequential bleaching is a sign of photo-damage; consider using fluorescent proteins that are more readily bleached than GFP (such as CFP, YFP, etc.). This will allow lower laser powers to be used during the bleaching step [20]. The mobile fraction however should be drastically reduced with sequential bleaching. To explain why assume 100 % of the fluorescence is bleached in the first FRAP measurement, after recovery the fluorescence observed is from that of the mobile fraction. After re-bleaching this region all the newly bleached molecules should be mobile so no immobile fraction should be detected. If you continue to get similar immobile fraction measurements you may be causing a photo-induced immobile fraction, or it may be due to bleaching a large proportion of the total fluorescence (*see* Subheading "The Gap Ratio").

- *Diffusion vs. binding*: If you wish to get an indication if the recovery is diffusion limited or reaction limited, it is advisable to try bleaching with different region sizes (*see* Subheading "Diffusion-Limited Recovery"). Bleaching cells with different expression levels can also be used to assess reaction-limited recovery (*see* Subheading "Reaction-Limited Recovery").

2.2 FRAP Analysis

2.2.1 Measuring Data

To extract the necessary data from a FRAP image sequence all that is required is software capable of reporting the mean intensity within a user specified region of interest over time. Most acquisition software is capable of doing this, but it is often more convenient to use free software such as ImageJ (http://rsbweb.nih.gov/ij/) to process this data away from the microscope.

Regions to Measure: (*see* Fig. 2)

- *FRAP area*: User-specified region of high-intensity bleach
- *Background*: Region outside of the cell where there should be no fluorescence.
- *Reference area*: Area within an adjacent structure/cell. This is used to assess undesirable bleaching due to image acquisition. This can also be measured in the same cell assuming the bleaching stage does not deplete a substantial proportion of the fluorescent molecules; though it is less suitable. *Whole cell*: Average of the entire cell to observe the proportion of molecules bleached.

The bleach area is the most fundamental measurement to obtain, and can be used by itself to measure the recovery kinetics. However, this is not ideal as the intensity within this region is not solely determined by fluorophore kinetics but is also affected by background within the imaging system and potentially corrupted by bleaching during acquisition. By measuring a background area, this value can be subtracted from the other traces improving reproducibility between experiments where background may vary. Measuring the mean intensity of the whole cell, or a reference area, to characterize the rate of bleaching is also highly recommended where possible, as bleaching during acquisition attenuates recovery, affecting curve fitting, and inflating the measured immobile fraction. Limiting bleaching is far more desirable and accurate than attempting to correct for it post-acquisition, so in the case of a high degree of photobleaching in your reference measurements, consider re-optimizing the acquisition settings.

2.2.2 Photobleaching Correction

Bleaching during image acquisition can make accurate interpretation of recovery curves difficult. If significant bleaching occurs it will attenuate the rate of recovery and also result in a lower plateau

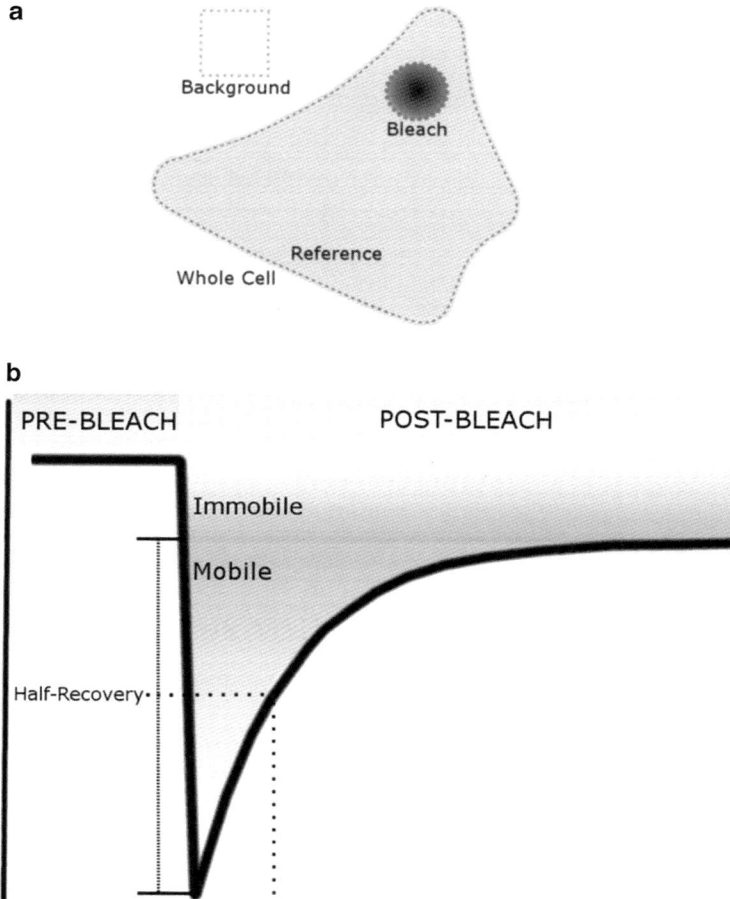

Fig. 2 (**a**) Cell illustration showing bleach, background and reference regions (**b**) recovery curve showing half-life recovery, mobile, and immobile fractions

value, which can incorrectly be interpreted as a higher proportion of immobile molecules. FRAP recovery curves are also typically normalized (*see* Subheading 2.2.3), and some normalisation methods incorporate photobleaching correction. For simplicity it is advised that these methods be used, where a suitable reference is available within the sample. However, it should be noted that in cases where this is not possible then mathematically describing the rate of photobleaching is an option. In a separate field of view acquire a time-lapse with identical acquisition settings to those used during the FRAP experiment. The rate of photobleaching should follow a single negative exponential decay:

$$\mathrm{sample}(t) = \mathrm{sample}(0) \cdot e^{-at}$$

where "sample(t)" is the intensity of the sample at time t, sample(0) is the initial intensity (t=0) and "a" is rate of photobleaching. Images collected under identical conditions can then be corrected by

$$\text{corrected}(t) = \text{measured}(t) \cdot e^{at}$$

It can be argued that modeling the properties of photobleaching to correct FRAP measurements is a preferable approach over correcting each time point with a measured reference. The measurement of a reference region contains unavoidable measurement error. This results in compounding the noise in the final decay curve, whereas the measurement error is avoided by using a modeled decay curve to account for the photobleaching correction.

2.2.3 Normalization

To aid analysis, recovery curves are normalized in one of the two ways. Firstly, the values get converted so that the pre-bleached intensity is equal to 1. This allows comparison between samples of varying brightness as it describes the data as a proportion of its original value. Secondly, to simplify model fitting the curve can be fully normalized so that the initial intensity is equal to 1 and the intensity at the time of bleaching is equal to 0. This represents the data as a proportion of recovery with 1 being equal to full recovery (100 %), 0 being no recovery (0 %).

If measurements are available for the bleach, background and a suitable reference region to characterize unintentional bleaching region, a typical approach is the double normalization [7]:

$$\text{Norm}(t) = \frac{\text{Ref}_{\text{pre-bleach}}}{\text{ref}(t)} \cdot \frac{\text{FRAP}(t)}{\text{FRAP}_{\text{pre-bleach}}}$$

Where $\text{Ref}_{\text{pre-bleach}}$ is the mean intensity pre-bleach, $\text{FRAP}_{\text{pre-bleach}}$ is the mean intensity of the FRAP region pre bleach, ref(t) the intensity of the reference region at time point t, FRAP(t) is the intensity of the FRAP region at time point t.

If suitable background levels and reference regions cannot be obtained, the data can still be normalized by removing the first fraction, however, when interpreting results always remain aware that acquisition photobleaching has not been corrected for unless the model used accounts for it, or it has been corrected for prior to normalization:

$$\text{Norm}(t) = \frac{\text{FRAP}(t)}{\text{FRAP}_{\text{pre}}}$$

For analysis that requires curve fitting that assumes point of bleaching equals 0 a further normalization step is required to fully normalize the measurements:

$$\text{Full norm}(t) = \frac{\text{FRAP}(t) - \text{FRAP}_{\text{bleach}}}{\text{FRAP}_{\text{pre}} - \text{FRAP}_{\text{bleach}}}$$

where $FRAP_{bleach}$ is the intensity of the FRAP region at the time of bleaching.

2.2.4 Further Manipulation

FRAP measurements can often be noisy due to the requirement of high temporal resolution. Curve fitting can be made easier by averaging in the range of 3–10 fully normalised recovery curves prior to fitting. Be aware that if the repeats you are performing genuinely are variable then this will report an average recovery curve that will not necessarily fit to any given model. Furthermore, if you are intending to measure binding kinetics, care should be taken when averaging normalized FRAP curves as different expression levels of GFP fusion products will contain different fractions of bound and free molecules within the cell [15].

If you have imaged at a constant time interval you may find that you have relatively few points at the early rapid recovery phase and a large number at the slower stage. This can result in biasing the curve fitting to weight it more heavily towards the slower part of recovery. In this case it is recommended that either several adjacent time points later in slower part of recovery are averaged or that certain points are removed to make the weighting more equal.

2.2.5 Qualitative Analysis

A qualitative analysis of FRAP recovery curves is relatively simple and perfectly suitable for a large variety of experimental conditions. It has the advantage of demonstrating a change between different experimental conditions without having to deal with the many assumptions that must be made when fitting various models in a quantitative analysis. In its most simple form, plotting the fully normalized recoveries can visually indicate whether altering some condition causes an increase or decrease in the recovery speed, and/or the fraction of immobile molecules (Fig. 3). These values can be easily quantified by fitting an exponential model to the normalized FRAP curve (*see* Appendix and **Note 4**):

$$y(t) = A \cdot \left(1 - e^{-\tau \cdot t}\right)$$

where "*A*" equals plateau intensity, τ = fitted parameter and t = time after bleach.

Depending on your experimental conditions, it is possible that a single exponential curve will not fit. This can be caused by a complex recovery due to mixed populations of molecules that have different kinetics, such as fast and slow exchanging populations. In this case, fitting a double exponential may be required.

The recovery time of FRAP curves are reported as $t_{1/2}$, which is the time to reach half the intensity of full recovery. Once an exponential recovery has been fit the $t_{1/2}$ is obtained by

$$t_{1/2} = \frac{\ln 0.5}{-\tau}$$

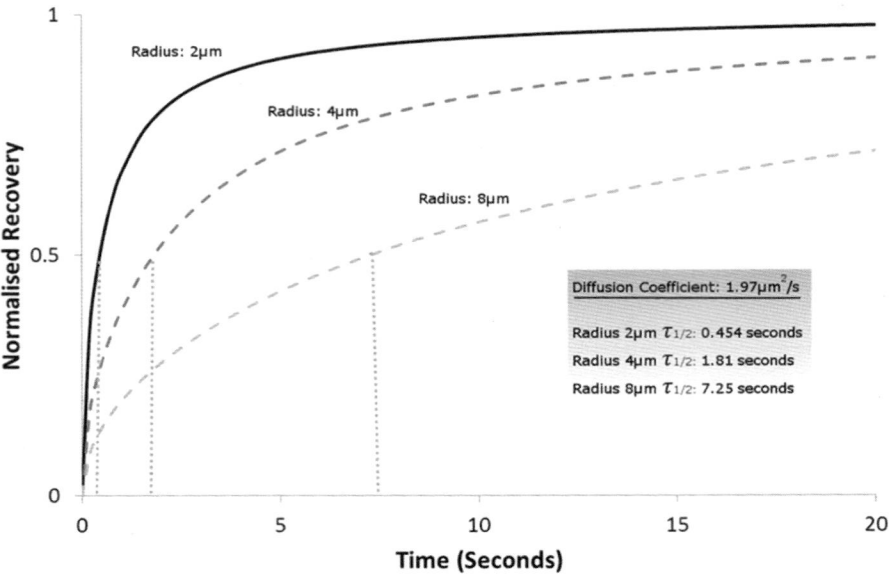

Fig. 3 Modeled recovery curves demonstrating the dependence of recovery on the radius of the bleaching area

Mobile/immobile fraction: From a fully normalized curve the mobile fraction is the value of the plateau, where the curve stabilizes. The immobile fraction is 1 – mobile fraction.

It should be noted that if the total fraction of molecules rendered non-fluorescent during bleaching is substantial, this can impact on the perceived mobile/immobile fraction. In the case of no immobile fraction being present, then the normalized recovery curve will plateau at a value of the original pre-bleach intensity minus the proportion of permanently bleached fluorophores. In the case of a tenth of the fluorophore being permanently bleached during the bleach step the final plateau value will be 0.9 for a normalized curve.

The Gap Ratio

The Gap ratio is a measure of the fluorescence lost at the point of bleaching. It is simply the mean intensity of the whole cell post-bleach divided by the mean intensity pre-bleach (after subtracting background and correcting for photobleaching). After measuring the mobile fraction from the plateau of a fully normalized recovery curve, it results in a more accurate prediction of mobile/immobile fractions by dividing the measured mobile fraction by the gap ratio:

$$\text{Gap ratio} = \frac{\text{Whole cell}_{\text{post-bleach}}}{\text{Whole cell}_{\text{pre-bleach}}}$$

where Whole cell$_{\text{post-bleach}}$ is the mean intensity of the whole cell post-bleach, Whole cell$_{\text{pre-bleach}}$ is the mean intensity of the whole cell pre-bleach.

$$\text{Corrected mobile fraction} = \frac{\text{Measured mobile fraction}}{\text{Gap ratio}}$$

2.2.6 Quantitative Analysis

In addition to describing the proportion of immobile molecules within a given region, and the time taken to reach half the final recovery value, it is possible to extract several properties such as diffusion coefficients, and binding kinetics. This is achieved by carefully selecting an appropriate model for both the system under investigation and the imaging set up. In some cases, extracting these values requires little extra effort over quantitative methods, simply fitting a different model to the full-normalized recovery curve. These, however, tend to suffer from assumptions that in practice are rarely upheld. Progress is being made in this area with newer models accounting for recovery during bleaching [21], using spatial information within the bleach area to overcome assumptions on the bleaching profile [22–24], models accounting for mixed populations and diffusion constants, and active transport and binding [15, 25–27].

Given there is currently no ideal model that does not rely on some assumptions and prior knowledge about the system under investigation, it is beyond the scope of this chapter to address all possible models. As such we will address the simplest generic models for diffusion and binding kinetics. It is suggested that prior to undertaking any quantitative FRAP experiments the researcher investigates the current literature for new models relating to their application, as well as considering a qualitative approach to the investigation.

Diffusion-Limited Recovery

On the assumption that the recovery of the FRAP curve is primarily due to the diffusion of molecules moving by Brownian motion, there are several models to which a fully-normalized recovery curve can be fit to estimate a diffusion coefficient. Here we will highlight two models that describe lateral diffusion suitable for membrane studies:

$$\text{Full norm}(t) = A \cdot e^{-\frac{2\tau_D}{t}}\left(I_0\!\left(\frac{2\tau_D}{t}\right) + I_1\!\left(\frac{2\tau_D}{t}\right)\right)$$

where

$$\tau_D = \frac{w^2}{4D}$$

where "A" is the plateau intensity, "D" is the diffusion coefficient, "w" bleach area radius, and $I_0()$ and $I_1()$ are modified Bessel functions [11].

Assumptions: Step function profile, instantaneous bleaching, recovery is diffusion-limited, single population with one diffusion constant, lateral diffusion, diffusion equal in all directions.

For those unfamiliar with the curve fitting procedure, and the daunting appearance of this model, we have provided an example in the appendix demonstrating how to use a nonspecialist piece of software, Microsoft Excel, to fit models using this model as an example.

Although this model was developed for studying lateral mobility, making it suitable for measuring diffusion in two dimensions within membranes, in theory it can be used in a 3D space if the bleaching volume is cylindrical.

The above model assumes a step function bleaching profile, if however the profile is Gaussian the diffusion can still be "estimated" using the following equation:

$$D = \frac{w^2}{4t_{1/2}} \bullet 0.88$$

where D is the estimated diffusion coefficient.

A model is available (*see* Axelrod et al. [1]) to be fit to a recovery curve although it requires multiple parameters including two parameters of the bleaching profile that must be experimental measured.

Both of these models assume diffusion dominated recovery. Confidence in this assumption can be increased by performing several FRAP measurements whilst varying the size of the bleaching region. In the case of diffusion, the rate of recovery will have a nonlinear relationship with the bleach size (Fig. 3). Specifically it scales with radius^{-2}, meaning if you double the radius of the bleach area the recovery will be four times slower. In principle this is not the case for reaction limited recovery, where its rate should not significantly change [28].

Reaction-Limited Recovery

Although the recovery of fluorescence is due to an exchange of molecules between the bleach region and the surrounding area, in some situations it is not the rate of diffusion or active transport that dictates the kinetics of recovery but rather the binding kinetics of the molecule. If the observed molecules are in an environment with fast diffusion, but the majority of the population are in a state of slow transient binding with stationary structures, the kinetics of recovery will be primarily dominated by the molecules dissociation constant [14]. In this ideal scenario the recovery curve can be fit to an exponential:

$$\text{Norm}(t) = A \cdot \left(1 - e^{-k_{off}t}\right)$$

where k_{off} is the fitted parameter.

Assumptions: The majority of fluorescent molecules are bound transiently to immobile structures, k_{off} is much slower than diffusion, it is a single population with the same dissociation constant [14].

Similar to the diffusion model above, there is a test to assess if recovery is reaction limited. If the sample is diffusion limited the rate of recovery should be independent of concentration. However, in a reaction limited scenario increasing the concentration of molecules is likely to lead to a higher concentration of unbound molecules; assuming no change in the number of binding sites. As such, faster recovery correlating with increased expression levels is indicative of a reaction limited sample [29, 30]. Consequently, it is imperative that during sample comparison, that the samples have similar concentrations or expression levels. Furthermore it should be noted that in reaction limited scenarios, unaccounted for diffusion can introduce errors.

Mixed Models

In previous sections we have outlined simple models that estimate either diffusion or binding kinetics. They both assume that the recovery of fluorescence is limited by the property being investigated. Not only would these estimates be more accurate using a model that takes both diffusion and binding into account, but this model would also be useful for scenarios where both diffusion and binding contribute equally to the recovery. Models that attempt to estimate more than one parameter are becoming increasing popular as an attempt to improve the accuracy of these estimations. It would be impractical to cover all these models in this protocol so instead we will direct you to several interesting examples.

In 2004, Sprague et al. developed a reaction-diffusion model accounting for single binding kinetics and incorporating a single diffusion coefficient [15]. Many models assume that binding interactions are transient interactions with immobile structures, but for the case where there are two populations of mobile molecules that transiently bind we direct you to Braga et al. [25]. When investigating proteins that can transiently bind to membranes recovery is affected by three factors, lateral diffusion when bound to the membrane, binding kinetics, and cytosolic diffusion, this has been addressed by Goehring et al. [26]. One of the more interesting recent developments is from Mai et al. [27]. Typically a suitable model is selected by the experimenter and then this is used to extract parameters from the recovery curve. Mai et al., provides a method for determining a suitable model from the recovery data itself, further reducing the number of assumptions being made [27].

3   Conclusions

FRAP provides an easy implemented method to investigate the mobility of proteins in their natural environment with minimal invasion. The characterization of diffusion provides relevant information of biological processes within the membrane as is evident by the number of publications using the FRAP technique.

However, quantification of diffusion requires careful selection of the correct model for analysis. In this chapter we have provided an overview for the application of FRAP to study membrane diffusion and binding, however each system will be uniquely different. The ability to utilize the technique on any microscope setup and the ability to analyses the data in nonspecialized software provides a very attractive proposition for studying membrane diffusion and binding.

4 Notes

1. At times it may be desirable to change your imaging interval during recovery. This will allow you to acquire quickly early in the recovery stage, and reduce bleaching by taking fewer points in the slower stage of recovery. However, be aware that if the photophysics of the dye is changed by the rapid excitation you should image at a continuous rate for the entire experiment.

2. After determining the time required to reach the plateau it is advisable to increase your imaging period by a further 30 % to account for slower recoveries within your samples.

3. In some cases, long-term experiments can result in protein synthesis or degradation occurring [31]. If long-term experiment are required then Photoactivatable GFP are available can be used as alternative. In this case only activated GFP molecules are fluorescent and newly synthesized molecules are non-fluorescent and will not affect the result [20].

4. The model was calculated over two columns C and D to try and simplify understanding and readability. Column C calculated the term $2\tau D/t$ for each time point which appears three times in the equation, which is then utilized in column D. However it could have been calculated in one column with the following formula:

   "=EXP(-(2*$H$3/A3)) * (BESSELI(2*$H$3/A3, 0) +BESSELI(2*$H$3/A3, 1))"

5 Appendix

Model fitting using Microsoft Excel 2010 (Fig. 4). Below is an example worksheet, cell formulas and instructions on how to use Excel's Solver tool. This example fits a fully normalized data set to a 2D diffusion model to obtain an estimated diffusion coefficient, but is easily adapted to other models. The principle is to produce a column that reflects the models values at the same time points as your measurements. Make sure that the coefficients that you want

	A	B	C	D	E	F	G	H
1	Time	Intensity	2τd/t	Model	Chi Squared			
2	0	0.000	0.000	0.000	0.000		Radius (ω) μm:	4
3	0.2	0.201	20.283	0.176	0.001		τD	2.028297
4	0.4	0.261	10.141	0.247	0.000		Plataue (A)	1.000
5	0.6	0.317	6.761	0.301	0.000			
6	0.8	0.331	5.071	0.345	0.000		Diffusion Coefficent (D) μm^2/s	1.972097
7	1	0.361	4.057	0.383	0.000		X^2 Sum:	0.006
8	1.5	0.436	2.704	0.461	0.001			
9	2	0.541	2.028	0.521	0.000		Model	
10	2.5	0.568	1.623	0.569	0.000			
11	3	0.630	1.352	0.609	0.000			
12	3.5	0.659	1.159	0.642	0.000			
13	4	0.669	1.014	0.671	0.000			
14	4.5	0.671	0.901	0.695	0.001			
15	5	0.741	0.811	0.716	0.001			
16	5.5	0.719	0.738	0.734	0.000			
17	6	0.763	0.676	0.750	0.000			
18	7	0.755	0.580	0.777	0.000			

$$f(t) = A \cdot e^{-2\tau D/t}\left(I_0\left(\frac{2\tau D}{t}\right) + I_0\left(\frac{2\tau D}{t}\right)\right)$$

$$\tau D = \omega^2/4D$$

τD [H3]: '=H2^2/4*H6)'

2τD/t [C3]: '=2*$H$3/A3' (see Note 1)

Model [D3]: '=H4*EXP(-C3) * (BESSELI(C3,0) + BESSELI(C3,1))'

Chi Squared [E3]: '=(B3-D3)^2'

X^2 Sum (H7): =SUM(E2:E22)

Fig. 4 Using solver to fit a model (*see* **Note 4**)

to determine are in separate cells which the model formula references. For example, in the excel sheet below the diffusion coefficient value is stored in cell "H6," this means by changing this cell all values in column D are updated. Next we use the Solver function of excel to try multiple values in "H6" and reveal which value gives the closest fit between our model and sampled data. If Solver is not visible under the "Data" tab you must enable it in the Ribbon: File → Options → Add-Ins. Select Excel Add-ins in the Manage drop-down box and click go. Select the Solver Add-in checkbox and click "OK". Click on "Solver" to open up the "Solver Parameters" dialogue. Here you get to specify what variables to change to achieve a certain objective, in this circumstance change the diffusion constant to that which gives the lowest X^2 sum. Set the field "Set Objective:" to point to the X^2 Sum field [H7] and select the "Min" radio button. This specifies that the solution should be that which gives the smallest X^2 sum. Set the "By Changing Variable Cells:" selection box to point to the diffusion Coefficient value [H6]. Click Solve. Plot the measured and modeled values to inspect how well the model fits your data. Due to noise the measured points should deviate above and below the modeled points at random.

Acknowledgements

The authors would like to thank the Mark Wainwright Analytical Centre, University of New South Wales for continuing support.

References

1. Axelrod D, Koppel D, Schlessinger J, Elson E, Webb W (1976) Mobility measurement by fluorescence photobleaching recovery kinetics. Biophys J 16:1055–1069

2. Magde D, Webb WW, Elson E (1972) Thermodynamic fluctuations in a reacting system - measurement by fluorescence correlation spectroscopy. Phys Rev Lett 29(11):705

3. Elson EL, Magde D (1974) Fluorescence correlation spectroscopy. 1. Conceptual basis and theory. Biopolymers 13(1):1–27

4. Gelles J, Schnapp BJ, Sheetz MP (1988) Tracking kinesin-driven movements with nanometre-scale precision. Nature 331(6155): 450–453

5. Lee GM, Ishihara A, Jacobson KA (1991) Direct observation of Brownian-motion of lipids in a membrane. Proc Natl Acad Sci U S A 88(14):6274–6278

6. Tsien RY (2010) Nobel lecture: constructing and exploiting the fluorescent protein paintbox. Integr Biol 2(2–3):77–93

7. Phair RD et al (2004) Global nature of dynamic protein-chromatin interactions in vivo: three-dimensional genome scanning and dynamic interaction networks of chromatin proteins. Mol Cell Biol 24(14):6393–6402

8. James PS, Hennessy C, Berge T, Jones R (2004) Compartmentalisation of the sperm plasma membrane: a FRAP, FLIP and SPFI analysis of putative diffusion barriers on the sperm head. J Cell Sci 117(26):6485–6495

9. Lai FPL et al (2008) Arp2/3 complex interactions and actin network turnover in lamellipodia. EMBO J 27(7):982–992

10. Hallen MA, Liang Z-Y, Endow SA (2008) Ncd motor binding and transport in the spindle. J Cell Sci 121(22):3834–3841

11. Soumpasis DM (1983) Theoretical analysis of fluorescence photobleaching recovery experiments. Biophys J 41(1):95–97

12. Braeckmans K, Peeters L, Sanders NN, De Smedt SC, Demeester J (2003) Three-dimensional fluorescence recovery after photobleaching with the confocal scanning laser microscope. Biophys J 85(4):2240–2252

13. Mazza D et al (2008) A New FRAP/FRAPa method for three-dimensional diffusion measurements based on multiphoton excitation microscopy. Biophys J 95(7):3457–3469

14. Bulinski J, Odde D, Howell B, Salmon E, Waterman-Storer C (2001) Rapid dynamics of the microtubule binding of ensconsin in vivo. J Cell Sci 114:3885–3897

15. Sprague B, Pego R, Stavreva D, McNally J (2004) Analysis of binding reactions by fluorescence recovery after photobleaching. Biophys J 86:3473–3495

16. Zadeh KS, Montas HJ, Shirmohammadi A (2006) Identification of biomolecule mass transport and binding rate parameters in living cells by inverse modeling. Theor Biol Med Model 3:36

17. Pederson T (2001) Protein mobility within the nucleus—What are the right moves? Cell 104(5):635–638

18. Houtsmuller AB, Vermeulen W (2001) Macromolecular dynamics in living cell nuclei revealed by fluorescence redistribution after photobleaching. Histochem Cell Biol 115(1): 13–21

19. Levin MH, Haggie PM, Vetrivel L, Verkman AS (2001) Diffusion in the endoplasmic reticulum of an aquaporin-2 mutant causing human nephrogenic diabetes insipidus. J Biol Chem 276(24):21331–21336

20. Patterson GH, Lippincott-Schwartz J (2002) A photoactivatable GFP for selective photolabeling of proteins and cells. Science 297(5588):1873–1877

21. Wu J, Shekhar N, Lele PP, Lele TP (2012) FRAP analysis: accounting for bleaching during image capture. PLoS One 7(8):e42854

22. Berk DA, Yuan F, Leunig M, Jain RK (1993) Fluorescence photobleaching with spatial Fourier analysis: measurement of diffusion in light-scattering media. Biophys J 65(6): 2428–2436

23. Deschout H et al (2010) Straightforward FRAP for quantitative diffusion measurements with a laser scanning microscope. Opt Express 18(22):22886–22905

24. Jönsson P, Jonsson MP, Tegenfeldt JO, Höök F (2008) A method improving the accuracy of fluorescence recovery after photobleaching analysis. Biophys J 95(11):5334–5348

25. Braga J, McNally JG, Carmo-Fonseca M (2007) A reaction-diffusion model to study RNA motion by quantitative fluorescence recovery after photobleaching. Biophys J 92(8):2694–2703

26. Goehring NW, Chowdhury D, Hyman AA, Grill SW (2010) FRAP analysis of membrane-associated proteins: lateral diffusion and membrane-cytoplasmic exchange. Biophys J 99(8):2443–2452

27. Mai J et al (2011) Are assumptions about the model type necessary in reaction-diffusion modeling? A FRAP application. Biophys J 100(5):1178–1188

28. Wu ES, Jacobson K, Szoka F, Portis A (1978) Lateral diffusion of a hydrophobic peptide, n-4-nitrobenz-2-oxa-1,3-diazole gramicidin-s, in phospholipid multibilayers. Biochemistry 17(25):5543–5550

29. Icenogle RD, Elson EL (1983) Fluorescence correlation spectroscopy and photobleaching recovery of multiple binding reactions. 2. FPR and FCS measurements at low and high DNA concentrations. Biopolymers 22(8): 1949–1966

30. Safranyos RG, Caveney S, Miller JG, Petersen NO (1987) Relative roles of gap junction channels and cytoplasm in cell-to-cell diffusion of fluorescent tracers. Proc Natl Acad Sci 84(8):2272–2276

31. Bancaud A, Huet S, Rabut G, Ellenberg J (2010) Fluorescence Perturbation Techniques to Study Mobility and Molecular Dynamics of Proteins in Live Cells: FRAP, Photoactivation, Photoconversion, and FLIP. Cold Spring Harbor Protocols 2010, pdb.top90, doi:10.1101/pdb.top90

Chapter 19

The Laurdan Spectral Phasor Method to Explore Membrane Micro-heterogeneity and Lipid Domains in Live Cells

Ottavia Golfetto, Elizabeth Hinde, and Enrico Gratton

Abstract

In this method paper we describe the spectral phasor analysis applied to Laurdan emission for the assessment of the fluidity of different membranes in live cells. We first introduce the general context and then we show how to obtain the spectral phasor from data acquired using a commercial microscope.

Key words Laurdan, Spectral phasor, Membrane fluidity, Lipid domains

1 Introduction

Lateral organization and lipid packing of cell membranes, both plasma and intracellular, allows for the regulation of numerous cellular processes essential to cell life and death. Membrane heterogeneity, which arises from variation in lipid composition, lipid order and diversified lipid-protein interaction, is essential for the correct functioning of biological membranes. It has been shown that lateral membrane heterogeneity influences and regulates cellular polarization, signaling events, transduction pathways, trafficking, and secretion of molecules [1–3]. Within this concept, two related aspects of membrane heterogeneity are particularly relevant: membrane fluidity (or order) and membrane domains.

Membrane fluidity plays a key role in the formation of membrane domains. A more rigid membrane environment influences the lateral diffusion and concentration of membrane proteins [4], which in turn regulates interactions among proteins and therefore the efficiency of signaling pathways through the formation of membrane domains. Some examples are viral entry/budding [5], cell polarization and adhesion [6], membrane trafficking [7, 8], and B cell signaling [9]. The membrane lipid composition defines the degree of lipid packing, the thickness of the bilayer, and the rotational freedom that the lipids have in the bilayer.

Dylan M. Owen (ed.), *Methods in Membrane Lipids*, Methods in Molecular Biology, vol. 1232,
DOI 10.1007/978-1-4939-1752-5_19, © Springer Science+Business Media New York 2015

All these parameters are included in the general concept of membrane fluidity or order. In order to evaluate membrane fluidity a key role is played by water molecules in the membrane. The degree of lipid packing can be quantified because more ordered packing excludes polar water molecules from the nonpolar lipid bilayer. This results in a change in the local environment polarity, which can be sensed by polarity environment-sensitive fluorescent probes, such as Laurdan.

Membrane domains are (micro) regions of the membrane, which are enriched in cholesterol and sphingolipids, such as gangliosides. According to their composition their intrinsic fluidity is different from the rest of the membrane. In response to signaling, membrane domains may fuse into larger and more stable structures, which could become efficient signaling platforms. In other words, they are thought to modulate cellular signaling and to initiate cellular processes before/after sequestering specific proteins from the rest of the membrane. A large effort has been spent to investigate the existence and nature of membrane domains or rafts, via several single-molecule high resolution techniques such as [10–14]. These techniques support the existence of the postulated domains and the importance of cholesterol in their composition and existence. In these studies the use of probes that bind to various membrane constituents is very common. One of the most widely used examples is Cholera Toxin subunit B (CT-B), from the bacterium *Vibrio cholerae*. This protein binds to the ganglioside GM1, which, similarly to other glycosphingolipids, has been found to be concentrated in ordered and cholesterol-enriched membrane domains.

As mentioned previously, fluorescent probes are extensively used to study membrane heterogeneity. In order to address questions oriented toward measuring subtle changes of either membrane fluidity and/or order in membrane domains, two different classes of fluorescent probes can be distinguished: environment sensitive and binding probes. The first class of probes (environment sensitive) is particularly suited to explore membrane fluidity and order. These probes have equal and high affinity for every region of the membrane. Some examples of these fluorescent membrane markers are Laurdan (with its derivatives) [15–21], di-4-ANEPPDHQ [22, 23], and other solvatochrome derivatives [24–26]. In this work we focus on the Laurdan probe and its capability to sense the polarity of its environment [27]. In the case of lipid bilayers, Laurdan senses the presence of water in the membrane, thus providing information about water penetration, which is directly related to membrane fluidity. When originally used in model membranes, it was shown that Laurdan distinguishes lipid domains in model systems according to at least two separate classes: gel (alias solid-ordered) and fluid (alias liquid-disordered) phases [28]. To quantify the contribution of these two phases in membranes, a normalized ratiometric method of analysis based on a two-channel detection was developed the Generalized Polarization (GP) function [29].

The second class of probe (binding) includes probes that have affinity for membrane ordered domains, and therefore are considered to stain selectively only the liquid-ordered phase in membranes. Some examples are probes that bind to most glyco-phosphatidylinositol-anchored (GPI-anchored) proteins as well as palmitoylated proteins. As we mentioned before, the fluorescently labeled Cholera Toxin subunit B (CT-B), which binds to the ganglioside GM1, belongs to this family of probes [30–32]. It is usually assumed that CT-B binds to specific membrane domains but there are some issues. First, the ganglioside GM1 is not expressed by all cells, therefore in some cases CT-B is not useful as a membrane domain probe. Moreover, a significant fraction of GM1 is found outside ordered regions, at least when judged by detergent resistance [33, 34], and the toxin may also bind to other molecules such as glycosylated proteins [35]. Therefore, binding of CT-B alone should not be taken as evidence of a microdomain, but instead an indication, in the instance it is confirmed by additional approaches.

In this work we aim to explore membrane lateral heterogeneity using a new detection scheme based on a dual fluorescent labeling and novel approach to spectral analysis. More precisely, we combine the fluorescent labels Laurdan and CT-B, with the aim to detect the fluidity of the regions which CT-B binds. The reasoning behind this is that, despite CT-B being useful for imaging ordered domains due to its binding properties, it is not able to provide information about the actual micro-environment of these domains and is not very specific in live cells. Thus since Laurdan labels all the membranes in a cell and is sensitive to the domain micro-environment, a dual labeling could determine the specific fluidity of the regions where CT-B binds. Then for image analysis of the combined Laurdan and CT-B signal, we propose a novel detection scheme which exploits the additive property of the phasor to identify the specific "local fluidity" of the membrane regions where CT-B binds.

Laurdan fluorescence has been used to study lateral membrane heterogeneity both in model systems and in live cells due to the dependence of the Laurdan emission on the packing and water content of the membrane. This spectral sensitivity originates from the dipolar relaxation effect in which few water molecules in the proximity of the Laurdan molecules reorient after excitation to lower the energy of the excited state in times comparable to the Laurdan fluorescence lifetime producing a gradual shift of the spectrum. Traditionally, this shift is captured by the Generalized Polarization function or GP which is essentially a normalized ratiometric measurement obtained by measuring the emission at two wavelengths. Since the shift is gradual, the measurement at two wavelengths only reflects the projection of the spectral shift on the wavelength axis as captured at these two wavelengths. Therefore a limitation of the GP function is that all the spectra that give equal projections at these two wavelengths appear indistinguishable. This limitation is particular severe for the analysis of the GP of microscopy images of live cells where many different membranes are presents, each with different lipid composition and packing.

Here we describe the Laurdan spectral phasor approach, a detection scheme and image analysis of Laurdan emission in model systems and in live cells, that has the potential to capture in greater detail the process of dipolar relaxation as it occurs in time and at different pixels of microscope images. Since the possible spectral combinations in the pixels of an image arise from a continuum of Laurdan environments, it is difficult to establish a priori the number and shape of the different spectra in a given image. In this respect, the spectral phasor analysis offers two important advantages. First, due to the linear properties of the phasor transformation it could identify spectral species occurring in each and every pixel of an image. Second, using higher harmonics of the phasor transformation it could identify subtle changes in spectral width and spectrally overlapping components.

In this work, we first present the phasor analysis in the case of two well-established lipid phases, gel and fluid (liquid-disordered) in model membrane systems. Then, we perform a comprehensive analysis of Laurdan spectral phasor in NIH3T3 and HEK293 live cells. We found that the spectral phasor approach has the potential to identify different membranes in a live cell on the basis of the Laurdan spectral emission. We use these spectral signatures and their linear combinations to investigate micro-heterogeneities in live cells' membrane. Treatment with cholesterol reducing agents shows that we can quantify the abundance of regions with a given spectral signature. Finally, by coupling two different fluorescent membrane probes (Laurdan and fluorescently labeled Cholera Toxin subunit B) we are able to characterize the fluidity of the membrane around different GM1-enriched domains identified by Cholera Toxin binding.

Laurdan [27] is a fluorescent dye used to measure dipolar relaxation effects on a membrane, which are related to the extent of water penetration in the bilayer. Laurdan is therefore capable of determining the degree of membrane fluidity. When lipids are highly packed, it has an emission maximum at 440 nm. When lipids are in a disordered fluid phase its emission is centered at 490 nm. The GP function uses two emission channels centered at 440 nm and 490 nm, respectively, to determine membrane water content. In this work we use instead 32 separate channels over the entire emission wavelength range (416–728 nm). This results in a much higher spectral sensitivity, which gives us the chance to detect subtle and smaller changes in membrane lateral packing when compared with the GP function.

As a means of enabling a simple and fit free analysis of spectral imaging data, in this work we applied the phasor transformation to Laurdan spectral images. In the case of Laurdan studies this feature of the phasor approach is extremely relevant. In order to fit the Laurdan fluorescence response many assumptions are needed and they can easily lead to artifacts and wrong interpretations. The phasor approach has been proved to be very valuable for frequency and time domain lifetime imaging data [36–39]. The phasor approach provides a global analysis of entire images and does not require a priori knowledge of the number or spectral shape of the components in the sample. Fereidouni et al. recently applied the phasor approach to spectral imaging data as a simple graphical method for spectral un-mixing [40]. Here we use the spectral phasor approach to distinguish Laurdan interactions with different environments in a pixel. We use spectral images acquired at 416–728 nm wavelength range with 32 separate channels (bandwidth of 9.7 nm). Each pixel in the spectral image contains the emission spectrum at that pixel. Therefore, an emission curve is associated with every pixel. The spectral phasor transformation calculates the Fourier sine and cosine transforms of a given spectrum. For each Fourier harmonic, two coordinates are calculated: g corresponds to the cosine transform (x coordinate in a polar plot) and s corresponds to the sine transform (y coordinate in a polar plot). We use the following equations:

$$g = \frac{\sum_{\lambda} I(\lambda)\cos(2\pi n\lambda / L)}{\sum_{\lambda} I(\lambda)} \tag{1}$$

$$s = \frac{\sum_{\lambda} I(\lambda)\sin(2\pi n\lambda / L)}{\sum_{\lambda} I(\lambda)} \tag{2}$$

A point at coordinates (g, s) is called a phasor and is represented in a polar plot. L in Eqs. 1 and 2 is the total wavelength range, $I(\lambda)$ represents the intensity at a given wavelength and given pixel, and n is the harmonic order.

Figure 1 illustrates the basic principles of the spectral phasor transformation. Gaussian spectra (Fig. 1a) are transformed into a spectral phasor (Fig. 1b). The angular position in the polar plot (the phasor angle) is proportional to the spectral center of mass. The distance from the origin (the phasor radius) is inversely proportional to the spectral width. Figure 1c, d show reference phasors obtained varying spectral center of mass and width of Gaussian spectra, using the first harmonic ($2\pi\lambda/L$) and the second harmonic ($4\pi\lambda/L$) in Eq. 1. Basically, it's easier to interpret phasor positions and phasor visual properties using the first harmonic when the spectral response extends over the entire wavelength range (416–728 nm) and the spectral components have emission wavelengths

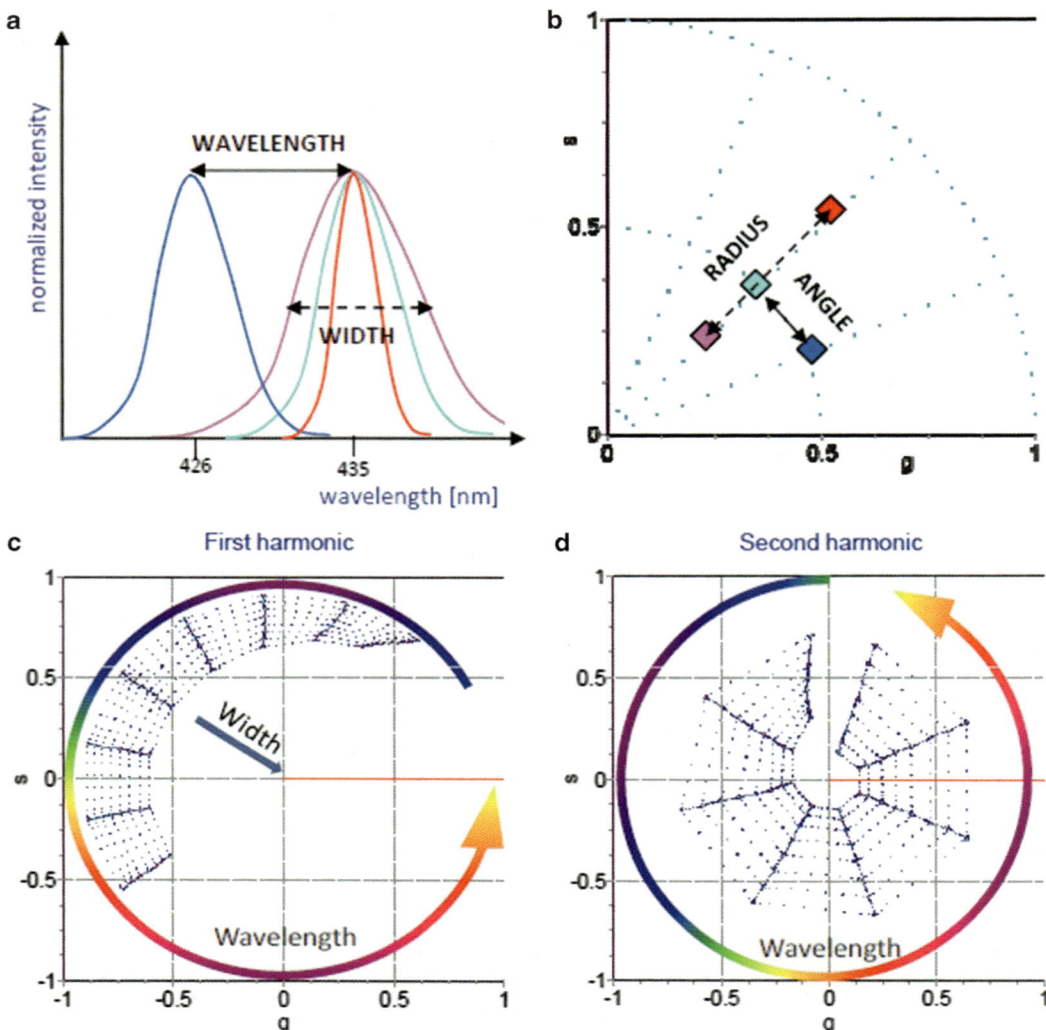

Fig. 1 Basic principles of the spectral phasor approach. Gaussian spectra in (**a**) are transformed into phasors and represented in (**b**). Changes in the spectral center of mass are translated into different phasor angle (from blue spectrum/phasor to turquoise spectrum/phasor). Changes in the spectral bandwidth correspond to different phasor radius (*red, turquoise* and *pink* spectra/phasors). (**c**) Spectral phasor simulation using the first harmonic: *blue* pixels in the plot corresponds to Gaussian spectra with varying center of mass and bandwidth. (**d**) Analogous phasor simulation using the second harmonic

distant from each other. However, phasors can be deformed at the shortest wavelengths (Fig. 1a) due to the lack of points in the initial part of the spectrum. The second harmonic allows resolving spectral species that have close emission wavelengths, and it is less affected by distortion at the shortest wavelengths. However when spectral components are distant along the wavelength range, it can be more complex to visually interpret results due to the wrap around effect in the polar representation of Fig. 1. In other words, different harmonics of the Fourier transform provide additional polar plots that reveal finer details of the spectral response. This capability is particular advantageous when considering that each harmonic can have an optimal sensitivity for different spectral components. In this way we can specifically tune our analysis to discriminate spectrally separated as well as spectrally overlapping components using a different harmonic transform/analysis.

We performed a series of simulations in order to understand the level of sensitivity we can reach using the spectral phasor approach. Essentially, by coupling a low resolution instrument for spectral image detection (only 32 channels, therefore a 32-points-spectrum for each pixel in the image) with the spectral phasor image analysis, we obtain discrimination in the spectral domain due to subtle changes in the spectral position and width. Figure 2 shows Gaussian spectra detected using 32-channels acquisition and the corresponding calculated phasors. In Fig. 2a, ten spectra with increasing spectral width are represented. We are able to identify and correctly transform spectra with up to 2 nm difference in the bandwidth. This can be done using both first and second harmonic, and the results are represented in Fig. 2b, c. In Fig. 2d, ten spectra with a 2 nm shift in the central emission wavelength are represented. The spectral phasor transformation allows us to discriminate the spectra and to correctly transform each pixel (alias each spectrum) into a phasor without distortions. This can be done both by using first harmonic and second harmonic Fourier transform, as represented in Fig. 2e, f. In Fig. 2g very narrow spectra were reconstructed using 32 channels/points; The corresponding phasor plots for the first and the second harmonic analysis are shown in Fig. 2h, i.

In this work, we used the second harmonic when we analyze Laurdan labeled cells to better resolve the small spectral shifts induced by differences in the membrane micro-environment. Conversely, we used the first harmonic when we analyze cells labeled with both Laurdan and CT-B-Alexa 594 because the emission spectra of the two probes are distant and using the first harmonic we can better visualize a wider spectral range.

1.1 Laurdan Spectral Phasor Analysis in Solutions

Preparation of Giant Unilamellar Vesicles was performed using the electroformation method, originally developed by Angelova et al. [41]. Phospholipids were diluted with chloroform to a final concentration of 0.2 mg/ml. Two platinum wires attached to a Teflon chamber were coated with 2 μl of the lipid mixture and dried under N_2 (g). The water jacketed chamber was sealed with a No. 1.5 cover slip and it was attached to a circulating water bath. Phospholipids were rehydrated with 1 mM Tris–HCl pH 7.4. The platinum wires were attached to a frequency generator with alternating current set to 10 Hz and 2 V. A thermocouple was used to monitor the temperature of the chamber. GUVs were grown for 30 min and then imaged.

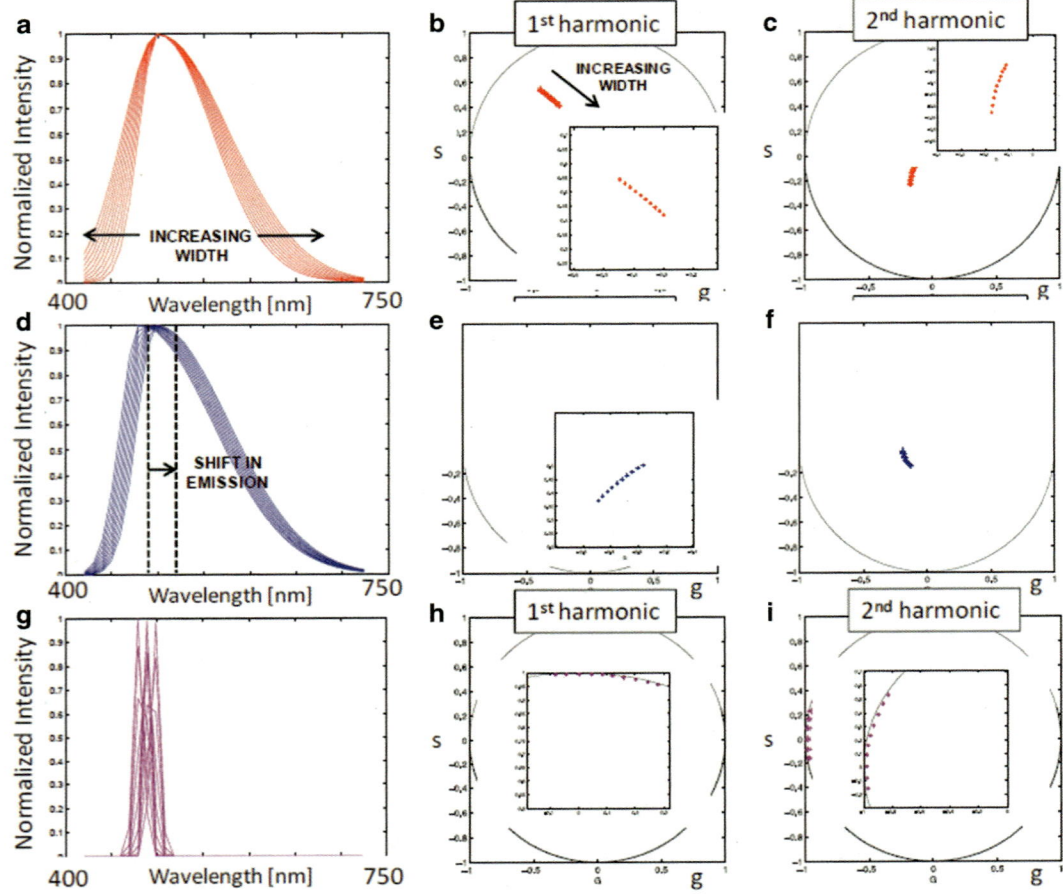

Fig. 2 Phasor simulations to test spectral sensitivity. (**a**) Spectra with increasing bandwidth are reconstructed using 32 channels/points from 416 nm to 728 nm; (**b, c**) corresponding spectral phasor plot for the first harmonic (**b**) and second harmonic (**c**). Each spectrum is transformed into a pixel in the phasor plot, it's possible to resolve and discriminate spectra which are 2 nm apart. (**d**) Spectra with shifted spectral center of mass were reconstructed using 32 channels/points; (**e, f**) spectral phasor plot for the first harmonic (**e**) and second harmonic (**f**). Each spectrum has been transformed into a phasor. Using the phasor transformation, it's possible to obtain a spectral resolution of 2 nm. (**g**) Narrow spectra were reconstructed using 32 channels/points; (**h, i**) corresponding phasor plot for the first and the second harmonic analysis. Considering 32 channels/points, the phasor transformation is capable of correctly calculate the phasor position using only three points of the spectrum

For cell culture NIH3T3 and HEK293 cells were grown at 37 °C and in 5 % CO_2 in Dulbecco modified Eagle medium (DMEM) from Invitrogen supplemented with 10 % of Fetal Bovine Serum, 5 mL of Pen-Strep and 2.5 ml of 1 M HEPES. Freshly split cells were plated onto 35-mm Mattek glass bottom dishes coated with fibronectin. The membrane probe Laurdan (6-dodecanoyl-2-dimethylamino naphthalene) was purchased from Invitrogen, it was dissolved in DMSO (dimethyl sulfoxide), 1.8 mM stock solution was prepared and added to the cell dishes at a dilution 1:1,000. Cells were incubated with Laurdan for 40 min at 37 °C before imaging.

For the chronic cholesterol depletion experiments, cells were incubated with lipoprotein-deficient serum (DMEM plus 5 % LPDS) for 2 days to deplete cholesterol. For the acute cholesterol depletion experiments, a 50 mM stock solution of methyl-β-cyclodextrin (Sigma-Aldrich) was prepared by dissolving in nano-pure water. The solution was added to the cell dishes with a final concentration of 2 mM and they were incubated for 1 h at 37 °C. Cells were then rinsed with PBS and new medium was added.

Spectral data were acquired with a Zeiss LSM710 META Laser scanning microscope, coupled to a 2-Photon Ti:Sapphire laser (Spectra-Physics Mai Tai, Newport Beach) producing 80 fs pulses at a repetition of 80 MHz. A 40× water immersion objective 1.2 N.A. (Zeiss) was used for all experiments. The Laurdan fluorophore was excited at 780 nm. Excitation at 780 nm induced negligible autofluorescence since 2-photon excitation of intracellular metabolites is centered at 740 nm [42]. For Laurdan and CTB-A594 co-localization measurements the excitation was set at 800 nm. Spectral images were acquired in 32 channels, each with 9.7 nm bandwidth, the first channel is centered at 421 nm and the last one at 723 nm. For image acquisition, the pixel frame size was set to 256×256 and the pixel dwell time was 177 μs/pixel. The average laser power at the sample was maintained at the mW level. Spectral data were processed by the SimFCS software developed at the Laboratory for Fluorescence Dynamics (www.lfd.uci.edu).

First, we performed measurements in different solvents to visualize Laurdan spectral phasor in solutions (Fig. 3a). We measured the Laurdan's spectral phasor in dimethyl-sulfoxide (DMSO), methanol (MeOH), glycerol, and glycerol with 10 % water. For the latter two samples, the shift caused by the change in polarity due to the addition of water is evident in the phasor plot. The images were taken using the Zeiss LSM710 microscope and 780 nm two-photon excitation is used in order to reduce photo-bleaching issues. The emission spectrum is detected from 416 nm to 728 nm in 32 wavelength channels, each having a 9.6 nm bandwidth. In Fig. 3a normalized fluorescence emission spectra of Laurdan in solution are shown. The emission spectrum at every pixel is Fourier transformed, the real and imaginary components of the Fourier transform are used to calculate the x and y coordinates in the spectral phasor plot, according to Eqs. 1 and 2. Here we use the second harmonic analysis, the corresponding phasor distributions are represented in Fig. 3b. The first major advantage of spectral phasor analysis is that for every pixel of the image we obtain a value identified by the two coordinate g and s, while conventional GP function calculates a single value. This means that instead of representing every pixel onto a graph with one dimension, we represent every pixel into a

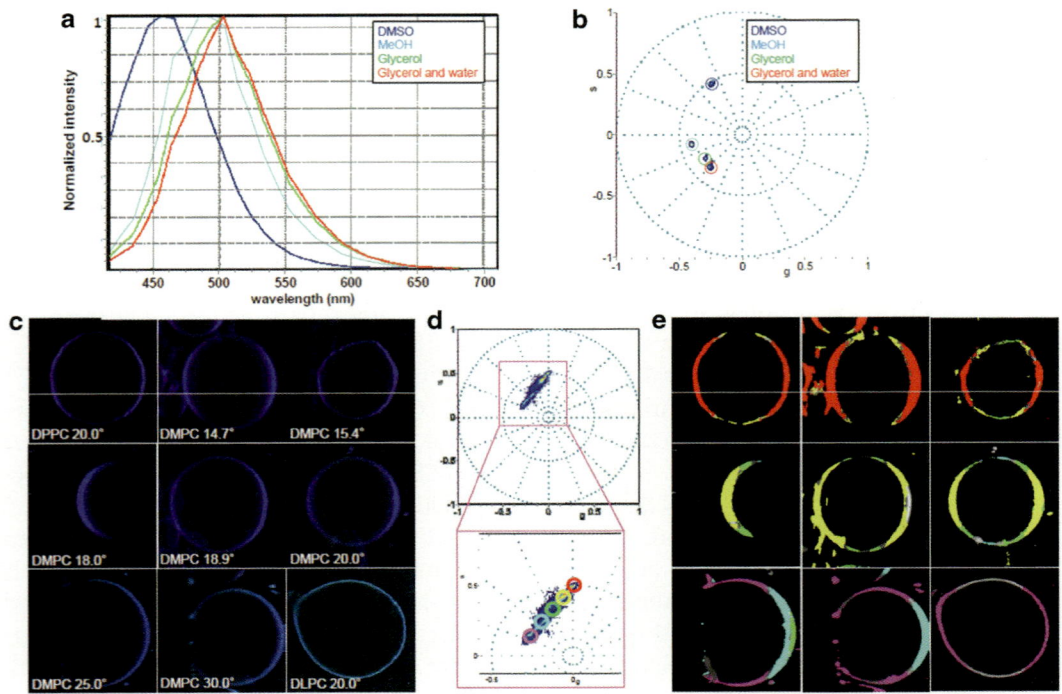

Fig. 3 Laurdan spectral phasors in solutions and GUVs. (**a**) Normalized fluorescence spectra of Laurdan in different solvents. (**b**) Spectral phasors of Laurdan in different solvents using the second harmonic analysis. All the phasor distributions are very well resolved in the phasor plot. (**c**) Spectral images of DPPC, DMPC (at different temperatures) and DLPC single lipid GUVs. The diameter of the GUVS varies from 20 to 50 μm. (**d**) Spectral phasor analysis using the second harmonic, the phasor distributions of all images are represented in the same phasor plot. (**e**) The images in (**c**) were pseudo-colored using the color map represented in the phasor plot. Each pixel in the spectral images has been colored depending on the phasor position, the color cursors that have been used are represented in (**d**)

phasor plot in two dimensions. Indeed a two-dimensional space is more informative than a one-dimensional scale which results in an improved resolution and a better capability of discrimination.

Next, we show Laurdan spectral phasor analysis in single lipid Giant Unilamellar Vesicles (GUVs) (Fig. 3c–e). To prepare GUVs we used the electroformation method, originally designed by Angelova et al. [16, 41, 43, 44]. We used phospholipids with different saturation (different rigidity) and at different temperatures. We imaged GUVs of dipalmitoyl phosphatidyl choline (DPPC) at 20°, GUVs of dimyristoyl phosphatidyl choline (DMPC) in a temperature range varying from 14.7° to 30.0°, and GUVs made of dilauroyl phosphatidyl choline (DLPC) at 20°. Figure 3c shows spectral images taken at the equatorial center of GUVs. Tor these measurements we used *linear* polarized excitation light so that in the equatorial region, only those molecules with an absorption transition moment parallel or nearly parallel to the plane of excitation light polarization are excited, and the efficiency is proportional to the cosine squared of the angle between the transition moment of the probe and the polarization plane of the excitation light [45]. The diameter of GUVs varies from 20 to 50 μm. In Fig. 3d the spectral phasor analysis using the second harmonic is shown, the phasor distributions of all the images are shown in the same polar plot. Using the second harmonic analysis, phasors of GUVs made of single lipids are distributed along two positions, corresponding to two Laurdan environments such as gel and liquid phases. GUVs made with different lipid composition and at different temperatures have different phasor positions. If we know the phasor position of these differently packed membrane environments, using the linear combination properties of phasors, we can determine the fractional contribution of liquid versus gel like phase in a given membrane. In Fig. 3d clusters of points in the phasor plot are selected by cursors of different colors. The points in the image that correspond to each specific cluster are colored with the same color of the cursor (Fig. 3e). Red cursor/color identifies the shortest wavelengths of emission, corresponding to a highly packed phospholipids phase (gel like). Pink cursor/color identifies the longest wavelengths, corresponding to a disordered lipid phase (liquid).

1.2 Laurdan Spectral Phasor in Live Cells

After determining Laurdan's response using spectral phasor image analysis in model systems, we performed Laurdan spectral imaging in live cells. In Fig. 4, we show images obtained at different z-values of a NIH3T3 and HEK293 cells, respectively. Live cells were labeled with Laurdan; spectral images were obtained at 1.2 μm and 1.4 μm distance, respectively, along the z-axis (Fig. 4a, d). First, for both cell types phasor plots show that the phasors fall on a different line with respect to the GUV's line and the total range of the phasor cluster is reduced in cells. This confirms that cellular membrane heterogeneity cannot be explained solely as a combination of gel and liquid phase. Moreover, the fluidity values obtained within a live cell membrane are not as extreme and separate as in model membranes. Using the four color cursors represented in the phasor plot (Fig. 4c, f) we can distinguish at least four groups of membranes, which are characterized by increasing fluidity values (or water content). When we color the pixels in the images according to their location in the phasor plot, we can identify small regions of relatively low fluidity along the plasma membrane (red pixels), plasma membrane (green pixels), and intracellular membranes (blue and pink pixels), including the nuclear envelope and other internal membranes, which could be associated to the Golgi, ER, and small organelles such as endosomes (Fig. 4b, e). Interestingly, the plasma membranes of both NIH3T3 and HEK293 cells display regions of different fluidity according to the Laurdan spectral phasor analysis. We cannot identify these small regions with membrane micro domains at the present stage of this research, since we

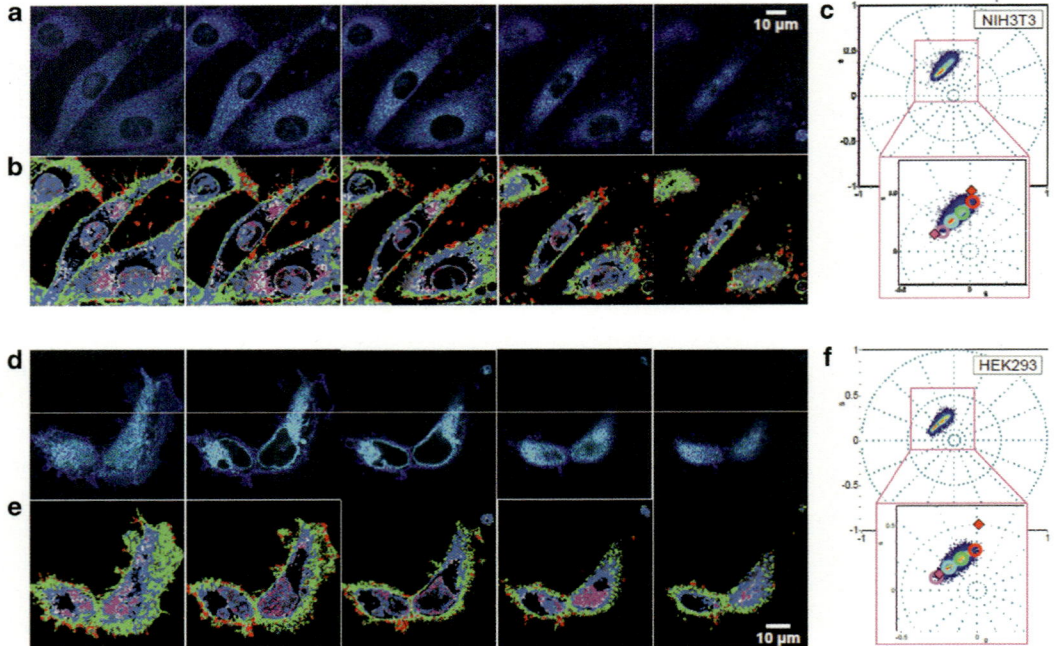

Fig. 4 Laurdan spectral phasor analysis in live NIH3T3 and HEK293 cells. (**a**) Spectral images of NIH3T3 live cells in different z-planes. The planes are 1.2 μm apart. (**b**) Phasor images: pixels of the images are colored using the color cursors for the phasor distribution in (**c**); *red* pixels are characterized by high rigidity, they are the pixels in which Laurdan emits at very short wavelengths (non polar environment). (**c**) Spectral phasor distribution of the images in (**a**) using the second harmonic: we used four cursors (*red, green, blue* and *pink*) to map cell membranes in order of increasing fluidity (therefore decreasing rigidity). (**d**) Spectral images of HEK293 live cells in different z-planes. The planes are 1.4 μm apart. (**e**) Phasor images: pixels of the images are colored using the phasor distribution in (**f**); *red* pixels are characterized by high rigidity according to the spectral phasor analysis. (**f**) Spectral phasor distribution of the images in (**d**) using the second harmonic: we used four cursors (*red, green, blue* and *pink*) to map cell membranes in order of increasing fluidity

cannot distinguish between membrane folds, small vesicles attached to the plasma membrane or proximity of membranes with different phasor values.

Since cholesterol plays a major role in influencing membrane fluidity/order and it is a major component of membrane domains, we monitored membrane fluidity changes during cholesterol removal using the Laurdan spectral phasor analysis. We induce chronic cholesterol depletion by cultivating the cells with lipoprotein-deficient serum (LPDS) for 44 h. During this time the cell should be able to partially compensate cholesterol depletion. We also induced acute cholesterol depletion by treating the cells with methyl-β-cyclodextrin (MβCD) for 1 h. Results of these two treatments are shown in Fig. 5. Intensity images of the control, and cells treated with LPDS and MβCD, respectively, are shown in Fig. 5a. A low intensity and high intensity threshold were applied

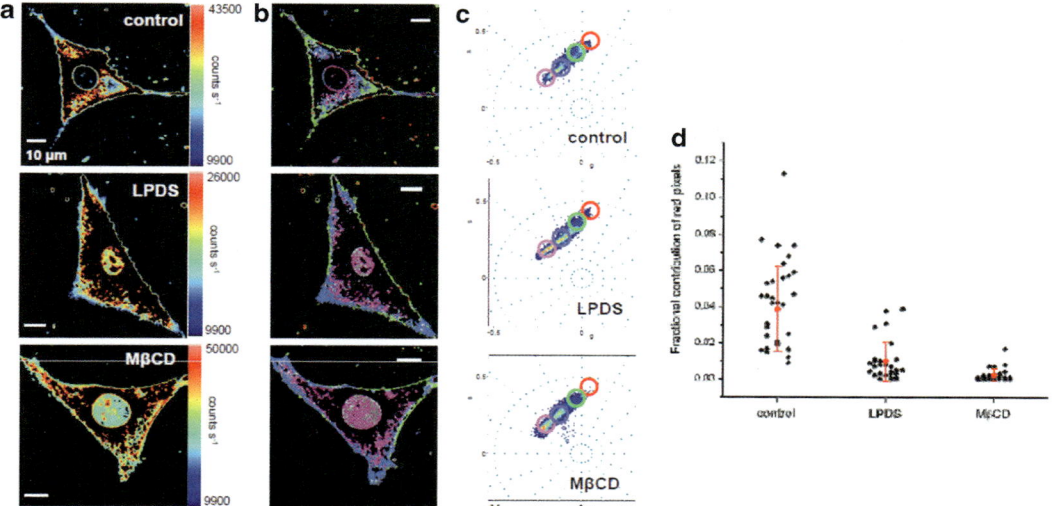

Fig. 5 Laurdan spectral phasor analysis show fluidity changes in cell membranes' fluidity upon chronic (LPDS) and acute (MβCD) cholesterol depletion. (**a**) Intensity images for control, LPDS, MβCD: pixels with very high or very low intensity were threshold, scale and thresholds are represented on the right. (**b**) Phasor images for control, LPDS, MβCD: pixels of each image are colored using the corresponding phasor distribution in (**c**); *red* pixels are characterized by high rigidity according to the spectral phasor analysis. (**c**) Spectral phasor distribution of the images in (**a**) using the second harmonic: we used four cursors (*red, green, blue* and *pink*) to map cell membranes in order of increasing fluidity. (**d**) Scatter plot representing the fractional contribution of red pixels (pixels characterized by a very rigid environment) in control images ($N=32$), LPDS images ($N=28$) and MβCD images ($N=24$). For each image, we plot the percentage of pixels for which their phasor is located in the area represented by the red cursor in the phasor plot

to the images in order to highlight changes in the plasma membrane. Figure 5b shows pseudo-colored intensity images obtained using the second harmonic analysis and the cursor location in the phasor plots shown in Fig. 5c. The color cursors' positions are the same for all the three images. Examining the phasor plots in Fig. 5c, we observe that the control (before cholesterol depletion) shows two defined large phasor clusters, which can be associated to pixels belonging to plasma (red and green cursors) and intracellular membranes (blue and pink cursors). Upon cholesterol removal, the number of pixels in the red cursor (very rigid domains) decreases. This decrease is prominent in the plasma membranes while the internal membranes are less affected by the treatment. Also, we are able to monitor the difference between the two treatments. Particularly, the number of red pixels (representing the most rigid regions in the plasma membrane) decreases differently for the two treatments, as can be seen in Fig. 5d. We used $N=32$ cells for the control, 28 cells for LPDS treatment and 24 cells for MβCD treatment. For each image, we plot the percentage of pixels having phasors highlighted by the red cluster. The control shows large variations among cells of the number pixels corresponding to the red cursor (rigid domains); the fractional contribution of red pixels has values between 0.12 and 0.01 (12–1 %). The LPDS population presents a fractional contribution of red pixels between 0.04 and 0 (4–0 %), while the MβCD population does not present heterogeneity anymore and the fractional contribution of red pixels is very close to 0 % for all cells. While for the control population the two groups of phasor (plasma and intracellular membranes) are quite separated and distant in the phasor plot, for LPDS treated cells it is difficult to separate the two groups (the phasors appear all along a line), and for MβCD treated cells the two phasor groups appear even closer.

1.3 Cholera-Toxin Co-labeling Experiments

To further illustrate the capability of the spectral phasor approach we co-labeled live cells with Laurdan and a putative marker of membrane domains, the Cholera Toxin subunit B (CT-B), fluorescently labeled with Alexa Fluor 594 which binds to ganglioside GM1. In performing these measurements we use a specific property of the phasor transformation, which is related to the linear combination of properties in the same pixel [38, 39]. Specifically, if in a pixel we have the linear combination of two fluorophores, the phasor distribution is found along a straight line joining the phasor distributions of the two independent fluorophores. The schematic of this concept is represented in Fig. 6b; as we mentioned in the introduction we use the first harmonic for this analysis. For Laurdan and CT-B-A594 this means we can visualize pixels having fluorescence emission due to a rigid Laurdan environment and the presence of CT-B-A594 in these pixels because they must lie along the line connecting the two separate phasor distributions (red line). Similarly, we can identify pixels characterized by the contributions of a fluid Laurdan environment and CT-B-A594 along the blue line. Spectral images are shown in Fig. 6a. Spectral phasor analysis is shown in Fig. 6c–h. Pixels of the spectral image have been colored according to cursors in the phasor plot. Yellow colored pixels correspond to pixels where CT-B-A594 is dominant; blue colored pixels show the co-localization of CT-B and Laurdan in fluid environment; red colored pixels show the co-localization of CT-B and Laurdan in rigid environment. In Fig. 6d, f, h we show spectral phasor distributions of the images in Fig. 6c, e, g, respectively. Using the phasor approach we can directly visualize the pixels where Laurdan and CTB-A594 are both emitting. Using the trajectory that represents the co-localization between CTB-A594 and the most rigid part of Laurdan emission, we can identify regions of the plasma membrane which are rigid and enriched in GM1 (red pixels).

Laurdan spectral phasor analysis allows us to investigate the heterogeneity of membrane fluidity in live cells. By coloring pixels according to the clusters in the phasor plots using four cursors to select these clusters, we identify four groups of membranes in the

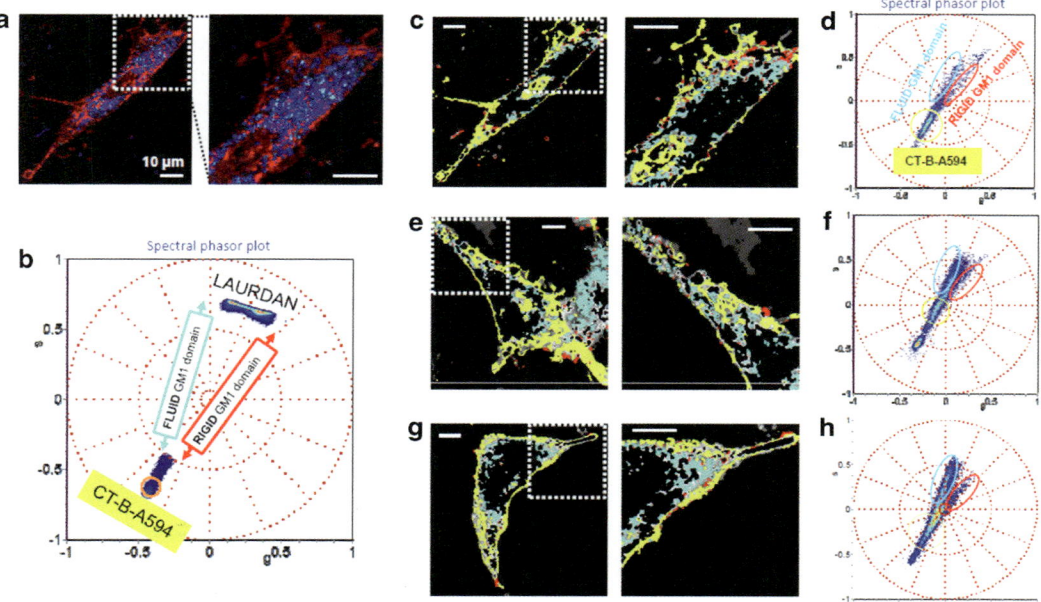

Fig. 6 Spectral phasor analysis of live NIH3T3 cells labeled with Laurdan and CT-B-Alexa 594. (**a**) Spectral image of NIH3T3 cells co-labeled with Laurdan and CT-B-A594; zoomed-in image on the right. Colors in the image correspond to the dominant emission wavelength. (**b**) Schematic representation of where the spectral phasor locates depending on the contribution of Laurdan and CT-B-A594. The phasor plot was calculated using the first harmonic analysis. Phasor distributions of Laurdan and CT-B-A594 alone are represented. Also, we represent where the pixel phasor is expected to be located when the emission spectra of Laurdan and CT-B-A594 are both present in the same pixel. The phasor position also changes depending on the Laurdan environment (fluid versus rigid). (**c**, **e**, **g**) Phasor image (zoomed-in on the right): pixels are colored according to the color cursors in the phasor plot in (**d**, **f**, **h**) respectively. *Yellow* colored pixels correspond to pixels where CT-B-A594 is dominant; *blue* colored pixels show the co-localization of CT-B and Laurdan in fluid environment; *red* colored pixels show the co-localization of CT-B and Laurdan in rigid environment. (**d**, **f**, **h**) Spectral phasor distribution of images in (**c**, **e**, **g**) respectively using the first harmonic analysis

imaged cells (Fig. 4). We can use fractional contribution analysis of the pixels in the image to quantify the contribution and the percentage of each group over the rest of cellular membranes. In this work we applied the fractional contribution analysis to quantify the percentage of pixels of the plasma membrane characterized by a rigid environment of the Laurdan probe (Fig. 5) when treated with different agents. We pseudo-colored these pixels in red and we monitored changes in their number after chronic and acute cholesterol depletion, respectively LPDS and MβCD treatments. Control cells are characterized by a number of rigid regions along the plasma membrane (Fig. 5d), while this number is reduced in cholesterol depleted cells. Different control cells show a large heterogeneity of the relative number of pixels corresponding to the red cursor. LPDS treated cells still maintain some heterogeneities, while rigid regions are basically absent after MβCD treatment. We anticipate that the level of detail revealed by the spectral phasor method will prove invaluable to measure cell membrane heterogeneity and for monitoring membrane compositional changes, upon drug treatment or pathogenesis.

Considering the intrinsic biological complexity of cell membranes, the spectral phasor analysis of Laurdan significantly improves the sensitivity to different spectral components with respect to GP analysis. The Laurdan spectral phasor is able to discriminate subtle changes in membrane fluidity, which shows significant and consistent differences between plasma and intracellular membranes. In particular, using higher harmonics, the phasor analysis can be tuned to obtain different degrees of spectral separation. In addition, it can easily identify the presence of different environments (such as for Laurdan itself) or different fluorophores (Laurdan and CTB-A-594) in the same image pixel. We took advantage of this phasor property when we co-labeled cells with Laurdan and CT-B-A594. The purpose of these experiments was to characterize the membrane fluidity around GM1 domains, labeled by CT-B; more specifically we identified domains that are enriched with GM1 and also characterized by high rigidity. Laurdan labels all membranes but has different spectral properties depending on the membrane fluidity. The combination of the two probes provides a way to characterize the environment of the CT-B in the different locations in the cell. The linear combination property of the phasor transformation gives a simple way to identify the fluidity of the CT-B environment. In Fig. 6 we show that only some of the domains stained with CT-B co-localize with rigid domains. In particular, by visualizing the pixels that correspond to the contribution of both Laurdan in a rigid environment and CT-B-A594, we are able to specifically select only some of the domains stained by CT-B. We identify these domains with rigid regions of the plasma membrane possibly enriched with GM1.

In conclusion, Laurdan spectral phasor analysis substantially improves sensitivity toward discrimination of different emission spectra as compared to conventional methods of membrane characterization such as GP. This increased sensitivity stems from the fact that we measure all wavelengths of a spectrum in parallel and the phasor transformation results in a two coordinate system (the g and s phasor value) for each pixel of the image. Using this analytical approach we can better discriminate changes in Laurdan's environment and interaction with other probes. In particular, we can tune our analysis to detect subtle shifts in Laurdan's emission due to micro-environment heterogeneity or significant differences in emission spectra when Laurdan is co-labeled with CT-B by using a higher or lower harmonic analysis, respectively. Considering the large number of questions that still need to be answered about membrane fluidity and conformation, lipid composition and domains, Laurdan's spectral phasor analysis has proven to be a valuable method that can be applied to many other biological problems concerning cell membranes.

Acknowledgments

We thank Milka Stakic for cultivating the cells. We also want to thank Michelle A. Digman for helping with the GUV's preparation, for the reagents for CT-B-A594 staining and valuable suggestions. We want to thank Susana A. Sanchez, Moshe Levi, and Luca Lanzano for important suggestions and help during cholesterol depletion experiments and William Rosales for his helpful assistance during GUV's preparation. This work is supported in part by NIH-P41-RR03155, P41 GM103540 and NIH P50-GM076516.

References

1. Jacobson K, Mouritsen OG, Anderson RG (2007) Lipid rafts: at a crossroad between cell biology and physics. Nat Cell Biol 9(1):7–14

2. Mayor S, Rao M (2004) Rafts: scale-dependent, active lipid organization at the cell surface. Traffic 5(4):231–240

3. Pike LJ (2004) Lipid rafts: heterogeneity on the high seas. Biochem J 378(Pt 2):281–292

4. Tanimura N et al (2003) Dynamic changes in the mobility of LAT in aggregated lipid rafts upon T cell activation. J Cell Biol 160(1): 125–135

5. Li K et al (2013) IFITM proteins restrict viral membrane hemifusion. PLoS Pathog 9(1): e1003124

6. Gaus K et al (2006) Integrin-mediated adhesion regulates membrane order. J Cell Biol 174(5):725–734

7. Ikonen E (2001) Roles of lipid rafts in membrane transport. Curr Opin Cell Biol 13(4): 470–477

8. Hanzal-Bayer MF, Hancock JF (2007) Lipid rafts and membrane traffic. FEBS Lett 581(11):2098–2104

9. Gupta N, DeFranco AL (2007) Lipid rafts and B cell signaling. Semin Cell Dev Biol 18(5):616–626

10. Hell SW, Wichmann J (1994) Breaking the diffraction resolution limit by stimulated emission: stimulated-emission-depletion fluorescence microscopy. Opt Lett 19(11):780–782

11. Gustafsson MGL (2000) Surpassing the lateral resolution limit by a factor of two using structured illumination microscopy. J Microsc 198: 82–87

12. Betzig E et al (2006) Imaging intracellular fluorescent proteins at nanometer resolution. Science 313(5793):1642–1645

13. Hess ST, Girirajan TP, Mason MD (2006) Ultra-high resolution imaging by fluorescence photoactivation localization microscopy. Biophys J 91(11):4258–4272

14. Rust MJ, Bates M, Zhuang X (2006) Sub-diffraction-limit imaging by stochastic optical reconstruction microscopy (STORM). Nat Methods 3(10):793–795

15. Parasassi T et al (1997) Two-photon fluorescence microscopy of Laurdan generalized polarization domains in model and natural membranes. Biophys J 72(6):2413–2429

16. Bagatolli LA, Gratton E (2000) Two photon fluorescence microscopy of coexisting lipid domains in giant unilamellar vesicles of binary phospholipid mixtures. Biophys J 78(1):290–305

17. Gaus K et al (2003) Visualizing lipid structure and raft domains in living cells with two-photon microscopy. Proc Natl Acad Sci U S A 100(26):15554–15559

18. Harris FM, Best KB, Bell JD (2002) Use of Laurdan fluorescence intensity and polarization to distinguish between changes in membrane fluidity and phospholipid order. Biochim Biophys Acta 1565(1):123–128

19. Romer W et al (2007) Shiga toxin induces tubular membrane invaginations for its uptake into cells. Nature 450(7170):670–675

20. Owen DM et al (2012) Quantitative imaging of membrane lipid order in cells and organisms. Nat Protoc 7(1):24–35

21. Sanchez SA, Tricerri MA, Gratton E (2012) Laurdan generalized polarization fluctuations measures membrane packing microheterogeneity in vivo. Proc Natl Acad Sci U S A 109(19):7314–7319

22. Jin L et al (2005) Cholesterol-enriched lipid domains can be visualized by di-4-ANEPPDHQ with linear and nonlinear optics. Biophys J 89(1):L04–L06

23. Jin L et al (2006) Characterization and application of a new optical probe for membrane lipid domains. Biophys J 90(7):2563–2575

24. Demchenko AP et al (2009) Monitoring biophysical properties of lipid membranes by environment-sensitive fluorescent probes. Biophys J 96(9):3461–3470

25. Klymchenko AS, Mely Y (2013) Fluorescent environment-sensitive dyes as reporters of biomolecular interactions. Prog Mol Biol Transl Sci 113:35–58

26. Kucherak OA et al (2012) Dipolar 3-methoxychromones as bright and highly solvatochromic fluorescent dyes. Phys Chem Chem Phys 14(7):2292–2300

27. Weber G, Farris FJ (1979) Synthesis and spectral properties of a hydrophobic fluorescent probe: 6-propionyl-2-(dimethylamino)naphthalene. Biochemistry 18(14):3075–3078

28. Parasassi T, Conti F, Gratton E (1986) Time-resolved fluorescence emission spectra of Laurdan in phospholipid vesicles by multifrequency phase and modulation fluorometry. Cell Mol Biol 32(1):103–108

29. Parasassi T et al (1990) Phase fluctuation in phospholipid membranes revealed by Laurdan fluorescence. Biophys J 57(6):1179–1186

30. Kenworthy AK, Petranova N, Edidin M (2000) High-resolution FRET microscopy of cholera toxin B-subunit and GPI-anchored proteins in cell plasma membranes. Mol Biol Cell 11(5):1645–1655

31. Dietrich C et al (2002) Relationship of lipid rafts to transient confinement zones detected by single particle tracking. Biophys J 82(1 Pt 1):274–284

32. Janes PW, Ley SC, Magee AI (1999) Aggregation of lipid rafts accompanies signaling via the T cell antigen receptor. J Cell Biol 147(2):447–461

33. Owen DM et al (2007) Optical techniques for imaging membrane lipid microdomains in living cells. Semin Cell Dev Biol 18(5):591–598

34. Parton RG (1994) Ultrastructural localization of gangliosides; GM1 is concentrated in caveolae. J Histochem Cytochem 42(2):155–166

35. Blank N et al (2007) Cholera toxin binds to lipid rafts but has a limited specificity for ganglioside GM1. Immunol Cell Biol 85(5):378–382

36. Jameson DM, Gratton E, Hall R (1984) The measurement and analysis of heterogeneous emissions by multifrequency phase and modulation fluorometry. Appl Spectrosc Rev 20(1):55–106

37. Verveer PJ, Squire A, Bastiaens PI (2000) Global analysis of fluorescence lifetime imaging microscopy data. Biophys J 78(4):2127–2137

38. Redford GI, Clegg RM (2005) Polar plot representation for frequency-domain analysis of fluorescence lifetimes. J Fluoresc 15(5):805–815

39. Digman MA et al (2008) The phasor approach to fluorescence lifetime imaging analysis. Biophys J 94(2):L14–L16

40. Fereidouni F, Bader AN, Gerritsen HC (2012) Spectral phasor analysis allows rapid and reliable unmixing of fluorescence microscopy spectral images. Opt Express 20(12):12729–12741

41. Angelova MI, Dimitrov DS (1986) Liposome electroformation. Faraday Discuss Chem Soc 81:303–311

42. Stringari C et al (2011) Phasor approach to fluorescence lifetime microscopy distinguishes different metabolic states of germ cells in a live tissue. Proc Natl Acad Sci U S A 108(33):13582–13587

43. Ruan Q et al (2004) Spatial-temporal studies of membrane dynamics: scanning fluorescence correlation spectroscopy (SFCS). Biophys J 87(2):1260–1267

44. Sanchez SA, Tricerri MA, Gratton E (2007) Interaction of high density lipoprotein particles with membranes containing cholesterol. J Lipid Res 48(8):1689–1700

45. Bagatolli LA (2006) To see or not to see: lateral organization of biological membranes and fluorescence microscopy. Biochim Biophys Acta 1758(10):1541–1556

Cholesterol Behavior in Asymmetric Lipid Bilayers: Insights from Molecular Dynamics Simulations

Semen O. Yesylevskyy and Alexander P. Demchenko

Abstract

Asymmetric lipid composition of the cell membranes plays an important role in the multitude of important biological functions. Much less is known, however, about the distribution and dynamics of cholesterol in asymmetric biological membranes. In this work we show how this issue could be addressed computationally by molecular dynamics simulations. The influence of the lipid head group charge, acyl chain saturation, spontaneous membrane curvature and the surface tension of the membrane on cholesterol distribution in asymmetric lipid bilayers is investigated. Four asymmetric bilayers containing DOPC, DOPS, DSPC, or DSPS lipids, were simulated on the time scale extended to tens of microseconds. We show that cholesterol strongly prefers anionic lipids to neutral and saturated lipid tails to unsaturated with distribution ratio ~0.4–0.6. Multiple flip-flop transitions of cholesterol were observed directly and their mean times range from 350 to 2,000 ns. It was shown that the distribution of cholesterol in the asymmetric bilayer depends not only on the type of lipids but also on the local membrane curvature and the surface tension. The geometric shape of spontaneously curved asymmetric bilayer changes dramatically in the presence of cholesterol. The membrane curvature becomes less homogeneous with large patches of flattened regions interleaved by rather sharp bends.

Key words Asymmetric membranes, Cholesterol distribution, Molecular dynamics, Coarse-grained

1 Introduction

One of the most intriguing aspects of the biological membranes is their asymmetry. In plasma membranes of eukaryotic cells the major part of phosphatidylcholine (PC) and sphingomyelin (SM) is located in its outer leaflet, whereas most of phosphatidylethanolamine (PE) and practically all phosphatidylserine (PS) and phosphoinositides face the cytosolic milieu [1, 2]. Much less is known about the distribution and dynamics of the other very important component of cell membranes—cholesterol [3]. Compared to phospholipids, it has much higher mobility and much lower free energy barrier of flip-flop transitions between the membrane leaflets [4, 5]. High mobility of cholesterol molecules makes their distribution sensitive to various external factors and to the changes of

Dylan M. Owen (ed.), *Methods in Membrane Lipids*, Methods in Molecular Biology, vol. 1232,
DOI 10.1007/978-1-4939-1752-5_20, © Springer Science+Business Media New York 2015

membrane structure and composition. However, the distribution and dynamics of cholesterol in cell membranes cannot be determined in present experiments with sufficient fidelity. Particularly the application of fluorescent analogs of cholesterol [6, 7] or cholesterol derivatives labeled with organic fluorophores [8] were discouragingly divergent [8, 9]. Earlier data suggested predominant cholesterol location at the outer leaflet [10], while recent studies based on microscopy of living cells suggest predominant cholesterol residence in cytoplasmic leaflet [7].

Structural modeling and free energy considerations suggest that cholesterol distribution in asymmetric membrane cannot be even [11–13]. It must possess stronger affinity to phospholipids with saturated chains [14]. Chain length of the lipid tails and the size of their heads may also influence cholesterol affinity [13], particularly in the case of unsaturated lipid chains [15]. Of special importance can be the mutual influence of the membrane curvature and cholesterol concentration [16]. The relative importance of these factors is a subject of controversy.

Even less certain are the results on the kinetics of redistribution of cholesterol molecules in the membranes. Reported flip-flop rates of cholesterol range extremely broadly, from milliseconds to several hours, depending on studied system and the method of observation. They could even be different in similar conditions in different labs [17]. Criticism is addressed to commonly used experimental techniques such as the usage of fluorescent cholesterol analogs [8], extraction of cholesterol from membranes by cyclodextrins [18], application of cholesterol oxidase [19], and small-angle neutron scattering experiments [17].

The facts that the results of experiments on cholesterol partitioning and dynamics in lipid bilayers are so challenging and contradicting stimulated numerous computational studies. The most popular method of simulating biological membranes is molecular dynamics (MD) [20, 21]. Classical Newton's equations of motion are solved iteratively in MD providing the trajectories of all atoms in time. Applying the concepts from statistical mechanics, the resulting trajectories can be used to evaluate various static and time-dependent structural, dynamic, and thermodynamic properties of the system. MD simulations provide an unique opportunity to study various aspects of the membrane functioning at the all-atom level of details, which is not accessible and will probably never be accessible for experimental techniques [20, 21].

Recently all-atom molecular dynamics (MD) simulations were used to study cholesterol interactions with the lipids [22], the condensing effect of cholesterol [23–25], and the properties of cholesterol-induced rafts [26]. There were attempts of obtaining the potentials of mean force (PMFs) for transferring cholesterol from lipid bilayers to water [27]. Despite significant insights obtained in these studies, the time scale of cholesterol redistribution between the monolayers was beyond the reach for all-atom simulations.

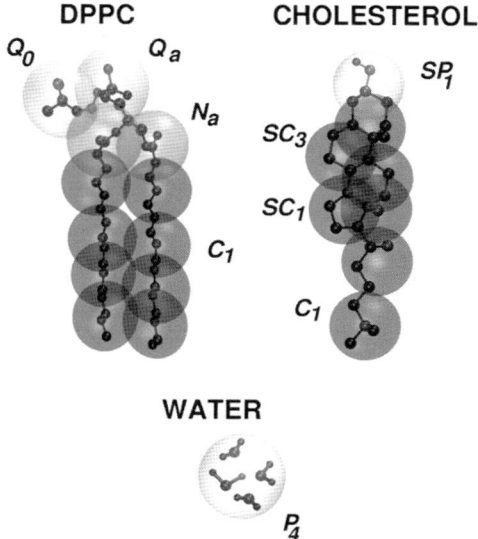

Fig. 1 The scheme of coarse graining used in MARTINI model [29, 46]. Coarse-grained particles are shown as *semitransparent spheres* and superimposed into the atomistic representations of the corresponding molecules. The types of coarse-grained particle, which are used internally in MARTINI force field, are indicated. The figure is adapted from MARTINI web site http://md.chem.rug.nl/cgmartini/index.php/about/martini

The coarse-grained (CG) models are widely used in MD studies of the membranes in order to increase the time and length scale of simulations. In these models the groups of adjacent atoms are combined into the "beads," which interact with each other by means of empirical potentials. Since the number of the beads is much smaller than the number of individual atoms, significant speedup of computations could be achieved. The coarse graining provides the nanoscopic description of the studied systems instead of purely atomistic treatment [28].

One of the most popular CG models for membrane simulations is MARTINI force field [29–32]. On average, in MARTINI one CG particle represents four heavy atoms with associated hydrogens (Fig. 1). MARTINI model was recently used in extensive study of cholesterol behavior in symmetric lipid membranes [33].

Despite the notable progress of MD simulations of lipid bilayers, surprisingly small attention is paid to membrane asymmetry and to the effects of strain and curvature associated with this asymmetry. The distribution of cholesterol in relation to its functional role in such membranes still did not attract the necessary attention. There are many experimentally observed cellular phenomena, which are likely to be governed or influenced by phospholipid asymmetry, membrane curvature and cholesterol distribution [34].

Such phenomena include formation of synaptic vesicles [35], formation of blebs and apoptotic bodies in the course of apoptosis [36, 37], the processes of membrane fusion [38, 39], budding of enveloped viruses from the plasma membrane [40], etc. These processes are of great practical importance for biomedical studies [41].

In this chapter we show how coarse-grained MD technique can address the problem of cholesterol behavior in asymmetric bilayers and provide significant insight into its distribution and dynamics. This chapter is based on the results of our recent MD simulations of asymmetric bilayers [42], which were reanalyzed with significantly improved techniques. New data of the large scale MD simulations of highly curved "meander-like" asymmetric bilayers are also included. We analyzed a number of factors, which influence the distribution and dynamics of cholesterol in asymmetric bilayers such as the phospholipid head group charge, the degree of hydrocarbon chain saturation, the membrane curvature, and the lateral strain in the membrane, caused by the unbalanced surface tension. Methodological issues in MD simulations of asymmetric membranes are discussed as well as special procedures of data analysis.

2 Materials

All simulations were performed with the GROMACS 4.5.5 software package [43]. The algorithm of analysis of the curved membrane was implemented using custom C++ software based on Pteros molecular modeling library [44]. VMD 1.9.1 is used for molecular visualization [45]. Origin 8.5.1 (OriginLab Corporation) is used for data analysis and plotting.

Preparation of the simulations and the analysis of trajectories were performed on the quad-core Intel workstation running Ubuntu Linux 12.04. Production simulations were performed on dedicated high-performance computational clusters using from 32 to 320 processor cores in parallel depending on the system size.

3 Methods

3.1 Simulation Setup

Different parts of real cell membranes can possess very different local curvatures, which originate from spontaneous curvatures of its monolayers with asymmetric lipid content, from the influence of the membrane proteins or from external mechanical stress. These factors can also produce significant mechanical strain in the plane of bilayer. The curvature and the strain may vary from one part of the real membrane to the other and influence the cholesterol distribution to different extent. Therefore we decided to consider two extreme cases: the non-planar membrane with optimal surface tension,

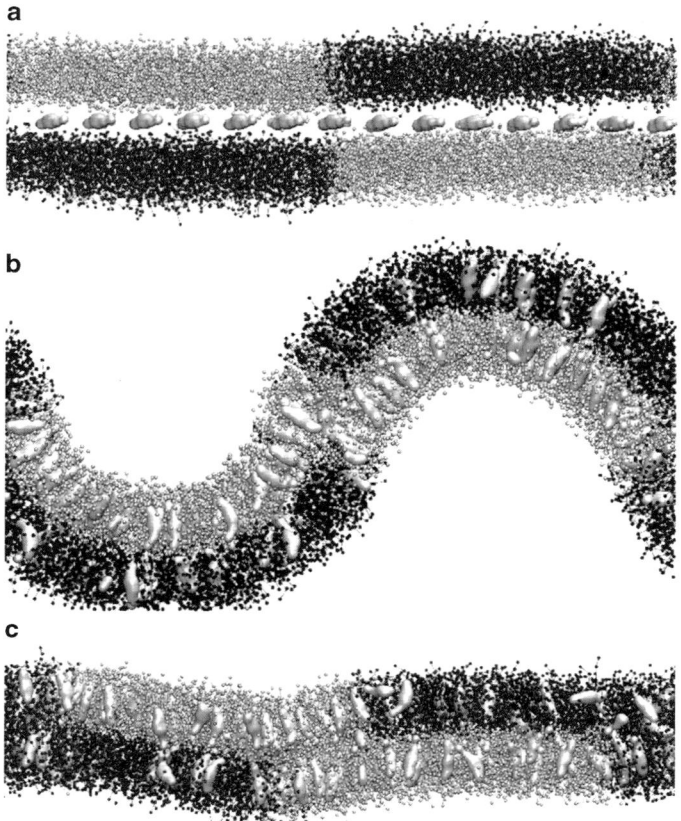

Fig. 2 Initial and equilibrated states of bilayers in setup 1. (**a**) Common initial state of all studied bilayers with cholesterol molecules located in the *XY* plane between the monolayers. (**b**) Equilibrated state of the DOPC/DOPS bilayer. DOPC lipids are *black*, DOPS lipids are *gray*. Similar picture is observed for DSPC/DSPS bilayer. (**c**) Equilibrated state of the DOPC/DSPC bilayer. DOPC lipids are *black*, DSPC lipids are *gray*. Very similar picture is observed for DOPS/DSPS bilayer. Lipids of different types are shown in the "balls and sticks" representation. Cholesterol molecules are shown in surface representation

which bends according to intrinsic curvatures of different lipids and the planar membrane subject to the surface tension stress caused by different areas per lipid for different lipid types.

In this work, lipid bilayers were modeled using the coarse-grained MARTINI force field version 2.1 [29]. Recommended simulation parameters for MARTINI force field were used [29] at the temperature of 320 K. In all simulations a Berendsen barostat with the relaxation constant of 5 ps was used for semi-isotropic pressure coupling. The pressure of 1 atm was maintained separately in *XY* plane and in *Z* direction.

In the first simulation setup (referred as "setup 1" hereafter) each monolayer consists of two rectangular blocks of lipids of

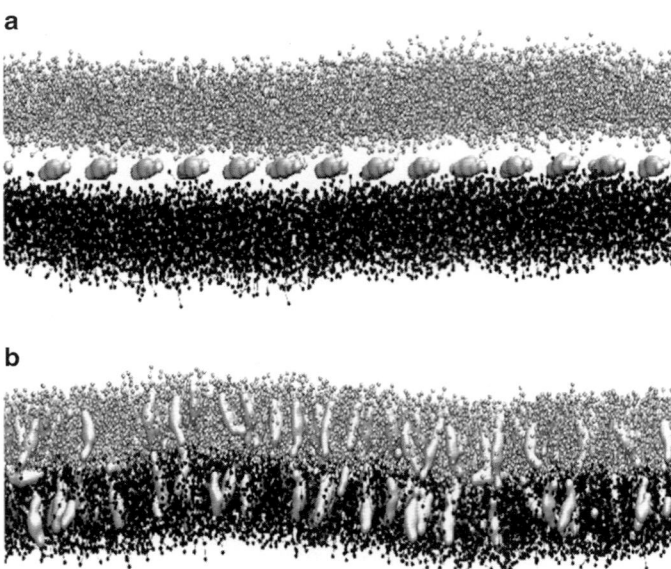

Fig. 3 Initial and final states of bilayers in setup 2. Colors and representations are the same as in Fig. 2. (**a**) Common initial state of all studied bilayers. (**b**) Final state of the DOPC/DOPS bilayer. DOPC lipids are *black*, DOPS lipids are *gray*. Very similar states without significant curvature are observed for all other bilayers

different type. Positions of the blocks in the second monolayer are inverted with respect to the first one. As a result, the whole system consists of two inverted asymmetric bilayer patches stacked side by side (Fig. 2a). The lateral dimensions of the blocks of different lipids changed in the course of simulations in order to optimize their areas per lipid. Thus no surface tension stress is developed in the monolayers. However, each monolayer becomes inhomogeneous and the membrane could develop significant curvature caused by different intrinsic curvatures of different lipid blocks (Fig. 2b).

In the second setup (referred as "setup 2" hereafter) two homogeneous monolayers containing different lipids were used (Fig. 3). The number of lipids was equal in both monolayers. Lateral homogeneity of the system ensures that the membrane remains approximately planar except for small random undulations. At the same time, the surface tension of two monolayers is different because the lipids, which form the monolayers, occupy different areas per lipid. Since both monolayers are confined in the same periodic box, they are subject to certain lateral strain.

The following lipids differing in head group charge and tail saturation were used: DOPC (dioleoylphosphatidylcholine, neutral, monounsaturated), DOPS (dioleoylphosphatidylserine, anionic, monounsaturated), DSPC (distearoylphosphatidylcholine, neutral, saturated), or DSPS (distearoylphosphatidylserine, anionic, saturated).

In the case of anionic lipids the coarse grained sodium ions were added to neutralize the system. Four initially asymmetric bilayers were formed: DOPC/DOPS (unsaturated, different charge), DSPC/DSPS (saturated, different charge), DOPC/DSPC (neutral, different saturation), and DOPS/DSPS (anionic, different saturation).

Initial system used in both setups contained 504 lipids (252 in each monolayer) arranged into square bilayer patch. The center of bilayer is placed at the point $Z = 0$, and the XY plane corresponds to the bilayer plane. In setup 1, initial bilayer was duplicated in X direction and its copy was inverted in Z direction. As a result, two square domains (compact patches, consisting of the lipids of one particular type) were formed in each monolayer (Fig. 2). In setup 2 no inversion was applied, and two homogeneous monolayers were formed (Fig. 3).

The final systems subject to production of MD runs contained 1,008 lipids (504 in each monolayer) solvated by ~24,000 coarse-grained water particles, which corresponds to ~24 CG water particles per lipid. Each CG water particle corresponds to 4 real water molecules leading to effective hydration of ~95 real water molecules per lipid, which is sufficient for full hydration of all studied lipid bilayers.

Additionally much larger DOPC/DOPS system was constructed using setup 1 to study the influence of cholesterol on the membrane curvature on larger time and length scale. It contained 4,032 lipids (2,016 lipids of each type), ~600,000 coarse grained water particle and 2,016 coarse grained sodium ions. The lipids were initially arranged into the planar elongated rectangular membrane patch with the dimensions 111.0×12.4 nm in the XY plane. This system is referred as "setup 3" hereafter (Fig. 4).

Artificial repulsive potential was introduced between the coarse-grained phosphate beads of the lipids from different blocks to prevent lateral mixing of the lipids of different kind. Lenard-Jones potential with the parameters $c_6 = 0$, $c_{12} = 2.581$ (GROMACS force field units) was used.

The CG model of cholesterol designed for MARTINI force field [46] was used. Cholesterol molecules were placed at the central plane of bilayer, oriented in XY plane and distributed evenly on rectangular grid with the account of periodic boundary conditions in XY plane. After that the monolayers were moved into opposite directions on Z scale for eliminating any steric overlaps of the lipid tails with cholesterol molecules. This forms a gap in the center of bilayer, which accommodates flat layer of aligned cholesterol molecules (Figs. 2a, 3a, and 4a). The number of cholesterol molecules was 196 in setups 1 and 2 and 784 in setup 3 (cholesterol to lipid ratio ~0.2 in all setups). In the course of subsequent equilibration the cholesterol molecules diffuse passively from the center of bilayer into both monolayers. In setup 3 the system without cholesterol was also simulated.

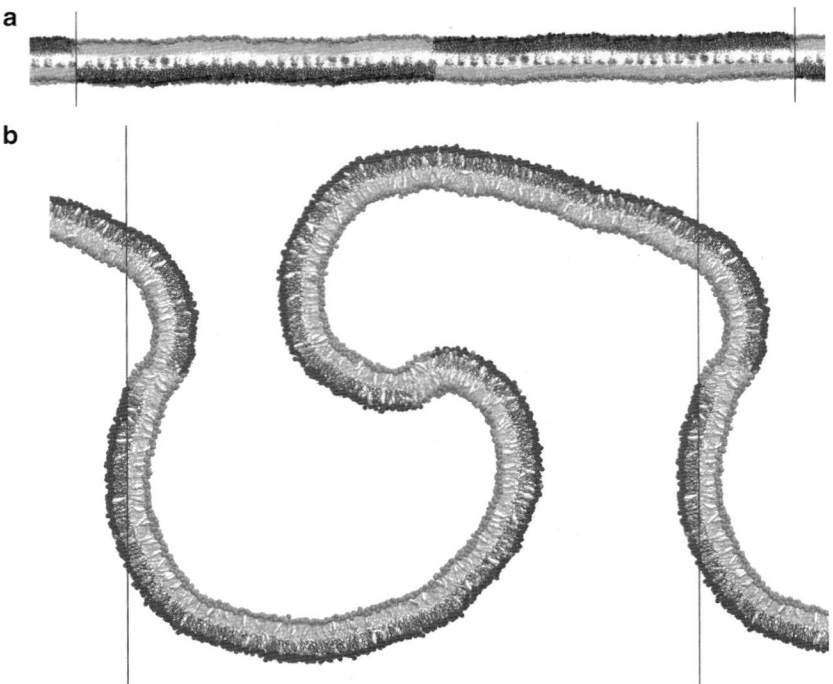

Fig. 4 Initial and final states of DOPC/DOPS bilayer in setup 3. Colors are the same as in Fig. 2. Lipid head groups are shown as spheres for clarity. (**a**) Initial state of bilayers. (**b**) Final meander-like shape of bilayer after 23 μs of simulation. *Vertical lines* show the boundaries of periodic simulation box in *X* dimension

The 12 μs MD trajectories for all systems from our previous work [42] were used for setups 1 and 2. The larger system in setup 3 was simulated for 23 μs. The effective time scale of CG simulations of lipids with MARTINI force field is proved to be four times longer than actual simulation time [29]. The effective simulation time is used hereon. Only equilibrated parts of trajectories were used for analysis.

3.2 Analysis

Significant large-scale curvature of bilayer was observed in setups 1 and 3 (Figs. 2b and 4b). This makes the analysis complicated because the properties of cholesterol molecules should be evaluated according to the normal to curved bilayer surface in the position of particular cholesterol molecule. The procedure proposed in our previous work [42] was used for analysis.

1. Positions of the PO4 particles (which represent the phosphate group in CG model) were determined for all lipids.

2. For each cholesterol molecule *i* the distances from its ROH particle (which represent the polar head group in CG model) to all PO4 particles of the lipids were computed.

3. Ten PO4 particles, which are closest to given ROH particle, were selected in monolayer 1 and their center of masses P_1 was found. This number was selected empirically to include all lipids, which are in contact with given cholesterol molecule.

4. Ten PO4 particles, which are closest to given ROH particle, were selected in monolayer 2 and their center of masses P_2 was found.

5. The local membrane normal was computed as $\vec{n} = \vec{P}_1 - \vec{P}_2$.

6. The position of local bilayer center was computed as $\vec{c} = \left(\vec{P}_1 + \vec{P}_2 \right) / 2$.

7. All distances were then computed in terms of $\vec{n}$ and $\vec{c}$. Cholesterol distribution ratio between the monolayers for a given trajectory frame was computed as

$$r(t_i) = N_1(i) / N_2(i), \qquad (1)$$

where $N_k(i)$ is the number of cholesterol molecules in the monolayer k for frame i.

Since the simulation time is long enough and the number of cholesterol molecules in the system is rather large, the flip-flop transitions in already equilibrated system can be observed in our simulations. We recorded the number of cholesterol molecules, which jump from one monolayer to another between two successive trajectory frames $i-1$ and i, K_i, considering only those jumps, which occur between the blocks consisting of different lipids. This allows disregarding those regions of the bilayer, where the lipids of the same type appear in both monolayers due to random lateral diffusion of the lipid domains and adjusting of their areas in the course of simulations. The number of cholesterol molecules, which remain in the same block of the same monolayer between trajectory frames $i-1$ and i, R_i, was also recorded. After that the mean flip-flop time τ_{flip} is computed as

$$\tau_{\mathrm{flip}} = \Delta t \Big/ \frac{1}{(M-s+1)} \sum_{i=s}^{M} \frac{K_i}{K_i + R_i}, \qquad (2)$$

where M is the number of trajectory frames, s is the first frame of equilibrated part of trajectory, Δt is the time interval between trajectory frames.

Much larger membrane size and simulation time in setup 3 allows analyzing the membrane curvature quantitatively. The principal scheme of the algorithm used to obtain the curvature profile of the membrane on each trajectory frame is the following.

1. The membrane midline in XZ plane is determined as a set of discrete points equally spaced along the membrane. Positions of the local bilayer centers and the normals, which were computed

as described above, are used on this step. As a result the shape of the membrane in XZ plane is reproduced accurately by the two-dimensional parametric curve $\{X(l_i), Z(l_i)\}$, where l is the length along the membrane, i is the number of discrete point.

2. The membrane midline is approximated by the sum of cosines using the fast Fourier transform (FFT). First and second analytical derivatives of the midline coordinates in respect to l $X'(l_i)$, $X''(l_i)$, $Z'(l_i)$, $Z''(l_i)$ are computed. Random noise on the derivatives is suppressed by applying the low-pass filter in Fourier space.

3. The curvature is computed in each discrete point along the midline as [47]

$$\kappa(l_i) = \frac{X'(l_i)Z''(l_i) - X''(l_i)Z'(l_i)}{\left(X'(l_i)^2 + Z'(l_i)^2\right)^{3/2}}.$$

It is more convenient to operate with the effective radius of curvature $R_{\text{curv}} = 1/\kappa$, which corresponds to the radius of the circle, which has the same curvature as the membrane in given point. The smaller is R_{curv}, the more curved is the membrane. For perfectly flat membrane R_{curv} is infinite.

In our previous work [42], for simplicity the analysis software did not account for periodic boundary conditions properly. As a result the cholesterol molecules, which were close to the boundaries of the periodic box, were excluded from consideration. In this study we improved the analysis algorithm significantly by accounting for periodic boundary conditions in all simulations. Although the qualitative trends in r and τ_{flip} remain the same, the absolute values of these parameters changed. It appears that in [42] the flip-flop times τ_{flip} were underestimated significantly, whereas the cholesterol distribution ratios r were computed quite accurately.

4 Notes

4.1 Cholesterol Distribution Ratio

The obtained results are summarized in Table 1. It is clearly seen that cholesterol strongly prefers anionic to neutral lipids, which was shown in all setups. This preference can be explained by less compact packing of the charged lipid head groups due to their mutual electrostatic repulsion, which makes the tendency of creating additional free volume in the membrane. This facilitates intercalation of cholesterol molecules between anionic lipids.

Cholesterol molecules strongly prefer saturated lipids to unsaturated. This trend is easily explained by better packing of straight saturated lipid tails, which facilitate easy intercalation of rigid cholesterol ring systems between them.

Table 1
Equilibrium cholesterol distribution, dynamics of its rearrangement between the monolayers, and the mean curvature in asymmetric bilayers

Bilayer		DOPC/DOPS	DSPC/DSPS	DOPC/DSPC	DOPS/DSPS
Charge		Neutral/anionic	Neutral/anionic	Neutral	Anionic
Saturation		Unsaturated	Saturated	Unsaturated/ saturated	Unsaturated/saturated
r	Setup 1	0.48 ± 0.06	0.43 ± 0.04	0.63 ± 0.08	0.46 ± 0.05
	Setup 2	0.74 ± 0.07	0.67 ± 0.08	0.44 ± 0.06	0.32 ± 0.04
τ_{flip}, ns	Setup 1	347.8 ± 165.3	$1{,}571.7 \pm 804.38$	645.16 ± 209.4	$2{,}005.01 \pm 1{,}073.1$
	Setup 2	267.5 ± 106.77	$1{,}403.5 \pm 654.9$	402.62 ± 202.4	$1{,}044.38 \pm 430.84$

r is distribution ratio of cholesterol between the monolayer. τ_{flip} is mean time of cholesterol flip-flop transitions (averaged over both monolayers)

In setup 1 the bilayers with different head group charges (DOPC/DOPS and DSPC/DSPS) have much smaller values of r than DOPC/DSPC bilayer with different tails saturation. In setup 2 the opposite situation is observed—both bilayers with different tails saturation (DOPC/DSPC and DOPS/DSPS) have larger r than the bilayers with different head group charges. This means that the membrane curvature enhances the effects of the lipid head groups on cholesterol distribution but somewhat decreases the effect of the tails saturation. In contrast, the presence of uncompensated surface tension decreases the effect of the head groups but emphasizes the effect of the lipid tails saturation.

If the leaflets have different net charge but the same tail saturation, the cholesterol distribution ratio is surprisingly insensitive to the saturation of lipid tails. Indeed, in setup 1 the values $r_{DOPC/DOPS} = 0.48 \pm 0.06$ and $r_{DSPC/DSPS} = 0.43 \pm 0.04$ are within the limits of statistical uncertainty. The same is true for setup 2 ($r_{DOPC/DOPS} = 0.74 \pm 0.07$, $r_{DSPC/DSPS} = 0.67 \pm 0.08$). In contrast, if the leaflets have different tails saturation but the same charge, then r becomes somewhat smaller for anionic lipids ($r_{DOPC/DSPC} = 0.63 \pm 0.08$, $r_{DOPS/DSPS} = 0.46 \pm 0.05$ in setup 1 and $r_{DOPC/DSPC} = 0.44 \pm 0.06$, $r_{DOPS/DSPS} = 0.32 \pm 0.04$ in setup 2). This suggests that the charge of lipid head groups plays a major role in distribution of cholesterol between the leaflets, while the influence of the saturation of the lipid tails becomes clearly visible only in the systems with the same charge of both monolayers.

4.2 Flip-Flop Transitions of Cholesterol

We observed a pronounced dependence of the mean flip-flop times on bilayer composition (Table 1). The fastest flip-flop transitions occur in completely unsaturated DOPC/DOPS bilayer (348 ns in setup 1 and 268 ns in setup 2). Neutral DOPC/DSPC bilayer with one saturated monolayer has larger τ_{flip} (645 ns in setup 1 and

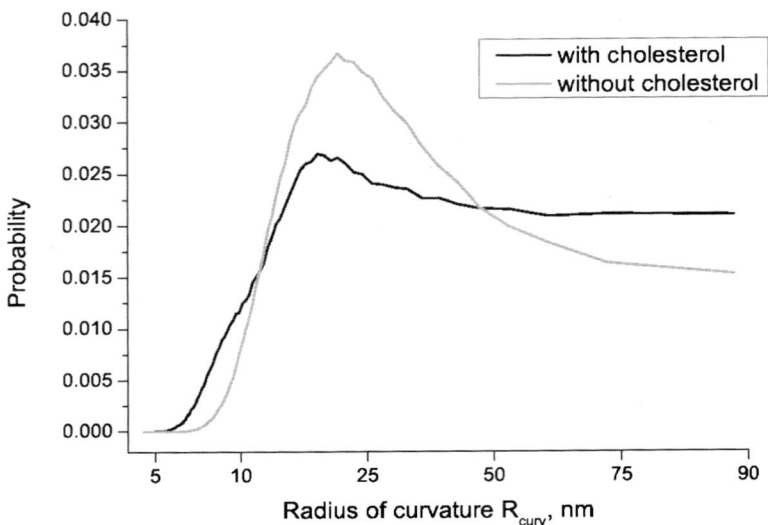

Fig. 5 Distributions of the membrane curvature in setup 3 in the simulations with and without cholesterol

403 ns in setup 2). The charged DOPS/DSPS bilayer with one saturated monolayer and the completely saturated DSPC/DSPS bilayer have slowest flip-flop transitions (up to 2005 ns in setup 1 and 1,404 ns in setup 2). We derive that both the tails saturation and the head group charge influence the flip-flop times significantly but the influence of the tails saturation is stronger.

The obtained flip-flop times are in perfect agreement with the data of Bennett et al. [33] in their study of symmetric coarse-grained DPPC bilayers. Out flip-flop times range from 0.27 to 2 μs, while the times reported for symmetric DPPC bilayer with 40 % cholesterol are between 0.32 and 4.34 μs [33].

4.3 Influence of Cholesterol on the Membrane Curvature

Our simulations allow answering very interesting question—what is the influence of cholesterol on the membrane curvature? Figure 5 clearly shows that the distributions of the membrane curvature are very different in the trajectories with and without cholesterol in setup 3. The peak of the distribution is observed at ~20.5 nm in the absence of cholesterol, and in its presence it becomes less pronounced, shifting to ~18 nm. In addition, the probability of observing both small radii of curvature (from ~5.5 nm to ~12 nm) and very large radii of curvature (larger than ~47 nm) increases at the expense of that corresponding to the maximum of distribution (from ~12 nm to ~47 nm).

These results could be interpreted in the following way. In the absence of cholesterol the membrane is bent rather uniformly. The most of the membrane has the radius of curvature around 20 nm while very sharp kinks (small radii of curvature) and flat regions (large radii of curvature) occupy only small parts of the membrane surface. In the presence of cholesterol the overall shape of the membrane changes dramatically. The amount of almost flat regions increases substantially. These regions are separated by rather sharp bends with very small radius of curvature, which results in the decrease of uniformly bent area.

4.4 Insight into Divergence of Experimental Results

The fact that experimental techniques perform so unreliably when measuring cholesterol distribution in asymmetric membranes gives us a hint about very high sensitivity of cholesterol distribution to some factors, which are hard to control even in otherwise pure experiments. The surface tension stress of the membrane and the membrane curvature could be among these factors. Indeed, the real membranes in experiments are rarely planar and completely homogeneous. Different parts of extended membrane may possess different curvatures or be subject to lateral mechanical stress. Our data show that the cholesterol distribution ratio between the monolayers consisting of different lipids depends strongly on the membrane curvature and the lateral mechanical stress. For example cholesterol distribution in DOPC/DOPS bilayer may range from 0.48 in the case of optimal curvature to 0.74 in the case of planar bilayer with the lateral stress. Therefore it is quite probable that divergent experimental data of cholesterol distribution in the asymmetric membranes are caused, at least partially, by the lack of control on membrane curvature and on lateral mechanical stress. Control of these parameters may not be possible in the conditions of experiment.

It is evident that the mutual correspondence exists between the membrane shape and cholesterol distribution. Instead of keeping monotonous curvature, cholesterol changes the overall shape of the curved membrane by promoting the appearance of flattened regions and sharp bends. In turn, the cholesterol distribution depends on the local curvature. Various self-organizing phenomena may also occur in such system due to feedback mechanisms between the cholesterol content and the membrane curvature. Investigation of the possibility of such mechanisms is planned in our future studies.

4.5 Perspectives

Cholesterol is the most abundant component of the vertebrate cell membranes after the lipids, thus there is little doubt that its concentration and distribution influences all major membrane properties. At the same time various factors, both external and internal to the membrane, lead to the changes of cholesterol distribution between the monolayers and inside the same monolayer. Such reciprocal dependencies lead to very complex self-consistent picture of cholesterol distribution and the membrane structure. Figure 6 is the scheme illustrating the connections between cholesterol distribution and various membrane properties.

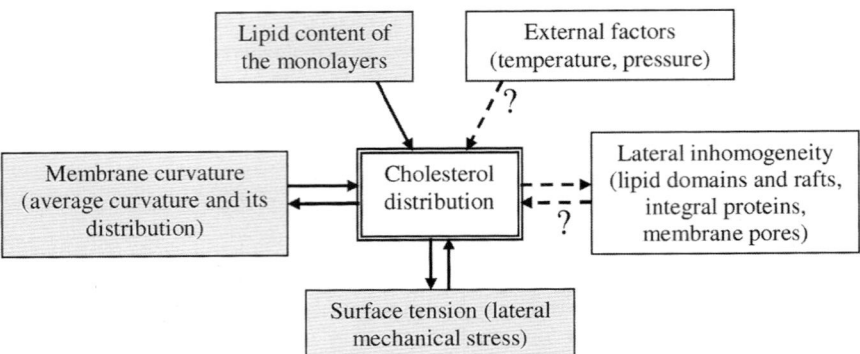

Fig. 6 The scheme illustrating the dependencies between cholesterol distribution, external physical factors and various membrane properties. *Gray boxes* and *solid lines* show the correlations established in this work. *White boxes* and *dashed lines* show hypothetical dependencies

Shaded boxes mark the dependencies, which are established in our work. Particularly there is mutual influence of cholesterol distribution on membrane curvature and surface tension. Two very important factors are not studied yet—external physical parameters (temperature and pressure) and the lateral membrane inhomogeneity (lipid domains and rafts, integral proteins, membrane pores, etc.). Present work establishes a solid methodological base for studying these factors. Future complementary computational and experimental studies may provide additional insight into role of cholesterol in various functions of real asymmetric membranes.

Acknowledgments

The calculations were partially performed on the HPC resources of the regional computational center Mesocentre of the University of Franche-Comté in the framework of "grand challenge" initiative in December 2012.

Software development is supported by the Ukrainian National Grid Technologies Program, project "Hardware and software complex for modeling and analysis of natural and artificial nanosystems in the Grid environment" and by the STCU grant 5525.

References

1. Kiessling V, Wan C, Tamm LK (2009) Domain coupling in asymmetric lipid bilayers. Biochim Biophys Acta 1788(1):64–71. doi:10.1016/j.bbamem.2008.09.003,S0005-2736(08)00294-0 [pii]

2. Demchenko AP, Yesylevskyy SO (2009) Nanoscopic description of biomembrane electrostatics: results of molecular dynamics simulations and fluorescence probing. Chem Phys Lipids 160(2):63–84

3. Ikonen E (2008) Cellular cholesterol trafficking and compartmentalization. Nat Rev Mol Cell Biol 9(2):125–138. doi:10.1038/nrm2336, nrm2336 [pii]

4. Hamilton JA (2003) Fast flip-flop of cholesterol and fatty acids in membranes: implications for

membrane transport proteins. Curr Opin Lipidol 14(3):263–271

5. van Meer G (2005) Cellular lipidomics. EMBO J 24(18):3159–3165. doi:10.1038/sj.emboj. 7600798, 7600798 [pii]

6. Hale JE, Schroeder F (1982) Asymmetric trans-bilayer distribution of sterol across plasma membranes determined by fluorescence quenching of dehydroergosterol. Eur J Biochem 122(3):649–661

7. Mondal M, Mesmin B, Mukherjee S, Maxfield FR (2009) Sterols are mainly in the cytoplasmic leaflet of the plasma membrane and the endocytic recycling compartment in CHO cells. Mol Biol Cell 20(2):581–588. doi:10.1091/mbc.E08-07-0785, E08-07-0785 [pii]

8. Gimpl G, Gehrig-Burger K (2011) Probes for studying cholesterol binding and cell biology. Steroids 76(3):216–231. doi:10.1016/j.steroids.2010.11.001, S0039-128X(10)00266-7 [pii]

9. Lange Y, Steck TL (2008) Cholesterol homeostasis and the escape tendency (activity) of plasma membrane cholesterol. Prog Lipid Res 47(5):319–332. doi:10.1016/j.plipres.2008. 03.001, S0163-7827(08)00025-8 [pii]

10. Fisher KA (1976) Analysis of membrane halves: cholesterol. Proc Natl Acad Sci U S A 73(1): 173–177

11. Ali MR, Cheng KH, Huang J (2007) Assess the nature of cholesterol-lipid interactions through the chemical potential of cholesterol in phosphatidylcholine bilayers. Proc Natl Acad Sci U S A 104(13):5372–5377. doi:10.1073/pnas.0611450104, 0611450104 [pii]

12. Radhakrishnan A, McConnell H (2005) Condensed complexes in vesicles containing cholesterol and phospholipids. Proc Natl Acad Sci U S A 102(36):12662–12666. doi:10.1073/pnas.0506043102, 0506043102 [pii]

13. Ohvo-Rekila H, Ramstedt B, Leppimaki P, Slotte JP (2002) Cholesterol interactions with phospholipids in membranes. Prog Lipid Res 41(1):66–97, S0163782701000200 [pii]

14. Róg T, Pasenkiewicz-Gierula M, Vattulainen I, Karttunen M (2009) Ordering effects of cholesterol and its analogues. Biochim Biophys Acta 1788(1):97–121. doi:10.1016/j.bbamem. 2008.08.022

15. Kučerka N, Perlmutter JD, Pan J, Tristram-Nagle S, Katsaras J, Sachs JN (2008) The effect of cholesterol on short- and long-chain monounsaturated lipid bilayers as determined by molecular dynamics simulations and X-ray scattering. Biophys J 95(6):2792–2805. doi:10.1529/biophysj.107.122465

16. Chen Z, Rand RP (1997) The influence of cholesterol on phospholipid membrane curvature and bending elasticity. Biophys J 73(1): 267–276

17. Garg S, Porcar L, Woodka AC, Butler PD, Perez-Salas U (2011) Noninvasive neutron scattering measurements reveal slower cholesterol transport in model lipid membranes. Biophys J 101(2):370–377. doi:10.1016/j.bpj.2011.06.014, S0006-3495(11)00711-9 [pii]

18. Steck TL, Ye J, Lange Y (2002) Probing red cell membrane cholesterol movement with cyclodextrin. Biophys J 83(4):2118–2125. doi:10.1016/s0006-3495(02)73972-6

19. Brasaemle DL, Robertson AD, Attie AD (1988) Transbilayer movement of cholesterol in the human erythrocyte membrane. J Lipid Res 29(4):481–489

20. Berendsen HJC (1996) Bio-molecular dynamics comes of age. Science 271:954–955

21. van Gunsteren WF, Bakowies D, Baron R, Chandrasekhar I, Christen I, Daura X, Gee P, Geerke DP, Glattli A, Hunenberger PH, Kastenholz MA, Oostenbrink C, Schenk M, Trzesniak D, van der Vegt NFA, Yu HB (2006) Biomolecular modeling: goals, problems, perspectives. Angew Chem Int Ed 45(25): 4064–4092

22. Berkowitz ML (2009) Detailed molecular dynamics simulations of model biological membranes containing cholesterol. Biochim Biophys Acta 1788(1):86–96

23. Hofsäß C, Lindahl E, Edholm O (2003) Molecular dynamics simulations of phospholipid bilayers with cholesterol. Biophys J 84(4): 2192–2206

24. Chiu SW, Jakobsson E, Mashl RJ, Scott HL (2002) Cholesterol-induced modifications in lipid bilayers: a simulation study. Biophys J 83(4):1842–1853

25. Falck E, Patra M, Karttunen M, Hyvönen MT, Vattulainen I (2004) Lessons of slicing membranes: interplay of packing, free area, and lateral diffusion in phospholipid/cholesterol bilayers. Biophys J 87(2):1076–1091

26. Niemelä PS, Ollila S, Hyvönen MT, Karttunen M, Vattulainen I (2007) Assessing the nature of lipid raft membranes. PLoS Comput Biol 3(2):e34. doi:10.1371/journal.pcbi.0030034

27. Zhang Z, Lu L, Berkowitz ML (2008) Energetics of cholesterol transfer between lipid bilayers. J Phys Chem B 112(12):3807–3811. doi:10.1021/jp077735b

28. Demchenko AP, Yesylevskyy SO (2009) Nanoscopic view on biomembrane electrostatics: molecular dynamics simulations and fluorescence probing. In: Vth International Symposium Supramolecular Systems in Chemistry and Biology, Kyiv. p 62

29. Marrink SJ, Risselada HJ, Yefimov S, Tieleman DP, de Vries AH (2007) The MARTINI force field: coarse grained model for biomolecular simulations. J Phys Chem B 111(27):7812–7824. doi:10.1021/jp071097f

30. Marrink SJ, de Vries AH, Mark AE (2004) Coarse grained model for semiquantitative lipid simulations. J Chem Phys 108(2):750–760

31. Monticelli L, Kandasamy SK, Periole X, Larson RG, Tieleman DP, Marrink SJ (2008) The MARTINI coarse-grained force field: extension to proteins. J Chem Theory Comput 4:819–834

32. Lopez CA, Rzepiela AJ, de Vries AH, Dijkhuizen L, Hunenberger PH, Marrink SJ (2009) Martini coarse-grained force field: extension to carbohydrates. J Chem Theory Comput 5(12):3195–3210

33. Bennett WFD, MacCallum JL, Hinner MJ, Marrink SJ, Tieleman DP (2009) Molecular view of cholesterol flip-flop and chemical potential in different membrane environments. J Am Chem Soc 131(35):12714–12720. doi:10.1021/ja903529f

34. Huttner WB, Zimmerberg J (2001) Implications of lipid microdomains for membrane curvature, budding and fission: commentary. Curr Opin Cell Biol 13(4):478–484

35. McMahon HT, Kozlov MM, Martens S (2010) Membrane curvature in synaptic vesicle fusion and beyond. Cell 140(5):601–605

36. Gores GJ, Herman B, Lemasters JJ (2005) Plasma membrane bleb formation and rupture: a common feature of hepatocellular injury. Hepatology 11(4):690–698

37. Taylor RC, Cullen SP, Martin SJ (2008) Apoptosis: controlled demolition at the cellular level. Nat Rev Mol Cell Biol 9(3):231–241

38. Jahn R, Lang T, Südhof TC (2003) Membrane fusion. Cell 112(4):519–533

39. Churchward MA, Rogasevskaia T, Höfgen J, Bau J, Coorssen JR (2005) Cholesterol facilitates the native mechanism of Ca^{2+}-triggered membrane fusion. J Cell Sci 118(20):4833–4848

40. Cadd TL, Skoging U, Liljeström P (2005) Budding of enveloped viruses from the plasma membrane. Bioessays 19(11):993–1000

41. Gondré-Lewis MC, Petrache HI, Wassif CA, Harries D, Parsegian A, Porter FD, Loh YP (2006) Abnormal sterols in cholesterol-deficiency diseases cause secretory granule malformation and decreased membrane curvature. J Cell Sci 119(9):1876–1885

42. Yesylevskyy SO, Demchenko AP (2012) How cholesterol is distributed between monolayers in asymmetric lipid membranes. Eur Biophys J 41(12):1043–1054

43. Hess B, Kutzner C, van der Spoel D, Lindahl E (2008) GROMACS 4: algorithms for highly efficient, load-balanced, and scalable molecular simulation. J Chem Theor Comput 4(3):435–447

44. Yesylevskyy SO (2012) Pteros: fast and easy to use open-source C++ library for molecular analysis. J Comput Chem 33(19):1632–1636

45. Humphrey W, Dalke A, Schulten K (1996) VMD—visual molecular dynamics. J Mol Graph 14(1):33–38

46. Marrink SJ, de Vries AH, Harroun TA, Katsaras J, Wassall SR (2008) Cholesterol shows preference for the interior of polyunsaturated lipid membranes. J Am Chem Soc 130(1):10–11

47. Coxeter HSM (1989) Introduction to geometry, 2nd edn. Wiley, New York

Chapter 21

Computer Simulations of Phase Separation in Lipid Bilayers and Monolayers

Svetlana Baoukina and D. Peter Tieleman

Abstract

Studying phase coexistence in lipid bilayers and monolayers is important for understanding lipid–lipid interactions underlying lateral organization in biological membranes. Computer simulations follow experimental approaches and use model lipid mixtures of simplified composition. Atomistic simulations give detailed information on the specificity of intermolecular interactions, while coarse-grained simulations achieve large time and length scales and provide a bridge towards state-of-the-art experimental techniques. Computer simulations allow characterizing the structure and composition of domains during phase transformations at Angstrom and picosecond resolution, and bring new insights into phase behavior of lipid membranes.

Key words Molecular dynamics, Domains, Rafts, Lung surfactant, Phase diagram

1 Introduction

Lipids refer to a broad class of compounds, and comprise fatty acids, glycerolipids, glycerophospholipids, sphingolipids, sterols, etc. [1]. Most lipid molecules are amphiphilic: they consist of a hydrophobic (apolar) tail and a hydrophilic (polar) headgroup [2]. Driven by the hydrophobic effect, they self-assemble into various phases in polar or apolar media, and act as surfactants at polar–apolar interfaces [3]. These phases include micellar, hexagonal, cubic, and lamellar morphologies [4]. The morphology of lipid aggregates depends on lipid and solvent concentrations, molecular shape (headgroup/tail ratio), and chemical details of lipids, and is modulated by temperature, pH, ion concentration, and other factors.

In this chapter we focus on flat morphologies: lipid bilayers in water and lipid monolayers formed at an air–water interface. Bilayers and monolayers are thin films as their thickness (of order of molecular length, i.e., ~nm) is much smaller than the in-plane size (~μm). They are ordered in the normal direction, having a preferred molecular position and orientation, and adopt various two-dimensional (2D) phases in the in-plane, or lateral, direction.

Dylan M. Owen (ed.), *Methods in Membrane Lipids*, Methods in Molecular Biology, vol. 1232,
DOI 10.1007/978-1-4939-1752-5_21, © Springer Science+Business Media New York 2015

The phase state of a bilayer is determined by its composition and temperature. In a monolayer, it also depends on the surface tension at the interface, γ, which varies with the surface density/area per lipid molecule, A_L: $\gamma(A_L) = \gamma_0 - \Pi(A_L)$. Here Π is the surface pressure arising from intermolecular interactions, and γ_0 is the surface tension of the bare interface (72 mN/m for the air–water interface at room temperature [5]). Upon reducing A_L, the surface tension decreases from γ_0 to near-zero values, and a monolayer can adopt phases ranging from gas to solid at a fixed temperature [6, 7]. In contrast, lipid bilayers are tensionless ($\gamma \to 0$) in the absence of external forces [8].

Lipid bilayers and monolayers constitute the structural basis for biological membranes. Biological membranes surround cells and their organelles, and cover biological interfaces such as in the lungs and in the eyes [2]. In cell membranes, a lipid bilayer separates their distinct chemical content and provides a platform for protein function [9]. In lung surfactant, a lipid monolayer at the gas exchange interface maintains low surface tensions during the breathing cycle [10, 11].

Biological membranes contain multiple lipid species; their composition varies between organisms, cells, and organelles [12]. Different lipids are organized in the membrane plane into regions of distinct properties, forming domains of coexisting phases [13]. Lipids are also distributed unevenly between the inner and outer leaflets of cell membranes. The phases formed by biological membranes are mainly liquid, such as a liquid-crystalline phase (Lα), which is similar to a liquid-expanded phase (LE) in monolayers. At high cholesterol concentration, a liquid-ordered (Lo) phase with increased orientational order of lipid molecules is formed; the phase depleted in cholesterol is called liquid-disordered (Ld). Gel (Lβ, also called So) phases in bilayers, or liquid-condensed (LC) phases in monolayers, are formed in some cases, for example in skin membranes [14], and possibly in lung surfactant.

Lateral organization of lipids in biological membranes plays an important role by modulating the membrane structure and properties, and optimizing the environment for protein function. In cell membranes, the raft hypothesis [15, 16] proposes the presence of nano-scale dynamic domains of Lo-like phase, enriched in saturated lipids and cholesterol, and incorporating selected proteins [17]. Rafts are implicated in membrane trafficking, signal transduction, including immune response, and entry of pathogens [18–21]. In lung surfactant, the lipid monolayer is separated into coexisting phases, which is necessary for its function [22–24]. It is conceivable that the squeeze-out of LE domains enriched in unsaturated lipids allows maintaining near-zero surface tensions [25, 26].

The nature of lateral organization in biological membranes, however, remains controversial [27–29]. In lung surfactant, the structure of more ordered domains, in particular on the nano-scale,

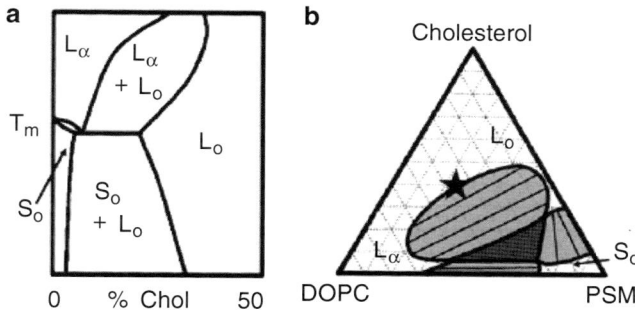

Fig. 1 Schematic binary phase diagram for a bilayer containing saturated phospholipid and cholesterol (**a**). Schematic ternary phase diagram of SM: DOPC: cholesterol mixture at a fixed temperature (**b**); three-phase coexistence is shown in *dark grey*, two-phase coexistence is shown in *light grey* with tie-lines and a critical point (*star*). Reproduced with permission from [35]. Copyright 2014

and their role in surface activity are not fully understood. Uncertainties remain regarding the small size and short life time of rafts in cell membranes. Current theories suggest several mechanisms underlying nano-scale heterogeneity in biological membranes, including non-equilibrium cellular processes, adhesion to cytoskeleton, lipid–protein interactions, or lipid–lipid interactions (*see* [27, 29, 30] for review).

Characterizing lipid–lipid interactions alone is complicated by the presence of thousands of different lipids [31], which implies multidimensional phase diagrams. This problem is approached by studying model systems of simplified composition. Many binary, ternary and some quaternary phase diagrams have been obtained in experimental studies (Fig. 1a, b), *see* [32–35] and references therein. It is important to note that discrepancies exist between experimental phase diagrams, which are related to both experimental methods and the quantities measured. Experimental observation of domains becomes challenging on the nano-scale as it requires simultaneously high spatial and temporal resolutions [28, 36]. Computer simulations have advanced significantly in recent years, reaching the time and length scales accessible by state-of-the-art experimental techniques, including super-resolution microscopy and secondary ion mass spectroscopy [37, 38]. Simulations allow characterizing interactions at atomic level, with Angstrom and picoseconds resolution, and complement experimental studies.

Here we review recent computer simulations of phase separation in model lipid membranes. We give a brief introduction to simulations methods and models, and describe typical lipid mixtures used as model systems in Subheading 2. We then discuss theoretical considerations and system setups to simulate phase transformations in Subheading 3, and conclude by covering several common simulation problems and providing possible solutions in Subheading 4.

2   Materials

2.1   Models

Standard particle-based computational approaches to study lipid membranes include Molecular Dynamics (MD) and Monte Carlo (MC) simulations. In simulations, a small membrane patch is represented as a system of many interacting particles. MC generates statistical ensembles of the system's configurations using trial moves combined with an acceptance criterion. MD generates the system's trajectory (coordinates of particles in time) by integrating numerically the classical equations of motion. In Dissipative Particle Dynamics (DPD), momentum-conserving friction and random noise are added to the forces acting on a particle.

The interactions between the particles are defined by a set of potentials for bonded and non-bonded interactions, which constitute a force field (FF). Based on the level of description, the FFs used for lipid simulations can be divided into two groups: atomistic, which represent (nearly) all atoms in the molecule as separate particles, and coarse-grained (CG), which group several (heavy) atoms into a particle. Atomistic FFs differ by treatment of non-polar hydrogen atoms, which are explicitly described in all-atom (AA) models, or grouped together with the neighboring heavy atom into one particle in united-atom (UA) models. CG FFs differ substantially by their resolution ("granularity"), the treatment of the solvent (explicit or implicit), and approaches to coarse-graining (top-down and bottom-up), which in turn influence their chemical specificity. In DPD simulations, soft-repulsive interactions between CG particles are typically applied.

The AA CHARMM FF [39] for lipids has been significantly improved in the recent version (C36) [40] compared to previous parameter sets (C27, C27R) [41, 42]. A number of lipid bilayer properties, including form factors (density profiles), area per lipid and orientational order parameters, have been validated against experimental data. The Stockholm lipids (Slipids) [43–45] is an AA FF based on C36, compatible with the AMBER FF family [46]; the bilayer properties were validated for various lipids in a range of temperatures. In a refined version of the AA AMBER lipid FF (Lipid14) [47], the structural and dynamic properties of tensionless bilayers compare well with experimental data. A recent UA GROMOS FF [48] (G53A6$_L$) [49] reproduces well the structural and hydration properties of phosphatidylcholine (PC) bilayers in the Lα phase [50]. The Berger modification [51] to GROMOS FF is an often used model which, however, has noticeable discrepancies with experimental data. As a result, further adjustments were incorporated to study phase separation [52, 53].

To date, atomistic simulations are limited to relatively small time and length scales insufficient to reproduce phase separation, due to the computational demand. A typical membrane lateral size is ~10–15 nm, with simulation times of hundreds of ns approaching

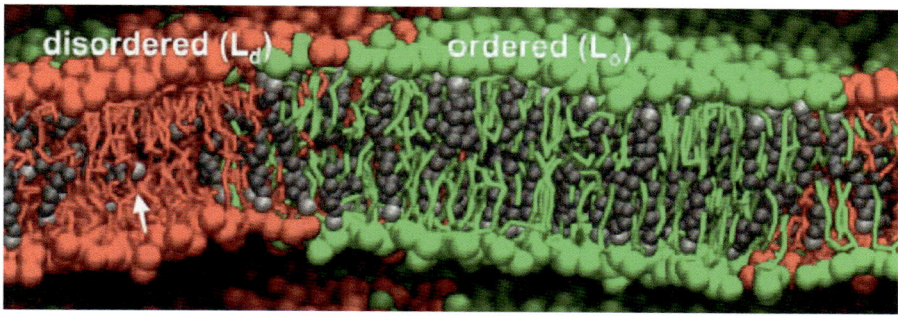

Fig. 2 Coexistence of Lo and Ld phase in the lipid bilayer of DPPC: DLiPC: cholesterol in the Martini FF, side view. DPPC is shown in *green*, DLiPC in *red*, cholesterol in *grey*. Reproduced with permission from [101]. Copyright 2014

10 μs in more recent simulations [54, 55]. Atomistic models have been mainly applied to study the properties of one of the phases in binary or ternary lipid mixtures. These simulations characterized the interactions between different lipid species, and a number of membrane properties including the area per lipid, chain order parameters, lipid diffusion coefficients, elastic moduli, and correlation functions. Early stages of domain formation on the nano-scale have been reproduced, and nano-scale lateral heterogeneity analyzed [54–56].

Coarse-grained (CG) models, on the other hand, have allowed significant computational speed-up, at the cost of atomic detail. The accessible time and length scales in CG models vary with their resolution, and the times required to sample compositional de-mixing also differ depending on the smoothness of the potentials. In the CG Martini model [57] approximately four heavy atoms are grouped into a particle. The model retains chemical specificity: the particles are assigned to different types based on their polarity, charge, and hydrogen bonding ability; non-bonded interactions between them are parameterized based on free energies of partitioning between various solvents. Using Martini, liquid–liquid and liquid–solid phase coexistence was reproduced in lipid bilayers and monolayers (Figs. 2 and 3), *see* for example [58] for review. The CG model of Shinoda [59] maps three heavy atoms into a particle. This model also distinguishes multiple polar and apolar particle types parameterized based on structural and thermodynamic data; unlike Martini, the interfacial properties of water were optimized [60]. The CG model of Lenz [61] developed for MC simulations is less detailed: lipid molecules are represented by a single chain, and the solvent is modeled by "phantom" molecules which only interact repulsively with lipid beads. The soft-repulsive-core model of Smit [62] used in DPD simulations has a mapping similar to Martini model, but distinguishes only hydrophilic head and hydrophobic tail particles, similar to the model of Lenz. For a comprehensive review of CG models used in lipid membrane simulations, *see* for example [62–66].

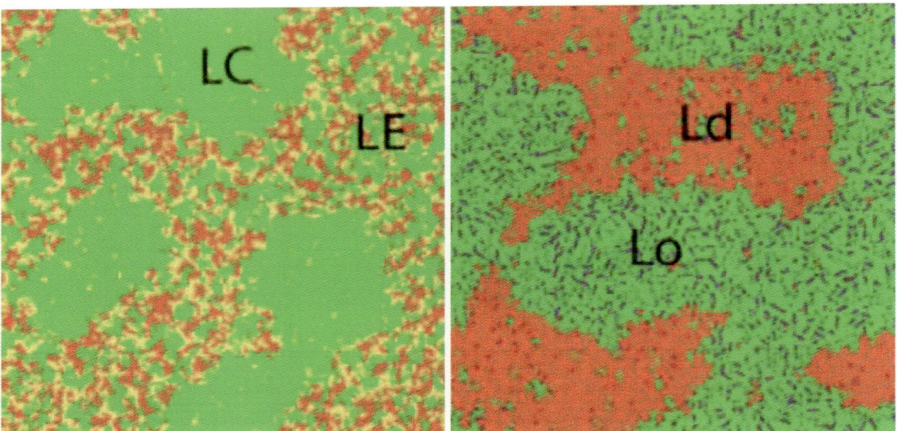

Fig. 3 Coexistence of LC/LE and Lo/Ld phases in lipid monolayers in the Martini FF, top view. DPPC is shown in *green*, DOPC in *orange*, POPG in *yellow*, cholesterol in *purple*. Reproduced with permission from [76]. Copyright 2014

2.2 Lipid Composition

Computer simulations generally focused on model membranes containing one to three common lipid components, although more complex mixtures have also been studied [67, 68].

Simulations of one-component membranes mainly used phosphatidylcholine (PC) lipids, in particular dipalmitoylphosphatidylcholine (DPPC). Extensive experimental data on DPPC bilayers and monolayers can be used for FF validation. Bilayer simulations investigated phase transformations between Lα and gel, as well as ripple and sub-gel phases, and determined the main phase transition temperature (Tm) for the FF studied [69–73]. The surface pressure (tension)–area isotherms of DPPC monolayers were simulated, and the coexisting LE/LC and LE/gas phases at the isotherm plateau were reproduced [59, 74, 75]. The LC/LE coexistence in a range of surface tensions was simulated for two and three component lipid monolayers containing DPPC and unsaturated lipids [76, 77].

For two-component membranes, usually two types of systems were simulated. The first is a mixture of a high Tm lipid and a low Tm lipid, which has a tendency to phase separate into Lα and gel phases at intermediate temperatures. Examples include a saturated and an unsaturated lipid, such as DPPC and dioleoylphosphatidylcholine (DOPC), or lipids with short and long saturated chains, such as dimyristoylphosphatidylcholine (DMPC) and distearoylphosphatidylcholine (DSPC) [78–83]. A more general case of a binary mixture with tunable repulsive interactions between the two lipid types was considered in DPD simulations of flat bilayers and vesicles [84–86] (Fig. 4).

The second type is a binary mixture of either PC or sphingomyelin (SM) with cholesterol, which can form Lo phases. Cholesterol has an ordering and condensing effect on lipids, forms

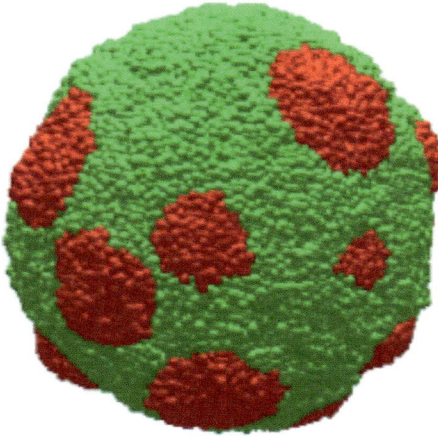

Fig. 4 Phase separation in a bilayer vesicle containing a binary lipid mixture in DPD simulations. Reproduced with permission from [84]. Copyright 2014

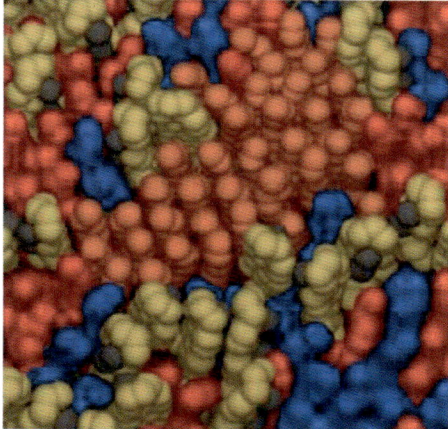

Fig. 5 Molecular structure of the Lo phase in the CHARMM FF. DPPC is shown in *red*, DOPC in *blue*, cholesterol in *yellow*; a sub-structure of DPPC with local hexagonal order is shown in *orange*. Reproduced with permission from [54]. Copyright 2014

rigid clusters/condensed complexes, and dramatically changes bilayer properties and phase behavior [87, 88]. Binary mixtures with cholesterol have been simulated in a large number of works [89–95], *see* also [96, 97] for review; the binary phase diagram of DMPC and cholesterol was calculated in CG simulations [98].

Lipid–lipid interactions in binary mixtures are important for understanding more complex phase behavior of ternary mixtures of a saturated lipid (e.g., DPPC, DSPC, or SM), an unsaturated lipid (e.g., DOPC or dilinoleoylphosphatidylcholine, DLiPC), and cholesterol. These ternary mixtures separate into Lo and Ld phases, and represent canonical model systems to study rafts in biological membranes. Atomistic simulations have provided details of the properties of the Lo and Ld phases [52, 54, 55, 99, 100] (Fig. 5), while CG simulations have reproduced large-scale phase separation in bilayers, monolayers, and vesicles [76, 101–103] (Figs. 2 and 3).

The difference in lipid composition of the inner and outer leaflets in biological membranes motivated a number of simulations of asymmetric bilayers [56, 68, 104, 105]. Domains in the two leaflets are coupled via inter-leaflet interactions; phase separation in one leaflet affects the phase behavior of the opposing leaflet [106, 107].

Biological membranes contain substantial fractions of the so-called hybrid lipids, i.e., lipids with a saturated and an unsaturated chain, such as palmitoyloleoylphosphatidylcholine (POPC), which could act as lineactants [108]. These lipids were simulated in quaternary [109] and ternary [110] mixtures forming Lo and Ld phases. These studies investigated the partitioning of hybrid lipids at domain boundaries and their effect on line tension and degree of phase separation.

To conclude this section, it is important to mention that in biological membranes, lipid–protein interactions play an active role in lateral organization. These interactions were investigated in a number of simulations, *see* for example [58] for review, which lie beyond the scope of this chapter.

3   Methods

Phase separation in lipid membranes implies lateral de-mixing of components and formation of regions of distinct properties. The phase state of the membrane is determined by the point in the phase diagram, which is given by the composition and temperature for bilayers, and additionally by the surface tension for monolayers. The location of the point within either binodal or spinodal regions in the phase diagram results in different mechanisms for phase transformation. Phase transformation proceeds via nucleation and growth in the former case, and via spinodal decomposition in the latter case. Nucleation is characterized by a free energy barrier; it is initiated by (compositional) fluctuations larger than a critical nucleus. Spinodal decomposition does not have a free energy barrier; it is spontaneous and uniform, spread over the entire membrane.

Simulations of phase separation in lipid membranes, even for one state point, rely on several important factors, outlined below.

3.1 Mapping of Phase Diagrams in the Simulation FF to Experimental Diagrams

This major question is often difficult to address directly, as it requires extensive simulations under varying temperature and composition. More feasible tests include the Tm for single-component bilayers, and the tendency of specific lipids to segregate, in the given FF [111].

Bilayer properties are typically parameterized in the Lα phase, and the FF parameters do not depend on temperature. In CG models, substitution of entropic interactions by enthalpic ones due to the reduced number of degrees of freedom leads to a weaker temperature dependence. Therefore, if Tm is not included in the FF validation, it can be somewhat underestimated [69, 72] (*see* **Note 4.1**).

In monolayers, phase behavior at a fixed temperature depends on the surface pressure–area isotherm in simulations. The latter, in turn, is affected by the surface tension at the bare air–water interface γ_0, which enters directly in the balance of interfacial forces and contributes to the monolayer lateral pressure profile [112, 113]. The γ_0 has correct values in the CG model by Shinoda [60], has improved values in the CG big multipole water (BMW) [114], and in TIP4P [115], is somewhat underestimated in SPC [48] and TIP3P [116] models, and noticeably underestimated in the CG Martini standard and polarizable [117] models. Underestimated surface tension offsets the surface pressure–area isotherm compared to experimental [74] (*see* **Note 4.2**).

Driving forces for lipid de-mixing determine the tendency of specific lipids to segregate. Formation of Lo or gel phases is accompanied by orientational ordering of the hydrocarbon chains and a decrease of their conformational entropy. Loss of intra-molecular degrees of freedom in CG models leads to lower entropy changes at phase transitions. Enthalpic contribution to de-mixing depends on the Lennard–Jones (LJ) interactions between different lipid types, and thus is model specific, (*see* **Note 4.3**). For example, the Lo/Ld phase separation in the DPPC:DOPC:cholesterol bilayer in the Martini model can be induced or suppressed by small changes of the particle types corresponding to unsaturated sites in the hydrocarbon chains [118, 119]. Therefore, it is essential to compare the composition of the coexisting phases in the considered FF to experimental tie-lines.

3.2 Correctness of the Structural and Dynamic Properties of the Phases

Such properties as area per lipid, chain orientational order parameters, and lipid lateral diffusion coefficient can serve as characteristics of the phase state of lipid membranes, as they change dramatically upon phase transitions. These properties are typically validated on parameterization of atomistic FF, as well as high-resolution CG models with chemical specificity. More generic or low-resolution CG models typically apply to several similar lipid types (e.g., saturated PC), and qualitatively reproduce the changes of membrane properties upon phase transitions. For phases not included explicitly in the FF validation, the properties may differ from real systems (*see* **Note 4.4**). For example, tilted gel and ripple phases in DPPC bilayers can be formed in the CG models of Lenz and Smit, but not in the CG Martini model (*see Subheading* 2.1) (*see* **Note 4.1**). If a CG FF is not parameterized to match the dynamic properties of lipid membranes and water, interpretation of the time scale is required. For example, in the Martini model, a scaling factor of 4 is applied to the simulation time, based on the diffusion of water and lipids, and other properties [57].

3.3 The Time Scales of Phase Separation

The times required to observe phase separation in simulations depend on several factors. Diffusion of individual lipids, characterized by the coefficient of (long-time) lateral diffusion, sets the time scale to sample compositional de-mixing in a membrane patch of a given size. For example, in a liquid phase with a typical diffusion coefficient of 10^{-7} cm^2/s, a lipid travels a distance of 20 nm during a 10 μs simulation. Nucleation of domains of a new phase requires overcoming a free energy barrier; it's characteristic time scales exponentially with the height of the barrier, which increases with the proximity to the de-mixing point (Tm). Lipid membranes represent nearly-2D films coupled to a three-dimensional (3D) solvent. Domain growth kinetics depends on the domain size, membrane and solvent viscosities, and other hydrodynamic parameters in simulations, as well as the mechanism of phase transformation [27, 120]. The so-called hydrodynamic length, given by the ratio of the membrane surface viscosity to the bulk solvent viscosity, separates the 2D and 3D domain growth regimes.

3.4 The Length Scales of Phase Separation

Phase transitions cannot be observed in simulations if the critical size of the nucleus of the new phase is larger than the simulation box size (*see* **Note 4.1**). The critical nucleus size increases in the vicinity of the phase transition point, as the difference in chemical potentials between the two phases become small. Once the domains of the new phase are formed, their characteristic size is also limited by the simulation box size. As the latter is typically of order of tens of nm, distinguishing between the macro- vs. nano-scale phase separation can be challenging in simulations.

3.5 System Setup

Typical setups to simulate phase separation include randomly mixed membranes, or membranes with pre-formed domains of the coexisting phases with the composition matching the experimental data. The latter approach is often used in atomistic simulations [54–56], to investigate the partitioning of selected components between the coexisting phases [109], or to bypass the slow nucleation stage [72]. Pre-formed atomistic domains can be obtained from equilibrated CG simulations using reverse transformations [121].

An alternative approach to investigate phase transitions is the semigrand canonical ensemble MD-MC simulations [52, 78, 90]. In this method, lipid types are exchanged using MC moves at a fixed difference of chemical potential (of two lipid components). This method allows overcoming slow diffusive de-mixing, characterizing non-ideal mixing, and comparing phase compositions to experimental tie-lines. Hybrid DPD-MC simulations also allow to enhance the sampling of conformations during phase transformations [71].

4 Notes

4.1 Gel Phases in Bilayers

(a) Problem: bilayer does not form a gel phase at the given temperature. Possible reasons: the Tm is different (underestimated) in the given FF and lies below the considered temperature (*see* Subheading 3.1); the simulation box size is too small to allow formation of the critical nucleus (*see* Subheading 3.4); the simulation time is too short to observe nucleation (*see* Subheading 3.3); the FF does not reproduce gel phases. Possible solution: to change the simulation conditions to move further into binodal region away from the transition point, e.g., reduce the temperature.

(b) Problem: bilayer does not form ripple/tilted gel phases. Possible reasons: the given FF does not reproduce molecular tilt (*see* Subheading 3.2). Tilting of the molecular axis takes place if the lipid headgroup area in the bilayer is larger compared hydrocarbon tail cross-section area. Possible solution: to increase the headgroup diameter/reduce the tail diameter by changing LJ interactions; however, care must be taken as it will alter the bilayer properties.

4.2 Monolayer Phases

Problem: phase separation does not occur inside the expected interval of surface tensions/pressures. Possible reasons: the surface tension at the bare air–water interface is underestimated in the given FF; as a result, the surface pressure–area isotherm is offset compared to experimental (*see* Subheading 3.1); the Tm of the lipid(s) in bilayers does not match the experimental value, which affects the surface tension-area dependence at a fixed temperature; the FF does not have the required accuracy; *see* also Note 4.1a. Possible solution: to change (reduce) the surface tension and/or the temperature.

4.3 Compositional De-mixing

(a) Lipid de-mixing is observed in experiments but does not occur in simulations at given conditions. Possible reasons: driving forces for phase separation are not correctly modeled in the FF used. Some FF validation is required, *see* Subheading 3.1 and 4.1a

(b) Problem: the composition of the coexisting phases does not match experimental data. Possible reasons: driving forces for phase separation (*see* above); the Tm of lipid components differs from experimental values which distorts the temperature axis in the phase diagram in simulations.

4.4 Temperature Dependence of Properties

Problem: variation of the properties of the coexisting phases, e.g., composition, with temperature, does not match experimental data. Possible reason: in CG models, substitution of entropic interactions by enthalpic leads to weaker temperature dependence of properties; in atomistic models, it is affected by the interval of temperatures and the (structural, thermodynamic) properties used for the FF parameterization.

Acknowledgements

This work was supported by the Natural Sciences and Engineering Research Council (Canada). DPT is an Alberta Innovates Health Solutions Scientist and Alberta Innovates Technology Futures Strategic Chair in (Bio)Molecular Simulation.

References

1. Fahy E, Subramaniam S, Brown HA, Glass CK, Merrill AH, Murphy RC, Raetz CRH, Russell DW, Seyama Y, Shaw W, Shimizu T, Spener F, van Meer G, VanNieuwenhze MS, White SH, Witztum JL, Dennis EA (2005) A comprehensive classification system for lipids. Eur J Lipid Sci Technol 107(5):337–364

2. Mouritsen OG (2005) Life - as a matter of fat, The frontiers collection. Springer, Heidelberg

3. Tanford C (1980) The hydrophobic effect. Wiley, New York, NY

4. Seddon JM, Templer RH (1995) Chapter 3: Polymorphism of lipid-water systems. In: Lipowsky R, Sackmann E (eds) Handbook of biological physics, vol 1. Elsevier, North-Holland, pp 97–160

5. Vargaftik NB, Volkov BN, Voljak LD (1983) International tables of the surface tension of water. J Phys Chem Ref Data 12(3):817–820

6. Knobler CM, Desai RC (1992) Phase-transitions in monolayers. Annu Rev Phys Chem 43:207–236

7. Kaganer VM, Mohwald H, Dutta P (1999) Structure and phase transitions in Langmuir monolayers. Rev Mod Phys 71(3):779–819

8. Jahnig F (1996) What is the surface tension of a lipid bilayer membrane? Biophys J 71(3): 1348–1349

9. Hames D, Hooper N (2006) Instant notes in biochemistry. Taylor & Francis, New York, NY

10. Clements JA (1957) Surface tension of lung extracts. Proc Soc Exp Biol Med 95(1): 170–172

11. Bachofen H, Schurch S (2001) Alveolar surface forces and lung architecture. Comp Biochem Physiol A Comp Physiol 129(1): 183–193

12. van Meer G, Voelker DR, Feigenson GW (2008) Membrane lipids: where they are and how they behave. Nat Rev Mol Cell Biol 9(2):112–124

13. Engelman DM (2005) Membranes are more mosaic than fluid. Nature 438(7068): 578–580

14. Plasencia I, Norlen L, Bagatolli LA (2007) Direct visualization of lipid domains in human skin stratum corneum's lipid membranes: effect of pH and temperature. Biophys J 93(9):3142–3155

15. Simons K, Ikonen E (1997) Functional rafts in cell membranes. Nature 387(6633): 569–572

16. Lingwood D, Simons K (2010) Lipid rafts as a membrane-organizing principle. Science 327(5961):46–50

17. Pike LJ (2006) Rafts defined: a report on the keystone symposium on lipid rafts and cell function. J Lipid Res 47(7):1597–1598

18. Simons K, Toomre D (2000) Lipid rafts and signal transduction. Nat Rev Mol Cell Biol 1(1):31–39

19. Brown DA, London E (1998) Functions of lipid rafts in biological membranes. Annu Rev Cell Dev Biol 14:111–136

20. Mayor S, Rao M (2004) Rafts: scale-dependent, active lipid organization at the cell surface. Traffic 5(4):231–240

21. Parton RG, Richards AA (2003) Lipid rafts and caveolae as portals for endocytosis: new insights and common mechanisms. Traffic 4(11):724–738

22. Bernardino de la Serna J, Perez-Gil J, Simonsen AC, Bagatolli LA (2004) Cholesterol rules: direct observation of the coexistence of two fluid phases in native pulmonary surfactant membranes at physiological temperatures. J Biol Chem 279(39): 40715–40722

23. Nag K, Perez-Gil J, Ruano MLF, Worthman LAD, Stewart J, Casals C, Keough KMW (1998) Phase transitions in films of lung surfactant at the air-water interface. Biophys J 74(6):2983–2995

24. Piknova B, Schram V, Hall SB (2002) Pulmonary surfactant: phase behavior and function. Curr Opin Struct Biol 12(4): 487–494

25. Zuo YY, Tadayyon SM, Keating E, Zhao L, Veldhuizen RAW, Petersen NO, Amrein MW, Possmayer F (2008) Atomic force microscopy

studies of functional and dysfunctional pulmonary surfactant films, II: albumin-inhibited pulmonary surfactant films and the effect of SP-A. Biophys J 95(6):2779–2791

26. Keating E, Zuo YY, Tadayyon SM, Petersen NO, Possmayer F, Veldhuizen RAW (2012) A modified squeeze-out mechanism for generating high surface pressures with pulmonary surfactant. Biochim Biophys Acta 1818(5): 1225–1234

27. Fan J, Sammalkorpi M, Haataja M (2010) Formation and regulation of lipid microdomains in cell membranes: theory, modeling, and speculation. FEBS Lett 584(9):1678–1684

28. Jacobson K, Mouritsen OG, Anderson RGW (2007) Lipid rafts: at a crossroad between cell biology and physics. Nat Cell Biol 9(1):7–14

29. Ziolkowska NE, Christiano R, Walther TC (2012) Organized living: formation mechanisms and functions of plasma membrane domains in yeast. Trends Cell Biol 22(3): 151–158

30. Honerkamp-Smith AR, Veatch SL, Keller SL (2009) An introduction to critical points for biophysicists; observations of compositional heterogeneity in lipid membranes. Biochim Biophys Acta Biomembr 1788(1):53–63

31. van Meer G (2005) Cellular lipidomics. EMBO J 24(18):3159–3165

32. Konyakhina TM, Wu J, Mastroianni JD, Heberle FA, Feigenson GW (2013) Phase diagram of a 4-component lipid mixture: DSPC/DOPC/POPC/chol. Biochim Biophys Acta Biomembr 1828(9):2204–2214

33. Davis JH, Clair JJ, Juhasz J (2009) Phase equilibria in DOPC/DPPC-d(62)/cholesterol mixtures. Biophys J 96(2):521–539

34. Feigenson GW (2009) Phase diagrams and lipid domains in multicomponent lipid bilayer mixtures. Biochim Biophys Acta Biomembr 1788(1):47–52

35. Veatch SL, Keller SL (2005) Seeing spots: complex phase behavior in simple membranes. Biochim Biophys Acta 1746(3): 172–185

36. Elson EL, Fried E, Dolbow JE, Genin GM (2010) Phase separation in biological membranes: integration of theory and experiment. Annu Rev Biophys 39:207–226

37. Eggeling C, Ringemann C, Medda R, Schwarzmann G, Sandhoff K, Polyakova S, Belov VN, Hein B, von Middendorff C, Schoenle A, Hell SW (2009) Direct observation of the nanoscale dynamics of membrane lipids in a living cell. Nature 457(7233): 1159–U1121

38. Frisz JF, Lou KY, Klitzing HA, Hanafin WP, Lizunov V, Wilson RL, Carpenter KJ, Kim R,

Hutcheon ID, Zimmerberg J, Weber PK, Kraft ML (2013) Direct chemical evidence for sphingolipid domains in the plasma membranes of fibroblasts. Proc Natl Acad Sci U S A 110(8):E613–E622

39. MacKerell AD, Bashford D, Bellott M, Dunbrack RL, Evanseck JD, Field MJ, Fischer S, Gao J, Guo H, Ha S, Joseph-McCarthy D, Kuchnir L, Kuczera K, Lau FTK, Mattos C, Michnick S, Ngo T, Nguyen DT, Prodhom B, Reiher WE, Roux B, Schlenkrich M, Smith JC, Stote R, Straub J, Watanabe M, Wiorkiewicz-Kuczera J, Yin D, Karplus M (1998) All-atom empirical potential for molecular modeling and dynamics studies of proteins. J Phys Chem B 102(18): 3586–3616

40. Klauda JB, Venable RM, Freites JA, O'Connor JW, Tobias DJ, Mondragon-Ramirez C, Vorobyov I, MacKerell AD Jr, Pastor RW (2010) Update of the CHARMM all-atom additive force field for lipids: validation on six lipid types. J Phys Chem B 114(23): 7830–7843

41. Feller SE, MacKerell AD (2000) An improved empirical potential energy function for molecular simulations of phospholipids. J Phys Chem B 104(31):7510–7515

42. Klauda JB, Brooks BR, MacKerell AD, Venable RM, Pastor RW (2005) An ab initio study on the torsional surface of alkanes and its effect on molecular simulations of alkanes and a DPPC bilayer. J Phys Chem B 109(11):5300–5311

43. Jambeck JPM, Lyubartsev AP (2012) An extension and further validation of an All-atomistic force field for biological membranes. J Chem Theor Comput 8(8):2938–2948

44. Jambeck JPM, Lyubartsev AP (2012) Derivation and systematic validation of a refined All-atom force field for phosphatidylcholine lipids. J Phys Chem B 116(10): 3164–3179

45. Jambeck JPM, Lyubartsev AP (2013) Another piece of the membrane puzzle: extending lipids further. J Chem Theor Comput 9(1):774–784

46. Case DA, Cheatham TE, Darden T, Gohlke H, Luo R, Merz KM, Onufriev A, Simmerling C, Wang B, Woods RJ (2005) The amber biomolecular simulation programs. J Comput Chem 26(16):1668–1688

47. Dickson CJ, Madej BD, Skjevik AA, Betz RM, Teigen K, Gould IR, Walker RC (2014) Lipid14: the amber lipid force field. J Chem Theor Comput 10(2):865–879

48. Hermans J, Berendsen HJC, Vangunsteren WF, Postma JPM (1984) A consistent empirical potential for water-protein interactions. Biopolymers 23(8):1513–1518

49. Poger D, Van Gunsteren WF, Mark AE (2010) A new force field for simulating phosphatidylcholine bilayers. J Comput Chem 31(6):1117–1125

50. Poger D, Mark AE (2010) On the validation of molecular dynamics simulations of saturated and cis-monounsaturated phosphatidylcholine lipid bilayers: a comparison with experiment. J Chem Theor Comput 6(1):325–336

51. Berger O, Edholm O, Jahnig F (1997) Molecular dynamics simulations of a fluid bilayer of dipalmitoylphosphatidylcholine at full hydration, constant pressure, and constant temperature. Biophys J 72(5):2002–2013

52. de Joannis J, Coppock PS, Yin F, Mori M, Zamorano A, Kindt JT (2011) Atomistic simulation of cholesterol effects on miscibility of saturated and unsaturated phospholipids: implications for liquid-ordered/liquid-disordered phase coexistence. J Am Chem Soc 133(10):3625–3634

53. Tjörnhammar R, Edholm O (2014) Atomistic simulations of gel and liquid crystalline lipid bilayers. Biophys J 106(2):403

54. Sodt AJ, Sandar ML, Gawrisch K, Pastor RW, Lyman E (2014) The molecular structure of the liquid-ordered phase of lipid bilayers. J Am Chem Soc 136(2):725–732

55. Hakobyan D, Heuer A (2013) Phase separation in a lipid/cholesterol system: comparison of coarse-grained and united-atom simulations. J Phys Chem B 117(14):3841–3851

56. Polley A, Vemparala S, Rao M (2012) Atomistic simulations of a multicomponent asymmetric lipid bilayer. J Phys Chem B 116(45):13403–13410

57. Marrink SJ, Risselada HJ, Yefimov S, Tieleman DP, Vries AH (2007) The MARTINI force field: coarse grained model for biomolecular simulations. J Phys Chem B 111(27):7812–7824

58. Bennett WFD, Tieleman DP (2013) Computer simulations of lipid membrane domains. Biochim Biophys Acta Biomembr 1828(8):1765–1776

59. Shinoda W, DeVane R, Klein ML (2010) Zwitterionic lipid assemblies: molecular dynamics studies of monolayers, bilayers, and vesicles using a new coarse grain force field. J Phys Chem B 114(20):6836–6849

60. Shinoda W, Devane R, Klein ML (2007) Multi-property fitting and parameterization of a coarse grained model for aqueous surfactants. Mol Simul 33(1–2):27–36

61. Lenz O, Schmid F (2005) A simple computer model for liquid lipid bilayers. J Mol Liq 117(1–3):147–152

62. Venturoli M, Sperotto MM, Kranenburg M, Smit B (2006) Mesoscopic models of biological membranes. Phys Rep Rev Sec Phys Lett 437(1–2):1–54

63. Marrink SJ, de Vries AH, Tieleman DP (2009) Lipids on the move: simulations of membrane pores, domains, stalks and curves. Biochim Biophys Acta Biomembr 1788(1):149–168

64. Mueller M, Katsov K, Schick M (2006) Biological and synthetic membranes: what can be learned from a coarse-grained description? Phys Rep Rev Sec Phys Lett 434(5–6):113–176

65. Brannigan G, Lin LCL, Brown FLH (2006) Implicit solvent simulation models for biomembranes. Eur Biophys J 35(2):104–124

66. Deserno M (2009) Mesoscopic membrane physics: concepts, simulations, and selected applications. Macromol Rapid Commun 30(9–10):752–771

67. Jo S, Lim JB, Klauda JB, Im W (2009) CHARMM-GUI membrane builder for mixed bilayers and its application to yeast membranes. Biophys J 97(1):50–58

68. Herrera FE, Pantano S (2012) Structure and dynamics of nano-sized raft-like domains on the plasma membrane. J Chem Phys 136(1):015103

69. Marrink SJ, Risselada J, Mark AE (2005) Simulation of gel phase formation and melting in lipid bilayers using a coarse grained model. Chem Phys Lipids 135(2):223–244

70. de Vries AH, Yefimov S, Mark AE, Marrink SJ (2005) Molecular structure of the lecithin ripple phase. Proc Natl Acad Sci U S A 102(15):5392–5396

71. Rodgers JM, Sorensen J, de Meyer FJM, Schiott B, Smit B (2012) Understanding the phase behavior of coarse-grained model lipid bilayers through computational calorimetry. J Phys Chem B 116(5):1551–1569

72. Coppock PS, Kindt JT (2010) Determination of phase transition temperatures for atomistic models of lipids from temperature-dependent stripe domain growth kinetics. J Phys Chem B 114(35):11468–11473

73. Revalee JD, Laradji M, Kumar PBS (2008) Implicit-solvent mesoscale model based on soft-core potentials for self-assembled lipid membranes. J Chem Phys 128(3):035102

74. Baoukina S, Monticelli L, Marrink SJ, Tieleman DP (2007) Pressure-area isotherm of a lipid monolayer from molecular dynamics simulations. Langmuir 23(25):12617–12623

75. Knecht V, Muller M, Bonn M, Marrink SJ, Mark AE (2005) Simulation studies of pore and domain formation in a phospholipid monolayer. J Chem Phys 122(2):024704

76. Baoukina S, Mendez-Villuendas E, Tieleman DP (2012) Molecular view of phase coexistence in lipid monolayers. J Am Chem Soc 134(42):17543–17553

77. Duncan SL, Dalal IS, Larson RG (2011) Molecular dynamics simulation of phase transitions in model lung surfactant monolayers. Biochim Biophys Acta Biomembr 1808(10): 2450–2465

78. Coppock PS, Kindt JT (2009) Atomistic simulations of mixed-lipid bilayers in gel and fluid phases. Langmuir 25(1):352–359

79. Stevens MJ (2005) Complementary matching in domain formation within lipid bilayers. J Am Chem Soc 127(44):15330–15331

80. Shi Q, Voth GA (2005) Multi-scale modeling of phase separation in mixed lipid bilayers. Biophys J 89(4):2385–2394

81. Faller R, Marrink SJ (2004) Simulation of domain formation in DLPC-DSPC mixed bilayers. Langmuir 20(18):7686–7693

82. Bennun SV, Longo M, Faller R (2007) Phase and mixing behavior in two-component lipid bilayers: a molecular dynamics study in DLPC/DSPC mixtures. J Phys Chem B 111(32):9504–9512

83. Muddana HS, Chiang HH, Butler PJ (2012) Tuning membrane phase separation using nonlipid amphiphiles. Biophys J 102(3): 489–497

84. Laradji M, Kumar PBS (2005) Domain growth, budding, and fission in phase-separating self-assembled fluid bilayers. J Chem Phys 123(22):224902

85. Illya G, Lipowsky R, Shillcock JC (2006) Two-component membrane material properties and domain formation from dissipative particle dynamics. J Chem Phys 125(11): 114710

86. Hong B, Qiu F, Zhang H, Yang Y (2007) Budding dynamics of individual domains in multicomponent membranes simulated by N-varied dissipative particle dynamics. J Phys Chem B 111(21):5837–5849

87. Sugar IP, Chong PLG (2012) A statistical mechanical model of cholesterol/phospholipid mixtures: linking condensed complexes, superlattices, and the phase diagram. J Am Chem Soc 134(2):1164–1171

88. McMullen TPW, Lewis R, McElhaney RN (2004) Cholesterol-phospholipid interactions, the liquid-ordered phase and lipid rafts in model and biological membranes. Curr Opin Colloid Interface Sci 8(6):459–468

89. Waheed Q, Tjornhammar R, Edholm O (2012) Phase transitions in coarse-grained lipid bilayers containing cholesterol by molecular dynamics simulations. Biophys J 103(10): 2125–2133

90. Meinhardt S, Vink RLC, Schmid F (2013) Monolayer curvature stabilizes nanoscale raft domains in mixed lipid bilayers. Proc Natl Acad Sci U S A 110(12):4476–4481

91. de Meyer FJM, Benjamini A, Rodgers JM, Misteli Y, Smit B (2010) Molecular simulation of the DMPC-cholesterol phase diagram. J Phys Chem B 114(32):10451–10461

92. Martinez-Seara H, Rog T, Karttunen M, Vattulainen I, Reigada R (2010) Cholesterol induces specific spatial and orientational order in cholesterol/phospholipid membranes. PLoS One 5(6):e11162

93. Khelashvili G, Pabst G, Harries D (2010) Cholesterol orientation and tilt modulus in DMPC bilayers. J Phys Chem B 114(22): 7524–7534

94. Mihailescu M, Vaswani RG, Jardon-Valadez E, Castro-Roman F, Freites JA, Worcester DL, Chamberlin AR, Tobias DJ, White SH (2011) Acyl-chain methyl distributions of liquid-ordered and -disordered membranes. Biophys J 100(6):1455–1462

95. Bennett WFD, MacCallum JL, Tieleman DP (2009) Thermodynamic analysis of the effect of cholesterol on dipalmitoylphosphatidylcholine lipid membranes. J Am Chem Soc 131(5):1972–1978

96. Rog T, Pasenkiewicz-Gierula M, Vattulainen I, Karttunen M (2009) Ordering effects of cholesterol and its analogues. Biochim Biophys Acta Biomembr 1788(1):97–121

97. Berkowitz ML (2009) Detailed molecular dynamics simulations of model biological membranes containing cholesterol. Biochim Biophys Acta Biomembr 1788(1):86–96

98. de Meyer F, Smit B (2009) Effect of cholesterol on the structure of a phospholipid bilayer. Proc Natl Acad Sci U S A 106(10): 3654–3658

99. Pandit SA, Jakobsson E, Scott HL (2004) Simulation of the early stages of nano-domain formation in mixed bilayers of sphingomyelin, cholesterol, and dioleylphosphatidylcholine. Biophys J 87(5):3312–3322

100. Niemela PS, Ollila S, Hyvonen MT, Karttunen M, Vattulainen I (2007) Assessing the nature of lipid raft membranes. PLoS Comput Biol 3(2):e34

101. Risselada HJ, Marrink SJ (2008) The molecular face of lipid rafts in model membranes. Proc Natl Acad Sci U S A 105(45):17367–17372

102. Risselada HJ, Marrink SJ, Muller M (2011) Curvature-dependent elastic properties of liquid-ordered domains result in inverted

domain sorting on uniaxially compressed vesicles. Phys Rev Lett 106(14):148102

103. Baoukina S, Mendez-Villuendas E, Bennett WFD, Tieleman DP (2013) Computer simulations of the phase separation in model membranes. Faraday Discuss 161:63–75

104. Perlmutter JD, Sachs JN (2011) Interleaflet interaction and asymmetry in phase separated lipid bilayers: molecular dynamics simulations. J Am Chem Soc 133(17):6563–6577

105. Piggot TJ, Holdbrook DA, Khalid S (2011) Electroporation of the E. coli and S. aureus membranes: molecular dynamics simulations of complex bacterial membranes. J Phys Chem B 115(45):13381–13388

106. Kiessling V, Wan C, Tamm LK (2009) Domain coupling in asymmetric lipid bilayers. Biochim Biophys Acta Biomembr 1788(1): 64–71

107. May S (2009) Trans-monolayer coupling of fluid domains in lipid bilayers. Soft Matter 5(17):3148–3156

108. Brewster R, Pincus PA, Safran SA (2009) Hybrid lipids as a biological surface-active component. Biophys J 97(4):1087–1094

109. Schäfer LV, Marrink SJ (2010) Partitioning of lipids at domain boundaries in model membranes. Biophys J 99(12):L91–L93

110. Rosetti C, Pastorino C (2012) Comparison of ternary bilayer mixtures with asymmetric or symmetric unsaturated phosphatidylcholine lipids by coarse grained molecular dynamics simulations. J Phys Chem B 116(11): 3525–3537

111. Hakobyan D, Heuer A (2014) Key molecular requirements for raft formation in lipid/cholesterol membranes. PLoS One 9(2):e87369

112. Marsh D (1996) Lateral pressure in membranes. Biochim Biophys Acta 1286(3): 183–223

113. Baoukina S, Marrink SJ, Tieleman DP (2010) Lateral pressure profiles in lipid monolayers. Faraday Discuss 144:393–409

114. Wu Z, Cui QA, Yethiraj A (2010) A new coarse-grained model for water: the importance of electrostatic interactions. J Phys Chem B 114(32):10524–10529

115. Abascal JLF, Vega C (2005) A general purpose model for the condensed phases of water: TIP4P/2005. J Chem Phys 123(23): 234505

116. Jorgensen WL, Chandrasekhar J, Madura JD, Impey RW, Klein ML (1983) Comparison of simple potential functions for simulating liquid water. J Chem Phys 79(2):926–935

117. Yesylevskyy SO, Schafer LV, Sengupta D, Marrink SJ (2010) Polarizable water model for the coarse-grained MARTINI force field. PLoS Comput Biol 6(6):e1000810

118. Davis RS, Kumar PBS, Sperotto MM, Laradji M (2013) Predictions of phase separation in three-component lipid membranes by the MARTINI force field. J Phys Chem B 117(15):4072–4080

119. Domanski J, Marrink SJ, Schäfer LV (2012) Transmembrane helices can induce domain formation in crowded model membranes. Biochim Biophys Acta 1818(4):984–994

120. Camley BA, Brown FLH (2011) Dynamic scaling in phase separation kinetics for quasi-two-dimensional membranes. J Chem Phys 135(22):225106

121. Wassenaar TA, Pluhackova K, Bockmann RA, Marrink SJ, Tieleman DP (2014) Going backward: a flexible geometric approach to reverse transformation from coarse grained to atomistic models. J Chem Theor Comput 10(2):676–690

INDEX

Dylan M. Owen (ed.), *Methods in Membrane Lipids*, Methods in Molecular Biology, vol. 1232,
DOI 10.1007/978-1-4939-1752-5, © Springer Science+Business Media New York 2015

Printed by Printforce, the Netherlands